Biological Paths to Self-Reliance
A Guide to Biological Solar Energy Conversion

Biological Paths to Self-Reliance

A Guide to Biological Solar Energy Conversion

RUSSELL E. ANDERSON, Ph.D.
International Federation of Institutes for Advanced Study
Ulriksdals Slott
Solna, Sweden

Van Nostrand Reinhold
Environmental Engineering Series

VAN NOSTRAND REINHOLD COMPANY
NEW YORK   CINCINNATI   ATLANTA   DALLAS   SAN FRANCISCO
LONDON   TORONTO   MELBOURNE

Van Nostrand Reinhold Company Regional Offices:
New York   Cincinnati   Atlanta   Dallas   San Francisco

Van Nostrand Reinhold Company International Offices:
London   Toronto   Melbourne

Copyright © 1979 by Litton Educational Publishing, Inc.

Library of Congress Catalog Card Number: 78-9847
ISBN: 0-442-20329-2

Manufactured in the United States of America

Published by Van Nostrand Reinhold Company
135 West 50th Street, New York, N.Y. 10020

Published simultaneously in Canada by Van Nostrand Reinhold Ltd.

15 14 13 12 11 10 9 8 7 6 5 4 3 2 1

Library of Congress Cataloging in Publication Data
Anderson, Russell E
  Biological paths to self-reliance.

  (Van Nostrand Reinhold environmental engineering series)
  Bibliography: p.
  Includes index.
  1. Biomass energy. 2. Solar energy. I. Title.
II. Title: Self-reliance.
TP360.A53   333.9'5   78-9847
ISBN 0-442-20329-2

To my parents

Van Nostrand Reinhold Environmental Engineering Series

THE VAN NOSTRAND REINHOLD ENVIRONMENTAL ENGINEERING SERIES is dedicated to the presentation of current and vital information relative to the engineering aspects of controlling man's physical environment. Systems and subsystems available to exercise control of both the indoor and outdoor environment continue to become more sophisticated and to involve a number of engineering disciplines. The aim of the series is to provide books which, though often concerned with the life cycle—design, installation, and operation and maintenance—of a specific system or subsystem, are complementary when viewed in their relationship to the total environment.

The Van Nostrand Reinhold Environmental Engineering Series includes books concerned with the engineering of mechanical systems designed (1) to control the environment within structures, including those in which manufacturing processes are carried out, and (2) to control the exterior environment through control of waste products expelled by inhabitants of structures and from manufacturing processes. The series includes books on heating, air conditioning and ventilation, control of air and water pollution, control of the acoustic environment, sanitary engineering and waste disposal, illumination, and piping systems for transporting media of all kinds.

Van Nostrand Reinhold Environmental Engineering Series

ADVANCED WASTEWATER TREATMENT, by Russell L. Culp and Gordon L. Culp

ARCHITECTURAL INTERIOR SYSTEMS—Lighting, Air Conditioning, Acoustics, John E. Flynn and Arthur W. Segil

SOLID WASTE MANAGEMENT, by D. Joseph Hagerty, Joseph L. Pavoni and John E. Heer, Jr.

THERMAL INSULATION, by John F. Malloy

AIR POLLUTION AND INDUSTRY, edited by Richard D. Ross

INDUSTRIAL WASTE DISPOSAL, edited by Richard D. Ross

MICROBIAL CONTAMINATION CONTROL FACILITIES, by Robert S. Rurkle and G. Briggs Phillips

SOUND, NOISE, AND VIBRATION CONTROL (Second Edition), by Lyle F. Yerges

NEW CONCEPTS IN WATER PURIFICATION, by Gordon L. Culp and Russell L. Culp

HANDBOOK OF SOLID WASTE DISPOSAL: MATERIALS AND ENERGY RECOVERY, by Joseph L. Pavoni, John E. Heer, Jr., and D. Joseph Hagerty

ENVIRONMENTAL ASSESSMENTS AND STATEMENTS, by John E. Heer, Jr. and D. Joseph Hagerty

ENVIRONMENTAL IMPACT ANALYSIS: A New Dimension in Decision Making, by R. K. Jain, L. V. Urban and G. S. Stacey

CONTROL SYSTEMS FOR HEATING, VENTILATING, AND AIR CONDITIONING (Second Edition), by Roger W. Haines

WATER QUALITY MANAGEMENT PLANNING, edited by Joseph L. Pavoni

HANDBOOK OF ADVANCED WASTEWATER TREATMENT (Second Edition), by Russell L. Culp, George Mack Wesner and Gordon L. Culp

HANDBOOK OF NOISE ASSESSMENT, edited by Daryl N. May

NOISE CONTROL: HANDBOOK OF PRINCIPLES AND PRACTICES, edited by David M. Lipscomb and Arthur C. Taylor

AIR POLLUTION CONTROL TECHNOLOGY, by Robert M. Bethea

POWER PLANT SITING, by John V. Winter and David A. Conner

DISINFECTION OF WASTEWATER AND WATER FOR REUSE, by Geo. Clifford White

LAND USE PLANNING: Techniques of Implementation, by T. William Patterson

BIOLOGICAL PATHS TO SELF-RELIANCE, by Russell E. Anderson

HANDBOOK OF INDUSTRIAL WASTE DISPOSAL, by Richard A. Conway and Richard D. Ross

HANDBOOK OF ORGANIC WASTE CONVERSION, by Michael W. Bewick

LAND APPLICATIONS OF WASTE (Volume 1), by Raymond C. Loehr, William J. Jewell, Joseph D. Novak, William W. Clarkson and Gerald S. Friedman

LAND APPLICATIONS OF WASTE (Volume 2), by Raymond C. Loehr, William J. Jewell, Joseph D. Novak, William W. Clarkson and Gerald S. Friedman

WIND MACHINES (Second Edition), by Frank R. Eldridge

Foreword

Johan Galtung
Professor
Institut Universitaire d'Etudes du Developpement
Geneva, Switzerland

When future historians look back at our epoch and try to arrive at some answer to the perplexing question "what went wrong?", maybe one answer will be along the following lines: arrogant Western Man thought he could do better than Nature, even to the point of working against Nature rather than with Her. For how can one otherwise interpret the intellectual-commercial complex known as industrialism, including industrial agriculture? It is based on knowledge of some of the simpler processes in Nature, studied systematically by "modern" science, which cuts out of extremely complex totalities a handful of variables, relating them to each other under artificial (laboratory) conditions, uncovers a relationship, and builds it into another artificial setting, "modern" industry. Little respect for the miracle that is Nature, for how a mature ecosystem manages to survive through diversity, complex cycles, and careful use of high-quality energy. Instead of diversity, standardization and efforts to arrive at *the* universal optimal solution (strain, breed, organizational structure); instead of balance, depletion of resources and pollution of what is left, with "wastes" piling up; instead of energy-saving devices, non-use, or wasteful use, of high-quality energy. The consequences are now dawning upon us.

Some of these consequences are ecological; others are political, economic, social—in short, human. Thus, it may be argued that if each little human unit—country, community, family, even individual person— behaved this way towards the Nature in their allotted part of the world only, then this becomes their problem. The country/community/family/

person who behaves unwisely towards Nature becomes the victim of that behavior, at the same time cause and effect, subject and object. But human societies are not organized that way. They are huge entities, designed among other things to deflect the negative consequences outward and downward, to put them at the feet of the peripheral and downtrodden. The garbage dumps of local, national and world societies are next to the poor rather than next to the rich—although often next to the poor in the rich societies. The cause is often a comfortable distance away from the effect, located at the opposite ends of vast economic cycles that at the same time are cycles of exploitation, dependence, and environmental deterioration.

Recently a new term has entered the development terminology and with increasing force: *self-reliance*. It means essentially this: whether at the local, national or regional levels—if you want to produce something, count on your own resources (energy, raw materials, capital, and above all the human resources) *first*. If that is insufficient, get things through exchange, but with others at roughly the same level; your neighbors in geographical and social space. Do not become dependent on others, do not give away raw materials and energy in exchange for capital and technology, because (1) you undermine the basis of your own existence—Nature, (2) you deprive yourself of the challenge to develop your own technology and (3) you fall into a vicious circle of dependency (moreover, you are likely to be cheated anyhow). However, the skeptic will say: you still need a technological basis—how can poor countries and communities feed themselves? Answer: through other types of food technologies, more advanced ones, more respectful of Nature, also less wasteful, more autonomous—in short more self-reliant. Thus, there is a biology of self-reliance, and this aspect of self-reliance is what Russell Anderson's book is about.

This important book combines the three facets of good, policy relevant research—empirical, critical, and constructive approaches—in a fascinating manner. The reader will find a host of empirical data that in a sense add up to one conclusion: how utterly irrational the present system is, except—perhaps—for a few located at its center, and even for them/us one may suspect that our material overabundance no longer tastes as good as before. There is a clearly critical approach, with a dedication to what, for the author, stands out as the goal of the whole human exercise with Nature—"(to) emphasize the man whom the technical system serves, the need to unite food and fuel with mental and physical health, education, and social life in a manner and on a scale appropriate to his potential and needs." And this book is indeed constructive, there is a wealth of rich ideas that will keep the reader captivated from beginning to end, with perhaps one basic thought in mind: how little the rest of us tend to know about energy and fuel, sunshine and rain, water and soil, food and fodder, waste and food chains. Thus, consider the small model presented by the author towards the end of the book, a three-point cycle using normal agriculture to produce food and "wastes;" putting the wastes from agricultural production and consumption into a biogas digester to produce methane for all kinds of energy use, fertilizer for agriculture and some carbon dioxide to feed the algal pond, the third element in the cycle. This algal pond can be used for food and fodder, to feed the biogas digester, *and* to provide for the famous biological fixation of nitrogen to be used as fertilizer for agriculture. It is such a fine example of putting first things—profound respect for Man-Nature symbiosis—first.

What does "Man-Nature symbiosis" mean? For one thing, it means a focus on natural processes. In Anderson's tale, the two heroes of Nature, sunshine and the photosynthetic process, are so obvious that they had to be pushed into the background by modern industry because they were too competitive. The former is described as our only true energy capital, one constantly renewed because it comes in abundance from the outside of the earth itself. The latter is described as a process infinitely more complex than industrial processes. So why not make optimal use of them—sunshine as high quality energy that can be converted into all kinds of energy, directly and indirectly, but with the care of a mature ecosystem; the photosynthetic process as the best factory in the world, yielding all we need when properly understood. And they cannot be patented (or at least so we assume)—they are truly part of Man's common heritage.

Anderson does not say so explicitly, and his style is not excessively imbued with Nature romanticism, but the reader will sense a great deal of identification with Nature and organic, biological processes, and a corresponding skepticism vis-à-vis colder, less organic, more mechanized, purely man-made operations. It may be said that this would be in the vested interest of the biologist as against the mechanical/physical/chemical engineers whose world views have dominated our economic relations to Nature for a period of a century or more. In that case, so be it: maybe the book should serve as a warning to the latter, that their time is up, that we need much richer world views with more respect for Nature, conducive to a greater ability to fuse the economic cycles on which our existence is based much better with the ecocycles that constitute Nature in its more mature stages. Actually, it is interesting to compare this with a corresponding phenomenon within the social sciences: the economists have shown a disrespect for the finer lines in the spectrum of the human psyche, and even for the most basic material needs of the majority, in their search for determinants of economic growth. They have brushed into very complex webs and totalities, with great faith in a handful of variables to be manipulated—and are now pitted against the "softer" social sciences in their effort to defend this complex thing called "human identity," at the individual level (psychology, social psychology) and at the social level (sociology, anthropology).

But this is not an occasion to search for heroes and villains; there are more important tasks. Biology and sociology, for example, do not constitute final answers to anything, but precisely because they are less mathematized and less axiomatic, they are more open as thought systems and hence more capable of adjusting to new insights. The price for approximation to pure mathematics (or to classical mechanics) as an intellectual model may be premature death as an ongoing intellectual enterprise—so let us hope those who want to formalize the alternative disciplines will not have too much success. At this point one might feel some worry in connection with the present book: I would be as afraid of a world clan of biologist high priests (or sociologists) as I am of the present alliance at the center of our highly centralized societies: the bureaucrats, the men who control the basic economic flows, and the researchers with those linear, nonholistic, often mathematically impressive thought models. In Russell Anderson's favorite world, decentralization is a key point, and his task in this book is to show—and I think rather convincingly—that it is possible to provide food and energy on a decentralized basis working *with* rather than *against* Nature. But one aspect of self-reliance is intellectual self-reliance: how do people advance their own ideas and concepts? Will the biologists be less inclined to try to make themselves indispensable than the physical scientists; the social scientists more than the economists?

There is probably an answer to this question hidden in these pages. By living in closer contact with Nature, making better use of sun and rain, using soil, photosynthesis and organic processes, self-interest will stimulate more creativity. But it should not be concealed in any way that what Anderson in fact is saying, when taken seriously, would be tantamount to a complete agricultural revolution, as encompassing and full of implications as the revolution that has led to what today is known as "industrial agriculture," with Justus Leibig as one of the forerunners. Instead of coupling industry and agriculture in such a way that the former is gradually encroaching one and transforming the latter, leading to enormous industrial commercial systems whose centers have been maintained up till now essentially on cheap oil—Anderson says to let agriculture penetrate industry, agriculturizing industry, rather than vice versa. For this to happen another mentality is probably needed, with more ability to listen to Nature and follow Her pointers—and one can only hope there are still enough people left with that kind of ability reasonably intact to understand what Nature says, and translate it into language understandable to all of us who have been systematically detrained in that direction. Let us only hope it happens before the last Western expert who still believes in the gospel of industrialization in the classical sense has managed to convert the last villager in the Third World.

One very important characteristic of this book is the absence of doomsday gloom. Anderson does not offer people the cheap kick of incipient catastrophe; he indicates both that there are ways out, that they seem feasible, and that there is much hard work to do. There is one particular feature to his cautiously worded optimism: the geography of sunshine correlates well with the geography of hunger, indicating that an economy more based on sunshine (and less on fossil fuels such as oil) might be in favor of our present world's underprivileged. In fact, the solar energy ratio is between two to one and three to one in favor of the tropics; and only in a very limited part of the poor countries of today does the need for water exceed the objective water supply. A world more based on solar energy and photosynthesis would require initial investment—and this is of course where the snag is located: certain forces in the Third World might see that as their chance of retaining the upper hand, not giving too much of a play to factors that might upset that "balance."

Biological Paths to Self-Reliance is an important book. It brings together old material in new ways, it pushes our thinking in these fields further, and it makes it very clear to the reader that to push further in action, in practice, research will not be sufficient; there is also a political job to do. This book, which combines pedagogy sufficient to make even the second law of thermodynamics comprehensible with highly technical elaborations, is in itself a part of that job, and no doubt the product of a self-reliant mind. May it serve to stimulate many others as it has stimulated,

JOHAN GALTUNG
Geneva

Preface

The International Federation of Institutes for Advanced Study (IFIAS, Ulriksdals Slott, Solna, Sweden) commissioned me in 1975 to undertake a small study, ''Photosynthesis: Its optimal use and impact in less-developed countries,'' as part of its Enzyme Engineering project. As originally intended by Unesco (the instigators and original supporters), this report was merely to summarize the potential of microbiology and enzyme engineering for solar energy conversion in developing countries. Interest from Swedish organizations and increased familiarity with the material have broadened the concept of ''developing countries'' to include the industrial world, while retaining an emphasis on the Third World.

A greater expansion of the study's scope has also resulted from the influence of several IFIAS workshops on both technical and developmental problems, plus visits to a wide range of research institutes, particularly the Energy Studies Unit of the University of Strathclyde (Glasgow, Scotland). After the study began, it became clear that biological solar energy technology could not be confined within the normal limit of a scientific discipline, as it was inseparable from its larger role in both technical and social development. Biological solar energy conversion cannot be separated from energy's role in ecology, ecology's role in agriculture, nor agriculture's role in population distribution and development.

To deal adequately with both the developmental and technological aspects of biosolar energy conversion, the two have been artificially separated. Part One, The Potential Role in Development of Biological Solar Energy Conversion, has therefore been written as a free-standing entity for those less interested in scientific details. It relates development conceptually to its energetic and ecological antecedents and limits. Chapters 6 and 7 of this first part briefly review the technology of Part Two. Part Two, The Technical Basis of Biological Solar Energy Conversion, deals exclusively with the scientific aspects, but does not require of the reader an extensive scientific education. The absence of a detailed description of physical solar energy devices, many of which would play a central role in the integrated solar systems described, is regrettably necessary due to lack of space; a good US National Academy of Sciences report is referenced instead (NAS, 1976a).

One function of this book is to provide the basis for a further "Self-Reliant Development" IFIAS program which will continue in collaboration with UNEP (the United Nations Environmental Program), Unesco (the United Nations Educational, Scientific and Cultural Organization), and ICRO (the International Cell Research Organization). More generally, this book is intended for three groups of readers: first, those scientists whose biological research can be related to solar energy conversion and who in Part One can begin to appreciate the broader context of their potential technological contributions. Second, those increasing numbers of officials responsible for technical development who in Part One can become acquainted with the limitations, potential and full implications of solar energy development; Part Two can for this group serve as reference material. The third group are those students who wish to dedicate their studies to Man's long-term survival and to appreciate the equally important, nontechnical aspects of their efforts. For them, Part One provides a context for their enthusiasm and Part Two the scientific and technical principles with abundant references for further reading.

It is the author's sincere intention to convey the excitement of social breadth with an honest foundation of scientific detail. Where the breadth has left detail wanting, it is hoped that the references will help to fill the void.

RUSSELL E. ANDERSON

Acknowledgments

Seldom is one given both the opportunity and the resources to write a book according to one's own taste. While this was not the original intention, the International Federation of Institutes for Advanced Study (IFIAS, Ulriksdals Slott, Solna, Sweden) has, in fact, provided just that and more. For this opportunity, and for the support, encouragement and, above all, patience received, I am indebted to IFIAS and its Director, Dr. Sam Nilsson.

Professor Carl-Göran Hedén (Department of Microbiological Engineering, Karolinska Institute, Stockholm), a man of many hats, has led for the past five years the IFIAS project on enzyme engineering and instigated this particular study as a part of that project. In addition to my appreciation for his official role, I am no less grateful to Göran for providing working space in his department, his personal generosity as a landlord, and as a tolerant fount of inspiration and encouragement.

Financial support during the two years of preparation has had a varied history: The initiating support for this study was provided by Unesco's Division of Scientific Research and Higher Education, where Dr. A.C.J. Burgers kindly allowed a generous reading of the contract and the resulting expansion into book form. This expansion was supported by Swedish organizations, both private and public: First, the Swedish Board for Technical Development provided a grant, after which the private Salén Foundation supported the study and a mid-course workshop. In the final months, support was provided by the new National Swedish Board for Energy Resource Development. Without this generous and above all liberal support, this study could not have been completed.

While the technical content is heavily referenced, much of what stands between the lines has resulted from innumerable scientists who have hospitably received me at their institutes or shared their thoughts at conferences and meetings. Those most deserving special mention for text contribution and criticisms are Drs. David Anderson, Rolf Eriksson, Tony

Gloyne, Alistair Neilson, Buck Nelson, and Lynn Williams. Special artwork was contributed by Rolf Eriksson and Inger Kühn.

Two influences from outside the biophysical sciences deserve special mention. Dr. Malcolm Slesser and his staff at the Energy Studies Unit at the University of Strathclyde (Glasgow, Scotland) extended their hybrid hospitality and imparted some of their knowledge of Energy Analysis during my one-month stay.

A number of persons have aided this natural scientist's fledgling steps into the dark forest of the social sciences, but Professor Johan Galtung was particularly instrumental for the theory of self-reliance and social development. I am, in addition, grateful for his contributing the book's Foreword which spotlights the politics of self-reliance, leaving the technical foundation to stand for itself.

Finally, a special word of gratitude to my co-worker, Barbara Adams. There is no doubt but that without her steady, substantial contribution to the structure and language of the text, plus her penchant for detail, it would not have reached this form. For her often inspired involvement in this book's preparation, I offer my inadequate thanks.

Contents

Biological Paths to Self-Reliance
A Guide to Biological Solar Energy Conversion

PART 1

POTENTIAL ROLE OF BIOLOGICAL SOLAR CONVERSION IN DEVELOPMENT

1

The Question
of Development

This book will try to show that self-reliance can be achieved based on biological solar-energy conversion. What makes biological systems and solar energy uniquely conducive to this form of development will become clearer later; but, at the outset, the goal—self-reliance—must be understood. Self-reliance does *not* mean isolation, nor is it equivalent to self-sufficiency. Self-reliance is development which stimulates the ability to satisfy basic needs locally; the *capacity* for self-sufficiency, but not self-sufficiency itself. Self-reliance represents a new balance, not a new absolute.

Self-reliance is both the goal of the proposed development plan and a consequence of the decentralized development required to realize solar energy's full potential. As a goal, it responds to the greater consideration now being paid to the social and technical limits of centralized industrialization and the fragmentation of life's basic processes to a sea of specialists. The primitive family farm provided one example of a well-integrated socio-economic production unit, but few would propose a return to the levels of nutrition, hygiene and suffering normally associated with that more restrictive concept of self-sufficiency. A modern reintegration of life processes must not only assure secure technical survival at a high level, it must also reunite Man with his traditional environment on a scale which both allows and demands greater participation.

Self-reliance is the concept which most accurately unites the need for a local reintegration, yet it has the flexibility to avoid isolationism and exploit the advantages of trade. It seeks "not to avoid interaction but to interact according to the criterion of self-reliance" (Galtung, 1978).

Basic needs, however, encompass little more than food, water, and energy, and in this sense self-reliance as an ideological goal is identical

3

with the technical result of decentralized development. Decentralized development (discussed later in this chapter) can also have ideological antecedents, but scientific necessity is more convincing: decentralized development is the logical result of optimizing the potential of this world's only free, renewable energy resource—solar energy.

Volumes have been written on development theory, and this book purports to be neither unique nor definitive. It is written to highlight both the ecological constraints within which Man must plan his future and the energy costs we incur for deviating from Nature's own path. Nature's path does not recognize Man's uniqueness as He himself does. Energy will be emphasized throughout, not as much in technical detail but more as a broad unifying concept. In the following chapter, Chapter 2, the physical concept of energy will be developed to unite the many diffusely-related everyday experiences of energy. Then Chapter 3 will discuss energy accounting, the economics of energy use. These energy tools will be applied first to the general problem of biological survival (ecology) and then specifically to Man and agriculture. With this conceptual framework, the unique potential of biological solar-energy conversion systems for Man's long-term survival can be understood along with their limitations and ecological constraints.

All energy forms have their limitations and solar energy is no exception, despite what some proponents claim. Sunlight is a variable, diffuse energy form whose use, when converted biologically, demands an understanding of Nature's laws. What might be viewed as limitations can be positively redefined as early warnings of abuse and used as guides to optimal development. Decentralization, characteristic of biological systems, shrinks the size of nutrient/waste cycles to a more local level and thus speeds awareness of all the effects—good and bad—of Man's efforts. Such feedback mechanisms would, for example, have limited birth rates at a more manageable population level had not oil reserves temporarily compensated for the excess by slowing down our awareness. This delay complicates adaptation after the optimum has been passed.

Both parts of this book emphasize the biological side of what later will be seen to be a more general integrated system of both biological and physical components. Achieving a new balanced view in a world that equates biological processes with the primitive past and regards physico-chemical industrialization as axiomatic to future modernization, requires this compensatory overemphasis of the biological world's potential and its neglected limits. A new balance must be one that assures long-term survival, a balance that requires a new blend of the traditional with a careful selection from the most sophisticated biological and industrial (physical) worlds. Just as the deification of science has lulled people into ultimate faith in the as-yet unknown "technological fixes" (the hope for a cancer cure and a new "unlimited" energy source, the wishing away of an excessive population or starvation), biology should not become the technological fix of the future. Technology can solve many problems, but its successes have raised expectations to irrational limits. Likewise exploitation, a word with both positive and negative connotations, is possible within limits, but the constraints must be understood by all for the responsibility of restraint to be shared by all.

BIOLOGICAL SOLAR-ENERGY USE

Integrated biological solar-energy conversion for decentralized development is proposed here as a means for achieving a high standard of life in both presently developed and developing countries. Developed, or as some say, overdeveloped countries hold themselves in an aloof position of envy during development discussions, but no development plan can seriously be proposed which does not both recognize the present world inequities between these two crude categories and strive for the same goal for each. The inequities eliminate any plans which call for identical paths to that common goal. For the Third World to follow the industrial world's path would always leave them one step behind, and is probably not possible.

It is the intention of this book to establish the technical potential, feasibility and steps needed for self-reliant development. The potential is based on:

1. *solar energy,* the only energy resource which is free, continually renewable, and already established within the ecological equilibrium;
2. *biological organisms,* their products or constituent parts (e.g., enzymes) due to their—
   a. self-construction of both "factory and product" from simple, continually recyclable building blocks—water and air—according to highly refined and carefully stored genetic information;
   b. efficiency, specificity, and complexity of biological processes which are incomparably more sophisticated than even the most advanced industrial processes;
   c. self-regulation, self-repair, and reusability of all which is beyond repair;
   d. mild temperature and pressure requirements;

    e. amenability to small-scale operation;
    f. mutability to desirable abnormal functions under controlled conditions.
3. *nonrenewable resources,* invested in processes or structures of catalytic importance for the long-term utilization of renewable resources;
4. *integration,* or the combination of complementary processes to give a result greater than the sum of the individual results. This includes integration—
    a. of biological elements to simulate the efficiencies of complex ecological food chains, yet specifically aimed—within ecological limits—at supporting Man;
    b. of biological elements with those physical processes consistent with and complementary to biological solar-energy conversion;
    c. of education based on the unity of basic life processes: food, fuel, and fertilizer production; water and waste treatment and reuse; health protection;
    d. of networks for communication, production and service functions.

This rather theoretical description will become clearer in the following chapters, but at this stage one can simply recall that plants, for example, capture and store solar energy in the forms of cellulose, carbohydrates (starch), protein, and oils. Sunlight is an energy flow, which, like a waterfall, must be stored as a fuel (e.g., food) to later be burned or eaten. Plants have a complex series of processes which can store this energy flow during the daytime for use at night in much the same manner as animals do through food digestion. Each of these innumerable steps is made possible by biological catalysts, enzymes, which are manufactured by plants as well as by animals. In contrast to chemical catalysts, which are often made from rare earth metals and require high temperatures and pressures for their use, this multitude of enzymes is made from little more than air and water according to a most intricate genetic design. Even more incredible than the design and construction of these enzymes is their specificity; each of several thousand enzymes has a highly specific function. Many of these have built-in regulatory functions which prevent inefficient or excessive enzyme production, or once produced, unnecessary use. This efficiency is but one important lesson from these highly integrated organisms.

Biological solar-energy conversion is thus in its simplest form (which is incomparably more complex than any industrial process) plant or tree growth, but it also includes processes that convert plant or tree matter to some other form (e.g., yeast fermentation of grains to alcohol), or the use of any organism's by-products or subcomponents (such as the enzymes mentioned above). Traditional farming and food processing is the first stage of biological solar-energy conversion. Full exploitation, though, is an open-ended development which encompasses genetics, microbiology and enzymology for food, fuel, fertilizer, and several other specific industrial syntheses.

Integration, used in the context of biological energy conversion, implies the tight interweaving of complementary processes to achieve maximum efficiency. Integration occurs to varying extents in all industrial processes, but integration in a well-tuned complex ecosystem has evolved to an incomparable extent to assure survival. This is not to ignore the occasional need for apparent or short-term inefficiency due to the unique value of a particular result, such as sperm or egg production. The cost must be accepted for any such inefficiency, and the ability to meet it is presupposed. Inefficiency in industry is compensated for with oil energy and may be eliminated by high fuel costs, but inefficiency in Nature usually ends in extinction.

In the biological world, integration occurs first on the molecular level, where the cells' energy storage, release and use reactions are closely coupled to avoid excess and to maximize by-product use. The occurrence of this same integration in complex food chains in Nature will be discussed in Part I, Chapter 4.

In the following discussion, biological and physical processes will be linked together in this manner and on a scale which allows full waste utilization: plant and animal waste processing to recover energy, fertilizer, and pure water, waste gas use for plant growth or combustion, and waste by-product heat for low-grade heat uses. Technically, integration results in increased efficiency, but at the same time integration enables process uses which are inconceivable out of context; algae, for instance, cannot be grown unless wastes are readily available. Scale is, of course, an ancillary factor which determines if plant, animal or process "waste" is indeed waste. If a "waste" requires more energy for its collection and use than the energy extractable from it, then it is indeed a waste. A change in scale can redefine waste as a valuable by-product.

Integration and scale also interact on a social or psychological level. The integration of one's life functions with society can, on a local scale, offer a security that on a larger scale is read as the "cog in the wheel" syndrome. The multilevel question of scale and its role in integration will be discussed in Part I, Chapter 8, but here it is of interest to consider how, at the development level, scale merges with the concepts of centralization and decentralization.

DECENTRALIZED DEVELOPMENT

Decentralization is an unspecific concept which has meaning only relative to its opposite—centralization—and is based on two subvariables, scale and degree. Consider the family as an example. A family is an organizational structure smaller than a nation but whose degree of power centralization may or may not exceed that of a nation. In addition to these relatively independent descriptive variables, there are also properties which are concentrated by centralization; examples include population, power, money, resources, industry, etc. A moment's thought reveals that the interdependency of several of the above examples originates not from inadequate factorization of the term centralization, but from its inherent nonreducibility: centralization and decentralization cannot be discussed purely from a single point of view or in terms of a single factor.

Industry provides the most familiar example. Centralized industry is as axiomatic to modern thinking (though I maintain not to *future* thinking) as economy of scale. This judgment has, however, resulted from the optimization of a single variable, namely economic profit, which is determined by the available resources and desired product(s). The resource base was reckoned as a purely economic function, i.e., as one resource became scarce its cost became less competitive relative to another, but its supply or that of the energy used was assumed to be unlimited. In this text, resource limitations will appear repeatedly as a general problem, but a detailed discussion is best found in other literature. Instead, full use of this example is being made to illustrate that centralization has noneconomic and even nontechnical elements, only some of which can even potentially be recalculated in economic terms.

Population distribution is one obvious consequence of centralized industry. A concentrated labor force supports centralized industry, and the resulting urbanization sets in motion a chain of secondary effects, such as agriculture, transportation, waste collection and treatment, and all support industry. To these technical and economic factors must be added the centralization of social support functions, such as education, health, entertainment, and politico-economic power. Optimization of economic reward is not without its Faustian backside: concentration of ecological perturbations (thermal, air, water, and noise pollution), loss of a sense of personal importance and freedom, a host of specifically urban social problems, and, most importantly, the subordination of the individual to the increasingly superordinate power invested in a few individuals, institutions, and the system inertia itself.

The ideological justness of the political ramifications from centralization is of less interest—or has less place here—than the realization that they occur, and that causal lines can be drawn from choice of technology to nontechnical factors, such as social order, political power, etc. But choice of technology does not occur in isolation. The potential for the choice rests in turn upon the elements it requires: energy and material resources. Here lies the point of division for biological and physico-chemical development:

	Physio-chemical	*Biological*
Energy Source:	diffuse sunlight	local concentrations of oil, coal, hydroelectric, and nuclear power
Material Resources:	air (N, CO_2), water, and a few trace minerals	mines with concentrated mineral reserves: iron ore, gold, bauxite, coal, oil, etc.

Ease and controllability resulting from the concentration of resources has permitted industrial development in the Western world, most commonly from resources in countries leading a "biological" life. Biological development uses an energy resource which is inherently non-conducive to control, capture or destruction. These properties result from solar energy being a diffuse energy *flow* (rather than *stored* energy), whose rate can neither be increased nor decreased and can only be concentrated to a limited degree. Solar energy is, along with air, the most decentralized resource: geographical variations in annual solar energy do occur, but do not exceed a factor of 2 to 3. Water is only relatively decentralized and uncontrollable; wars have been fought for it and civilizations have centralized around particularly good water sources. While it may ultimately prove to be a limiting resource, relative to oil or mineral deposits, water is, however, a decentralized resource.

Solar energy, and its biological storage in green plants, is by nature decentralized. Its exploitation, in turn, both requires and results in a *relatively* decentralized social organization. Its relative uncontrollability offers less opportunity for concentration of political or economic power and hence has not been the method of previous development. To the extent modern agriculture is energized by controlled resources (see Part I, Chapter 5), such as oil for machinery or fertilizer, its dependence is lost to the power center in a stereotype centerperiphery power relation. Such is the position of developing countries whose biological industry—farming—has via cash cropping invited exploitation by those controlling markets and resources; self-reliance has been traded for the periphery's subservience to the power center. To the extent that solar energy is judged in the context of a social-technical organization optimized for oil, it will remain an impotent alternative.

These arguments seem to equate biological solar-energy development with national suicide. One might falsely conclude that only through Western-style industrialization can the periphery—the less-developed countries—hope to compete and have living standards comparable to the center—the industrialized countries, or that without Western-style industrial development a nation can only choose between subservience or isolation.

Both conclusions are false, and the hope for an alternative lies in the nature of the limits to development. As Mahatma Gandhi is said to have asked: If Britain with a population of 30 million had to exploit half the world to give its people a high standard of living, how many worlds would India need to exploit to give its population the same standard? In all probability, the poor countries of the world will never have the luxury of wasting resources with the abandon of the industrial world. While this injustice is irreparable, its certainty requires the validity of only one of the following limits to growth:

1. Stored fossil fuel will be consumed until the reserves in the ground have no net energy value, i.e., the energy costs to locate, mine, or drill and process the fuels exceed their gross energy value (see Part I, Chapter 3). At the present production rates, no oil will remain in 30 years, while if the world consumed oil at the present US rate, the reserves would be exhausted in about seven years. *Gross* coal reserves are said to meet world fuel needs at US rates for a few centuries, but *net* fuel reserves based on this coal are believed by some to be a mere fraction of that (Odum and Odum 1976); the ecological effects of strip mining coal can only partially be counted in economic or energy terms, but the profound climatic changes resulting from coal combustion may prove to be an earlier technical limit (see point 3 below).

2. Nuclear breeder or fusion reactors, if developed, seem to cause a series of environmental, security and social risks of catastrophic magnitude. Since 1949 the US has misplaced 1490 kg plutonium, sufficient for the construction of several hundred atomic bombs. This loss in a single country indicates the potential for accidental or planned losses as the number of nuclear-fuel-reprocessing countries increases and plutonium stockpiles double every three to four years (Barnaby 1977). Accident or sabotage free storage of radioactive wastes in ever-increasing quantities lasting for thousands of years is a technological fix whose safety requirements can never match the magnitude of the potential catastrophe.

3. Climate changes, local or global:
   a. Increased heat production: The present world's population, if it consumed energy at the US rate, could increase the average earth temperature by 0.03°C; the population at year 2035 (16 billion), consuming at conservative projected US rates, would heat the earth 1.2°C. Temperature is only the first-order change, to be followed by altered rain, wind, and ultimately flooding from melted ice caps. Recall that temperature changes since the ice age are about 4°C in Europe and nearly unchanged in tropical regions (Bolin 1978, Schneider and Mesirow 1976).
   b. Decreased heat loss: The above effects are true for any energy form, but fossil-fuel combustion increases the atmospheric carbon-dioxide concentration which, in turn, slows the earth's heat loss to the atmosphere and increases its air temperature, the so-called greenhouse effect. A special advisory panel to the US Academy of Sciences concluded "the primary limiting factor on energy production from fossil fuels over the next few centuries may prove to be the climatic effects of the release of carbon dioxide" (NAS 1976). Normally the seas and forests store much of the produced carbon dioxide, but the ocean is reaching its limit and forests are being cut down. Coal could increase the carbon dioxide (CO_2) concentrations four to eightfold, while only a twofold increase would increase global temperatures 2°C with the greatest effect at the poles. Pure heating effects mentioned above are additional to and exacerbated by this greenhouse effect.
4. Population growth, if unabated in the face of limited resources, will lead to social, environmental and ultimately food limitations; half the world's population will be in urban centers by the year 2000 (Schumacher 1976), many of these centers will be less-developed but none will be less vulnerable to power loss than New York City in July, 1977.
5. Mineral reserves will be depleted. Natural lead and zinc reserves will run out by the year 2000 while other minerals in short supply will be asbestos, fluorine, gold, mercury, silver, tin, and tungsten (Carter, Leontief, and Petri 1976). Not only the cost but the energy required for use of marginal and diluted resources will increase.

Time is crucial, but hope can be based on pessimism, on the limitations of the present type of growth, and a need to use the remaining resources for the mutual survival of all peoples. But in each of the above cases, delay raises the costs and decreases the choices: developing a solar energy potential requires material investment, but delay diminishes the energy and resource capital left to invest. Delay in reducing energy consumption raises expectations, decreases reserves, and worsens the period of adaptation. Delays in environmental protection require much costlier, and not necessarily successful, technological fixes. Delays in stemming population growth distract attention from future planning. The immediacy of emergency feeding takes priority until even those efforts reach their maximum and the inevitable starvation begins, but at a higher rate. For each starving child saved today, three will starve by the year 2000 (Mesarovic and Pestel 1974). Is there a moral choice? Heroism may increase the suffering and leave no resources for long-term survival.

The prophets of gloom's heyday peaked with the two Club of Rome reports on growth and resource limits, but this initial period of alarm has now settled onto a bit more optimistic plateau. Whether the increased optimism—or decreased pessimism—results from numbed fatigue or from careful analysis remains to be seen. The most optimistic prediction is a stable world population of less than 15 billion by the year 2075 (Carter, Leontief, and Petri 1976).

One can hope that the world's finiteness will shock its residents into realizing the most optimistic predictions, but none can doubt that several of the pessimists' fears must be dealt with in time:

- Resources and the environment will command much more preventive attention than in the past, and waste as a concept will become obsolete. Energy and resource accounting (see Part I, Chapter 3) will stand side by side with economics.
- Oil and coal as net energy resources will eventually run out.
- Options decrease with delay and bring costs which far outweigh those of immediate prevention.
- Present development increases catastrophe risk: the absence of recent historical catastrophes or the successful heroic technological fixes provide no insurance for the future, rather they lull society into trust and complacency.
- Urban centers have an optimal size, exceeded in a number of cases due to unbalanced centralized development.

Each of the above has an obvious technical side plus an enormous range of hidden social ramifications. Clearly, population and catastrophic risk have social and mental health aspects and are related to urbanization and optimal human community size. Community size can be technically optimized, but does Man, the social animal, have the same optimum? Technical matters are amenable to rational discourse, but the social side of the same coin is generally avoided in futures studies or leads to discomfited thrashings and handwaving, or nebulous mutterings of "it can't go on like this." This trend will not end with this book as alternatives lie beyond this author's realm of competence, but the realized causal link between technology and its social side-effects leads to guidelines for technological development and its social goals:

- equalization of opportunity to develop, achieve and participate;
- technology whose labor requirements are conducive to Man's needs for individual significance and satisfaction;
- technology whose components are man-sized, convivial tools (Illich 1973);
- technology integrated with Man's basic needs;
- technologies "developed endogenously from the local context . . . " (Reddy 1976);
- "technologies which increase rather than diminish the possibility and effectiveness of social participation and control" (Reddy 1976);
- "technologies which facilitate the devaluation of power . . ." (Reddy 1976), or which inherently cannot lead to the amassing of power;
- technology whose educational support complements Man's understanding of himself;
- technology which permits diversity;
- technology which involves small risks to Man and his environment.

These criteria, a few among many for "appropriate technology," support the energetic, ecological, and technical discussions which follow, all of which are consistent with the concept of self-reliance through development of biological solar-energy conversion. While self-reliance can be a chosen goal, the emphasis here is placed on its being the result of the optimization of solar energy use through the decentralized development it requires. The practical achievement of this development is a question of personal and political will, not technology. The remainder of this discussion assumes there is a will to survive and instead deals with the fundamental biophysical constraints and technical potentials.

2

Understanding Energy

Considering its central role in survival and development, energy is a much maligned and poorly understood concept. It is one of those all encompassing concepts whose profound significance is rarely fully grasped. To say that without energy nothing lives is an insufficiently exclusive definition. To remind ourselves that the absence of energy—unachievable, but theoretically a system at rest at a temperature of absolute zero ($-273°C$) is characterized by a collapsed atomic structure, more likely gives a headache than understanding. Energy is what gives structure to this book, or to a banana, or to a tractor. Energy can describe structure, position and motion and as such is one of Nature's most unifying concepts.

Energy exists in innumerable different forms, many of which are seemingly unrelated, but all can be defined as a measure of the ability to do work. Work in a colloquial sense refers mostly to human effort, but while such examples are useful for understanding, the word's physical meaning is far broader. As a scientific word, work is performed wherever an object is moved some distance against a resisting force. For example, pushing against a brick wall produces no work in the strict physical sense if the wall does not move; energy *is* converted to waste heat (note the sweat), but recall that the definition referred to the potential or ability to do work; the wall *might* have moved. Further examples of work being done against the force of gravity include any actions which lift an object. Pumping water into a water tower, for example, results from work being done against the force of gravity and represents stored energy—work can be done when the water is released. A useful rule of thumb is that if conversion via any process can physically lift an object, then that which was converted was an energy source. Falling water can lift an object, as can dynamite, a man, light (shining on a photocell to drive a motor), steam (via a steam engine), any heat source (via a heat engine), or, as mentioned above, the humble banana.

ENERGY EQUIVALENCE

The banana is used not entirely as a joke but rather to emphasize the equivalence of all energy forms, including food. Food will only occasionally be discussed here as a substance different from any other energy source, as it is distinguished from other fuels only by its palatability and digestibility. Similarly, fuel is a word applied by tradition to those energy-rich materials which burn readily in common motors or burners. However, the distinctions among foods, fuels, and any energy-containing material is one of semantics alone. One can, of course, eat the banana for metabolic energy and then go out and carry water to the top of a hill (i.e., do work and store energy in the elevated quantity of water), or one can:

- burn the banana to drive a steam engine which pumps water;
- convert the banana's energy microbiologically (e.g., via yeast or bacteria) to alcohol or natural gas (methane), which in turn can be burned in a motor to pump water or generate electricity, or
- cover one's body or home with banana leaves as insulation to decrease the energy required to heat the body metabolically.

Food is therefore directly convertible to a fuel, but is the reverse true? In principle any energy form can also be used as food, but just as petroleum must first be converted to electricity before an electric motor will run, so must falling water, heat, electricity, oil, dynamite, or light be converted before it suits our bodies' machinery. Each of these energy forms is converted to food more or less directly via modern agriculture (especially light most directly), but other mechanical and chemical alternatives exist for converting fuels to food. Our conversion devices would be considerably more clumsy than Nature's, but the fundamental processes in automobile storage batteries, photosynthetic plants and human beings differ only in detail, not in principle.

All the above examples simply demonstrate that energy is the unifying concept for all seemingly disparate food, fuel, and other stored or moving energy forms. The lack of unity in popular concepts about energy forms is exacerbated by the variety of energy units used:

heat energy	- British thermal units, calories;
food energy	- calories (but means kilocalories!);
electrical energy	- kilowatt-hours;
mechanical energy	- horsepower-hours, foot-pounds;
atomic energy	- TNT equivalents;
electromagnetic energy	- ergs.

In addition to these units, the popular and scientific literature is plagued with tons (long, short, and metric) of coal or oil equivalents, barrels of oil, and food crop outputs measured in bushels, bags, pecks, plus various weights which can be dry, wet or somewhere in between.

Avoidance of this nightmare of measurement units requires the sacrifice of tradition and occasional short-range convenience in exchange for the International System of Units. For energy this unit is the joule (defined as one watt-second, 0.239 calorie, or 0.000949 Btu) whether one speaks of oil, gasoline, electricity, wheat, a water reservoir, or cow manure. A list of conversions is found at the end of this volume, and common equivalents will be given occasionally for convenience.

ENERGY CONSERVATION

The preceding discussion of equivalence of energy forms is a restatement of the familiar law of conservation of energy. This law can be stated several ways:

- energy can be neither created nor destroyed;
- energy flowing into a system must either be stored there or flow out at an equal rate;
- energy forms are interconvertible, albeit with losses.

Energy forms cannot be declared equivalent or interconvertible without some qualification regarding the quality of various fuels. Intuitively the law of energy conservation is incomplete or irrelevant. That energy can be neither created nor destroyed is of little interest to the motorist with an empty gas tank or the recipient of a bill for electricity consumption. Some quality factor has decreased, where quality means:

- a fuel's ability to do work;
- the maximum temperature a fuel can create;
- the flexibility with which a fuel form can be used.

Fuel quality is most easily understood in relation to heat energy. Mechanical or electrical energy can heat water with 100% efficiency, but that heated water cannot be reconverted to as much mechanical or electrical energy. An irreversible process has occurred in spite of the fact that no energy is lost—some energy unavoidably remains as low grade heat. The efficiency of converting heat to mechanical energy is less than 100% and depends on the temperature of that heat energy. More work can be removed from a body of water at 100°C than from a larger body which contains the same amount of energy at 30°C; heat below atmospheric temperature is therefore a waste in our environment. Mechanical and electrical energy are the energy forms having the highest quality, followed by chemical energy (fuels) and finally heat energy; the latter's quality increases with its temperature.

Intuition leads us to a more useful principle (the second law of thermodynamics) which describes a quality that *is* consumed. Two useful statements of this principle are:

- in any conversion process, energy loses some of its ability to do work via degradation to heat, or again,
- work (mechanical energy) can be completely converted to heat, but heat can only partially be converted to work.

Stated in this form, the law tells us that when mechanical or fuel energy is converted to heat, something—but not energy—is lost. The practical implication is that the number of conversion steps should be minimized, and when possible direct conversions to heat should be avoided.

Visualization of this concept is easier with Figures I.2.1 and I.2.2 in which the quantity of energy form, given by the area of the box(es), is unchanged (first law), but its position on a vertical "quality" axis decreases (Holmgren 1976). Interconversions of high-quality energy, electricity and mechanical motion, for example, occur with little loss, but steam-type power generator plants are inefficient because the conversion involves an intermediate step (steam, i.e., heat), which is lower quality than either the starting fuel (oil or coal) or product (electricity). These two

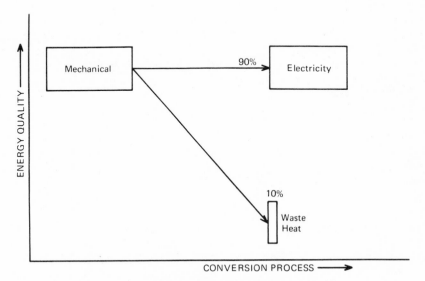

Figure I.2.1. Conversion of mechanical energy to electricity.

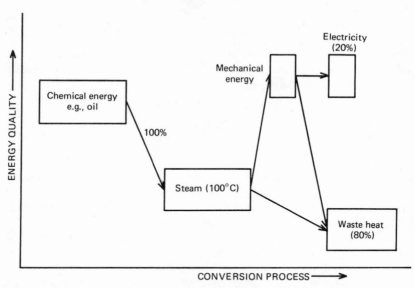

Figure I.2.2. Conversion of chemical energy (fuel) to electricity via an intermediate heat stage.

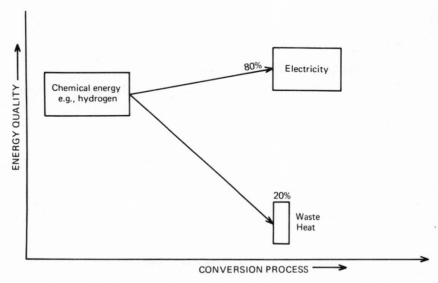

Figure I.2.3. Conversion of chemical energy (fuel) to electricity via a fuel cell.

examples are shown in Figures I.2.1 and I.2.2, while in Figure I.2.3 is shown the fuel cell (see page 294), a device which efficiently converts chemical energy to electricity by eliminating the heat step.

Photosynthesis, the process of particular interest here, converts solar energy to chemical energy, sugars. Sunlight is a form of heat energy whose quality is extremely high because of the sun's high temperature (6000°C). Its quality is not as high as, for instance, electricity, and as a result it can at best be converted with about 78% efficiency (Porter and Archer 1976). That absolute limit is lowered by the inefficiencies of the solar device used. The theoretical limit for photosynthesis, for example, is about 13% and in practice is at best a few percent (see page 185). The low efficiency in this case is not demended by the fundamental laws of physics as the examples above, but results due to a variety of reasons that are discussed later.

ENTROPY

Although energy is never produced nor consumed, quality is lost as discussed above. Quality, is, however, a rather vague concept and is better replaced by terms like order or organization. A fuel's quality is then determined by its degree of order; mechanical energy being completely ordered while heat is disordered.

A more correct, scientific statement of this "second law" builds on the word entropy, or disorder. Higher entropy then is associated with greater *dis*order, and this second law can be scientifically stated as:

- The disorder (entropy) of a closed system always increases during any irreversible process.

Freer translations, but ones which are at least as profound and far-reaching, might restate this law as:

- There is no such thing as a free lunch; or
- The world is going to hell.

Each of these statements again refers to a natural tendency toward degradation, disorder, or more technically, toward increased entropy. The concept is introduced here because of the need to understand the natural disordering or the increased entropy which occurs as a result of energy conversions; fuel quality, as drawn in Figures I.2.1, I.2.2, and I.2.3, is related to both high energy content and low entropy.

Entropy and energy are fundamental concepts whose principles affect not only fuel conversions, but as the lay translations above imply, include disordering in a universal sense. The tendency toward increased disorder (entropy) is fundamental to each of the following:

- energy lost to heat when electricity is transmitted;
- energy lost to heat when a ball is thrown and eventually stops, or when water is dropped and lies in a puddle;
- energy lost when an ink drop diffuses in a body of water;
- tendency for all ordered structures to fall (a standing tree, sand castles, and governments), for a garden to be taken over by weeds, and for the death of all organisms;
- tendency for information to become lost as random noise through genetic mutations, the rotting of a book or newspaper, or making a long series of Xerox copies and noting the last to be worse than the first.

To restore the order lost in each of the above, order (negative entropy) must be provided from another source. For instance, negative entropy is provided by the conversion of a high-quality fuel to low-quality heat. The degradation of fuel energy can restore order.

Entropy, the tendency toward disorder, is often called the arrow of time and as such is the force against which all energies and efforts are directed. The word "tendency" is essential as it is not the *accomplishment* of disorder, but rather the necessary expenditure of energy to *prevent* its accomplishment.

The success or failure of this struggle takes on special significance in the section on ecology (Part I, Chapter 4) in which the survival of biological organisms is seen to depend on their ability to combat this disordering tendency by maximizing the amount of energy available and the efficiency of its use. Survival of one individual species will be seen to depend not just on its maximizing its own energy budget, but also on the survivability of the system of which it is a part.

The species of particular interest, Man, is also a part of a system and stands not alone in his fight against disorder. Reorganization of the earth's well-balanced processes for the unique purpose of his comfort constitutes an extra-imposed ordering. Extra energy is required to support altered ecosystems (see Part I, Chapter 4), to support governments, armies, cities, and transport. High-quality fossil fuels have previously been abundantly available, but scarcity requires rationed (and *rational*) use. Evolution and survival will be the final cutting edge of efficiency, but a new accounting system, energy analysis, gives us a chance to predict the results and to alter our paths accordingly.

3

Energy Analysis
Accounting for Energy Use

Societies of all ages have had the problem of allocating scarce goods and services. Consider the problems. How does one compare:

- human labor (life)—of ultimate value to the individual but of widely varying economic "use" to the society; nonrenewable to the individual, but a renewable resource for the society; in limiting quantities in some areas, while in excess in others;
- natural resources—some have fixed quantity (e.g., land) but divided unequally among the people, while some have fixed quantity but are dispersed via use from high to extremely low concentration densities (mineral resources); some are essentially nonrenewable (e.g., petroleum), some are constantly renewed at a somewhat variable rate (e.g., biomass), and some flow constantly at a fixed rate (e.g., solar energy);
- produced goods—including food produced in distant lands, locally, or provided to dense urban centers, and manufactured goods which in turn require both natural and human resources;
- services—requiring primarily human resources.

Whereas barter systems functioned quite efficiently for local exchanges of renewable resources, modern economics has expanded the complexity of exchanges possible without adjusting to an age when the use of limited resources reaches critical limits. Economics remain limited to reducing all productive costs or exchange values to little more than labor and capital. Resources have been treated as other scarce goods whose use is determined by purely economic factors, which in turn reflects their relative abundance plus the capital and labor required to supply the resource. Nonrenewable resources have received similar treatment according to the theory of substitution, by which absolute scarcity (depletion of nonrenewable reserves) and relative scarcity (excessive labor and capital costs required) are treated identically. As the cost increases for one resource, a substitute waiting in the wings gradually becomes economically competitive and eventually preferable. By such mechanisms has plastic replaced many applications of steel or aluminum. Similarly, energy resources have been assumed to be replaceable with adequate investments of labor and capital. Historical precedent has previously been adequate justification for this substitution theory whereby oil replaced coal and now coal may again replace oil. But unfortunately for economics and Man, the implicit assumption of unlimited energy—whatever its form—is unjustified. Poorer reserves of natural resources, mineral or energy, cannot be utilized without the investment of not only labor and capital but also energy. One energy form can be substituted for another or used to produce another form, but energy as a resource class cannot be replaced. This realization has forced the recent birth of energy analysis, not as a replacement for economics but rather as a supplement.

Energy analysis has been officially defined as "the study of the energy, free energy, or availability of any other thermodynamic quantity sequestered in the provision of goods or services" (IFIAS 1975). Alternatively expressed by one of the leaders in the field, "If economics is the science of scarcity and substitution, energy analysis is no more than the application of economic theory to the one commodity in ultimate limitation on Earth—thermodynamic potential (energy) . . . Economics set up a boundary around the system of interest; in energy analysis the system boundary is the whole planet" (Slesser 1975). The close link with economics implied in the above statements is neither accidental nor as farfetched as perhaps thought. What will become clear is that economics will converge with the conclusions of energy analysis as time goes on. That is, as the price of energy increases, its contribution to the economic cost of goods and services will more nearly reflect the true energy costs and will agree with the energy-analysis evaluations. Economics and energy analysis for the present are not at all exclusory, but rather complement one another. Neither energy analysis, nor economics can provide an adequate criterion for comparing "apples and oranges," human life and natural resources.

In slightly less formal terms, the above energy-analysis definition can be restated as the accounting for all energy inputs and outputs involved in a given process, both direct and indirect. Figure I.3.1, for example, shows Odum and Odum's (1976) interpretation of the energy costs for transportation; fuel is the major energy expense as intuition predicts, but indeed it constitutes only 57% of the total energy consumed (or 73% of the total, if energy supports for the driver are excluded). Direct fuel costs are the only energy costs one typically considers for either home or industrial processes, for example, the electricity, oil, or coal that is directly consumed on site. When the criteria for energy inputs are wid-

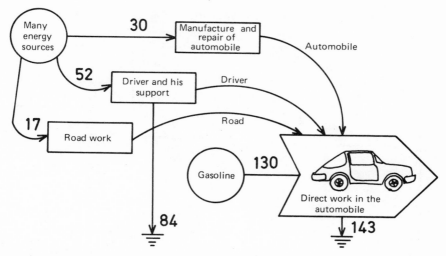

Figure I.3.1. Direct and indirect energy costs (in gigajoules) for a car driven ca. 16,000 kilometers per year (Odum and Odum 1976).

ened, however, there is seemingly no end: the fuel itself had a delivery cost (the coal truck's fuel) and a manufacturing cost (electrical power station inefficiency, coal-mining costs), in addition to the energy requirements for manufacturing the machinery involved in fuel production and delivery. The same is, of course, true for each of the materials or machineries used in the manufacture process. The processes that make these machines will have both direct and indirect energy costs, and so on. As summarized in Figure I.3.2, the analysis can be extended backwards until the headaches involved exceed the information gained. Level 1 for any process includes only the direct energy or fuel costs for that process or transport directly attributable to it. Level 2 includes the energy costs for producing, transferring and storing the direct energy costs of Level 1, plus the direct energy costs for providing and transporting process materials for Level 1. Level 3 includes energy acquisition costs for Level 2's direct energy costs, plus the direct energy of transport and processes involved in the final stage of capital-equipment manufacture. Level 4 accounts for the capital equipment producing the process capital equipment and for producing the material inputs, while higher levels can extend further in an obvious way.

Analysis can regress as far as one likes, but the increased uncertainty and effort rarely justify the little information gained. First level analysis is rarely sufficient and in fact often is more misleading than informative. Energy analysis limited to this level would, for example, mistakenly

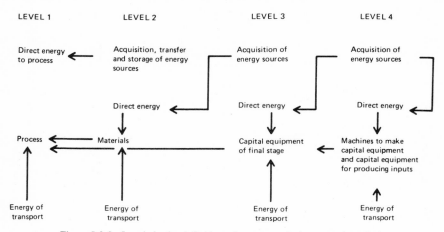

Figure I.3.2. Levels in the definition of energy analysis system boundaries.

exaggerate the number of coal mines or oil wells which are net energy yielders. Energy analysis to the second level is often the feasible limit and frequently accounts for 90-95% of the energy accounted for in fourth level analysis. In cases when the regression is limited to only the second level, care can easily be taken to at least estimate the errors incurred by not doing the complete analysis.

ENERGY AND AGRICULTURE

Modern maize agriculture is an example of central importance in this report and points out the value, complexity, and limitations of energy analysis. Anyone questioned as to the energy requirements for maize growth might first say that only sunlight is required, but farmers were reminded in the early 1970s of the dependence on fossil fuels for all tractor and other field operations. Another moment's thought would include the electric bill. All other inputs would probably be called materials, capital or labor and hence would be lacking in any energy component. Specific figures for the errors involved will be discussed later, but a brief glance at Figure I.3.3 shows that, in this case, analysis to Level 1, i.e., those obvious energy costs normally considered, includes less than half the real total. The most easily observed error is for electrical energy: electrical energy requires at least three times as much energy (fossil fuels, water power, etc.) input to the electrical generator process than is ever recovered. At higher regression levels, the energy costs to acquire the initial fuel must also be included.

Foremost among the neglected energy components of modern agricul-

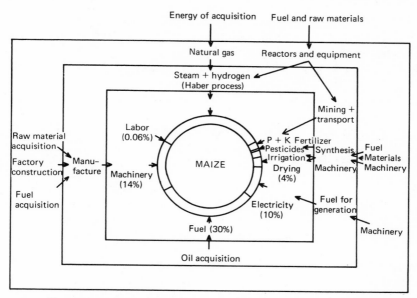

Figure I.3.3. Level boundaries for the energy analysis of maize production.

ture are the nitrogenous fertilizers (as urea or ammonia) essential for all protein production. Ammonia production will reappear several times in later discussions, but for the moment one must merely note that the final step of the so-called Haber process, the catalytic conversion of hydrogen and nitrogen gases to ammonia, requires high temperature and pressure steam. Steam, of course, requires energy, but the reaction material, hydrogen gas, is also a high-quality fuel. If treated as a material, the pressure steam. Steam, of course, requires energy, but the reaction material, hydrogen gas, is also a high-quality fuel. If treated as a material, the hydrogen gas energy content would not be included until Level 3 and ceasing the regression at Level 2 would conceal a 15% error due to the hydrogen alone.

Several other elements demand that the regression for agricultural analysis extend at least to Level 3. All of the capital equipment on the farm, and of course the capital equipment required for the former's manufacture, contain large energy components. These costs have been estimated and constitute nearly 15% of the maize production energy costs (Leach 1976).

Ultimately the analysis extends back to the material source which most commonly is the mine or quarry. Where exactly this contribution is placed depends upon its role in the overall production scheme. Phosphates for fertilizer use will, of course, play a much more direct role in maize production costs than, for example, the mining of iron ore for the irrigation pumps. The estimate for its mining must also be more accurate as, consequently, must the costs for the capital equipment employed in the mine. Each production step must be known in detail before its contribution can be neglected or roughly approximated.

ECONOMICS AND ENERGY ANALYSIS

A useful simple calculation provides an approximate conversion factor from economic to energetic capital costs. This figure is simply the ratio of a country's economic and energy fluxes: For the US in 1973, $1.4-trillion ($10^{12}$) circulated per year within the economy; this is *not* the same as the cash in the economic system as money circulates about four times per year (Odum and Odum 1976). Energy consumption during the same year was 35-million-billion calories (3.5×10^{16} cal or 1.46×10^{20} J). The conversion factor, the ratio of these two figures is simply 0.1 GJ (10^8 J or 25,000 cal) per dollar. That is each dollar of the GNP required on average 0.1 GJ. Such a factor allows a rough calculation of, for example, the per capita energy consumption as a function of annual income. Similarly, one can calculate the energy costs for agricultural inputs. The usefulness and dangers of such a conversion are apparent from the example of nitrogenous fertilizers (urea) before and after the oil price increase as shown in Figure I.3.4. This example demonstrates that if a process is highly energy

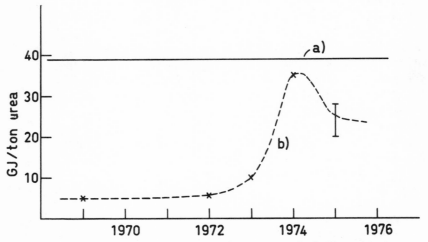

Figure I.3.4. True (a) and apparent (b) energy costs for fertilizer production.

intensive, e.g., nitrogen fixation for urea production, the low cost of the energy component will obscure the true energy cost if the simple energy per dollar conversion is used. An easy and significant improvement can be made by calculating a conversion factor for each of several industrial sectors which will—to a first approximation—compensate for variations in energy intensity.

PARAMETERS FOR ENERGY ANALYSIS

Maize production illustrates several other important principles of energy analysis, but now the system sketched in Figure I.3.3 can be simplified to an input-output black box as in Figure I.3.5. The output in this case is clearly maize, but of course one must carefully state if the output is the total biomass produced (i.e., grain, cobs, and stalks) or just grain; dry weights or energy contents of the respective portions are the only useful figures in this analysis, although economic transactions use volumes and wet weights. What inputs are assumed depends on the regression level reached in the energy analysis process. Useful analysis requires clear statement of the assumptions made, but all production and supply energies should be included as energy inputs. The following factors can then be calculated:

- Process Energy Requirement (PER) includes the fuel energy for all processes involved, including those material production costs specified by analysis level. Solar energy is by convention *not* included as it is a flow which cannot be altered;
- Gross Energy Requirement (GER) includes the process energy (PER) and the energy released if the inputs were combusted. For the maize example, inclusion of regression Level 3 would make GER and PER nearly equivalent. At Level 2, however, the energy content of am-

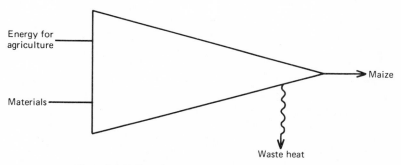

Figure I.3.5. Input/output process for maize production.

monia, for example, would be added to PER to calculate GER. GER is the essential factor for determining what energy resources must be devoted to, for example, maize production;
• Net Energy Requirement (NER) is the gross energy minus the heats of combustion of the products, in this case maize. NER must be negative for any fuel crop, i.e., an agricultural crop grown to produce energy.

This terminology is particularly useful for parameterizing the energy analysis of fuel conversion processes. For example, if maize is considered as a typical process input for microbiological conversion to ethanol, then its input-output black box is as shown in Figure I.3.6. In this case:

PER = all energy costs for the conversion process including capital equipment;
GER = PER plus the combustion energy of maize, i.e., the energy released if maize is used directly as a fuel;
NER = GER minus the combustion energy of ethanol.

All processes cost energy, and hence NER will almost always be positive; a negative NER is meaningful only for photosynthetic processes. As sketched above, conversion of maize to ethanol is a net energy consuming process, but if the system boundaries were redefined to describe an "energy plantation" whose output was ethanol, then NER must be negative. The system boundary required to evaluate whether an energy plantation is a net yielder of energy can be sketched in simplified input-output form as in Figure I.3.7. Note, however, that only solar energy can be justifiably excluded from an energy analysis and serves to remind us that

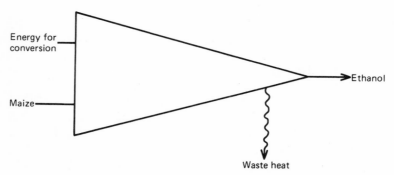

Figure I.3.6. Input/output process for maize-to-ethanol conversion.

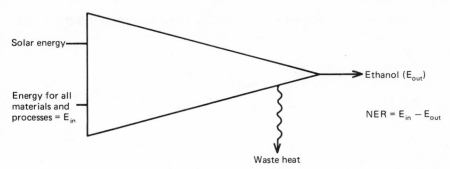

Figure I.3.7. Input/output process for solar energy to ethanol.

solar processes are our *only* energy "income," that is an energy flow which does not deplete our fossil fuel "capital."

Each of the terms used above cannot only have the average values as defined, but their marginal values can also be defined. The marginal process energy requirement is, for example, the extra process energy required to increase the output. Such a value can be useful for determining the optimal energy use or a type of efficiency figure.

SPECIAL CONSIDERATIONS

A word should be added for the purists about which thermodynamic quantity is used for energy analysis. Several alternatives are possible (Gibbs free energy, enthalpy, Helmholtz free energy, or internal energy) but the differences, except for special cases, are small (less than 10%). Most commonly used is the gross heat of combustion (i.e., the enthalpy of organic material combustion to gaseous CO_2 and water at 0°C, IFIAS 1975).

The illustrations discussed here have consistently been oversimplified, but only one of the many neglected elements will be mentioned. Each step of Figure I.3.3 produced a single product; the farm produced only maize, the factories produced only ammonia, machinery, or fuels. In fact, however, each process produces wastes which are nonnegligible in many cases. Little work has been done in these areas, but pure water has an energy value relative to polluted water. Although the former has not heat of combustion, the latter may be combustible as when the Ohio River in the US caught fire; but this is normally not viewed as a credit. The relative thermodynamic energies of pure and polluted water (24 KJ/liter for the change from sweet to sea water) are less intuitive than the energy re-

Table I.3.1   Efficiencies of UK Energy Supply Industries
(Leach 1976).

| | EFFICIENCY | | |
Energy Industry	1963 (%)	1968 (%)	1971–72 (%)
Coal	95.5	96.0	95.5
Coke	75.5	84.7	88.0
Gas	64.7	71.9	81.1
Oil	80.8	88.2	89.6
Electricity	22.0	23.85	25.2

quirement to purify polluted water. Various techniques are available—from distillation to modern sewage treatment plants—but each requires energy. Similar remarks apply to air pollution and point out the need to look at the total process output. In other cases, valuable by-products are available, or "wastes" whose energy contents justify reclassification and exploitation as products. Sewage and farm wastes are two such examples to be discussed later. Clever and careful integration on the proper scale is necessary if "wastes" are to be redefined as useful by-products.

Energy analysis can be extended further to express various waste and efficiency factors in direct analogy to corresponding economic terms. Such efficiency calculations will eliminate several processes often considered to be net energy producers. Most of these calculations are both controversial and preliminary, but among the net energy losers according to Odum and Odum (1976) are: some photoelectric cells, some commercial solar water heaters, many offshore oil or deep coal reserves, probably much of the shale-oil reserves, and the modern food production chain. Even nuclear energy is a positive but nominal producer as is the production of electricity from coal. Good coal mines, on the other hand, yield 30 units of energy for each unit invested while the yields from oil and gas wells are even far higher. Present fuel production efficiencies are shown in Table I.3.1, but as the easily tapped resources diminish, more energy will be required for digging or drilling deep until finally the zero net energy return (i.e., PER is greater than the product's combustion energy) level is crossed. At this point the reserves are of no value, and no increase in economic price or demand will make their exploitation possible. Similar arguments apply to all mineral reserves. Energy controls the accessibility of energy and other materials, and careful analysis will determine the long-term supply.

4

Ecology and Nature's Energy Balance

If disorder increases relentlessly on the average within any system (e.g., one's home, the world, the universe), how did the world as we know it now evolve? How did the highly ordered biological world—and our special interest Man—become formed from the "prebiological soup?" To answer these questions and to insure the future of Man's survival requires first the humility to forego our anthropocentric world view and then to perceive the profound implications of the highly interdependent terms—energy, entropy (disorder), and ecology.

SOLAR ENERGY

"On the average" is the key phrase in the first sentence above, one of the many restatements of the second law of thermodynamics discussed in Part I, Chapter 2. If the earth were isolated, we would indeed be on our way "downhill," and all energy stored in various chemical, biological and physical forms would be degraded to heat and randomized disorder. The heat could not be radiated if our earth were assumed isolated, and the temperature would rise to some new equilibrium. As heat can only do work if a temperature difference between two reservoirs exists, non-random, work-producing processes could not occur.

Our assumption is, of course, false; the earth is *not* an isolated system, but rather is an element of the universe which has one single input—solar energy. Only the solar energy input saves the earth from disorder. Figure I.4.1 conveys this useful "space ship" earth image in which a small fraction of the sun's radiation impinges the earth's surface. A large proportion of this irradiation is reflected by the earth's atmosphere, but all energy absorbed must eventually be re-irradiated if the earth's surface temperature is not to increase. If this loss did not occur, the earth would be like a perfectly insulated electric oven with the element on. Just as with the oven, however, the earth is not perfectly insulated and does—fortunately—re-irradiate energy at the same rate as it is absorbed.

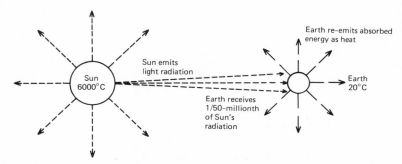

Figure I.4.1. "Space Ship Earth's" absorption of sunlight (6000°C, high-quality energy) and its total re-emission as waste heat (20°C, low-quality energy).

If no net energy is gained by the earth, how are disordering processes counteracted? How do we live? The earth's beautiful answer is sketched in Figure I.4.2: high-quality (low entropy) energy from the sun is used in an ordering or entropy-lowering machine (photosynthesis in green plants) which produces an ordered structure plus waste heat. The collapse of the ordered structure (plant matter breakdown via metabolism, decomposition or burning) also results in an amount of waste heat, which, when summed, equals the solar-energy input. The sun in effect provides only negative entropy to the earth; the energy inputs and outputs are equal, but the entropy of the emitted heat is far greater than that of the received sunlight. It is this difference which prevents disorder and allows net work to be done on earth.

An alternative view of this conceptual problem looks at the earth as a heat engine for which the relative entropy of the incoming and outgoing radiation is characterized by the temperatures of the sun and earth, respectively. In Figure I.4.1 are the two so-called black body radiators: the sun at a temperature of 6000°C, while the earth is at a temperature of ca. 20°C. All the solar energy absorbed is emitted as heat with only a change in quality or entropy characterized by this difference in temperature. As mentioned earlier, it is this temperature difference which powers the heat engine earth with a maximum 78% efficiency. A similar everyday example of a heat engine is the steam engine: high quality electricity or gas produces high-temperature steam which is reduced to low-quality (i.e., low temperature but equal quantity) heat as it produces its mechanical work. The ordering on earth is similarly possible because the solar energy is degraded to heat; the entropy of the solar system increased but on the average remains unchanged on earth. The qualifier "on the average" conceals most of the

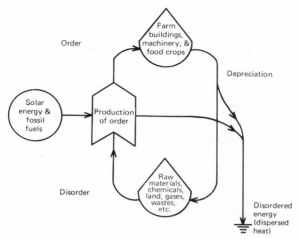

Figure I.4.2. Solar energy driven order-disorder cycle on a farm (Odum and Odum 1976).

processes on which life depends. Ordering and disordering may balance out "on the average," but all physical and biological structures represent small local imbalances in these fluxes.

Abstract concepts of order, entropy and structures are useful for organizing diverse observations, but understanding rests on the most trivial concrete examples. Consider for instance a maize seed planted in soil containing randomly dispersed minerals and water molecules. High-quality order (energy) exists in the seed, but it can only provide the genetic information for guiding the organization of dispersed nutrients for the desired maize plant's construction. Such organization from randomness requires a high-quality energy source whose flow is modulated by the seed's even more highly ordered informational energy. As shown in Figure I.4.2, solar energy powers this so-called photosynthetic process in which randomly distributed nutrient molecules (carbon dioxide, water, nitrogen, potassium, phosphorus, sulfur, iron, magnesium, molybdenum, etc.) are organized into highly ordered, high-energy macromolecules (sugars, cellulose, DNA, enzymes, etc.) which in turn constitute the macrostructure we know as maize.

Maize continually requires energy inputs to grow first and then to struggle against death and decay of its cellular mass, i.e., the fight against the disordering processes. Repair of the organism at some point (called senescence) can no longer compete against the decay processes, and the organism dies: growth, repair, senescence, and death are true not only of biological organisms but also of cars, houses and many bureaucratic organizations.

FOOD CHAINS

If the whole world consisted only of maize, all the nutrients would become bound up in maize plants, living, dying and dead: growing maize plants cannot metabolize dead ones, but instead require nutrients in a free form. Other plants and animals exist which not only can free bound nutrients for future maize plants, but which can also exploit the solar energy stored in maize. Man is one such organism that can, of course, eat maize seeds directly, but the bulk of the plant is a fuel form undigestible by Man. The return of free nutrients from the whole plant to the soil is impossible without the mediation of several more organisms. Maize can, of course, be used as silage and fed to meat and milk producing animals, but even the excrement from these animals and Man contains minerals bound in energy-rich form. As seen in Figure I.4.3, organic wastes are exploited as a rich energy source for microorganisms which constitute the principal mechanism freeing minerals for the sole net energy storage process on earth, the photosynthetic storage of solar energy in green plants like maize.

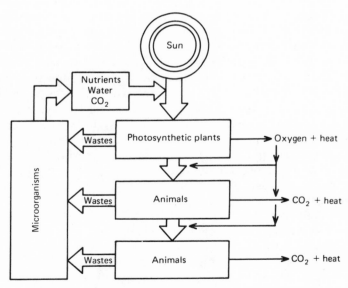

Figure I.4.3. Nutrient and energy cycles in nature.

Each organism in Figure I.4.3 is a coequal member of energy and nutrient cycles which originate with the photosynthetic process. This process "pumps up" low-quality materials to a more energy-rich form, or stated another way, organizes randomly dispersed nutrients into highly ordered forms. Each of the at least five following stages in the food chain metabolizes that organism's material which is suitable to its nutritional needs; each stage degrades large amounts of the feed material to wastes and upgrades some fraction as its own cell mass. Both the organism's own cell mass (e.g., a cow) and its wastes (manure) in turn provide freed material for two additional classes of organisms: the degraded portion is metabolized to its basic constituents by microorganisms, while the small portion upgraded to milk or meat can of course be metabolized by microorganisms, but it normally is exploited by higher animals. "Higher" here not only implies those organisms more nearly resembling egotistical Man, but those with similarly complex, restricted dietary requirements. Amino acids (protein) and vitamins are among Man's most restrictive nutritional needs.

Food chain development is an ecological subtopic in itself, but the essential point is that no organism stands alone. The survival of each organism depends not only on its ability to compete but on its compatibility with the intact cycle and its other member organisms. One can argue fairly convincingly that the removal of Man would constitute no great system loss, compared for example with the loss of the microbiological portion, but for the purposes of this anthropocentric discussion, that solution is uninteresting. Man, incidentally, is the only organism which stands outside the food chain except on the input end; all others consume from not only one or more sources, but are consumed themselves.

ECOLOGICAL NICHE

Each organism has a niche or optimal place in the food chain, and each niche is occupied by some organism(s). The competition for each niche is determined in part by the Darwinian survival theory, but Lotka extended the concept to the system level: that system survives which maximizes the availability and use efficiency of power from all sources (Odum 1971). This optimization typically involves:

- storage of high-quality energy for winter or drought periods;
- feedback loops to stimulate increased inputs, including reseeding;
- total recycling of all nutrient materials;
- built-in control mechanisms to increase adaptability and stability;
- exchanges with other systems to provide special and otherwise unavailable energy needs.

Survival of the whole system determines the optimum, not merely the survival of individual species.

Lotka's principles distinguish between two interesting cases. When energy is in excess, the successful system will be characterized by maximized power consumption, rapid growth and probably waste. "Growth or perish" concepts are exemplified in new plant growth in disturbed areas or modern civilizations with newly tapped fossil-fuel reserves.

Growth periods are, however, always transitional, so the more interesting case is for development trends when energy is limiting. An undisturbed ecosystem, one such example, will develop or mature to a level of complexity in which energy-use efficiency is maximized and a steady or "climax" state is achieved in which no *net* production occurs, i.e., the total bio(logical)-mass does not change. In such a mature system, the solar energy stored by photosynthesis in green plants and trees is exactly balanced by the plant's metabolic requirements (its fight against disorder), plant decomposition (plants which lost the fight), and some animal grazing (for their fight against disorder). Net storage occurs during the day to survive the darkness and during the summer to survive the winter, but on the average no net storage occurs. This climax status, characterized by great complexity, redundancy and diversity in the food chain, represents the ecosystem with the greatest energy capture and use efficiency.

In mature systems, high-quality energy is used with great care to control and construct ordered structures which collect and process lower-quality energy; energy is invested in, for example, plant structures, not consumed as in growth systems. Wasted high-quality energy, such as organic material, is virtually nonexistent. It is interesting to note that such mature systems do not contribute to our oil and coal reserves. The occurrence of fossil fuels requires of course fossils of nondecomposed organisms, but nondecomposed organisms represent an under-exploited energy resource, an unfilled niche characteristic of an immature, inefficient ecosystem. This sole exception is pictured in Figure I.4.4 where the constant flow of solar energy is likened to water flowing past a water wheel doing work, i.e., processes occurring on earth. The flow continues—energy radiating from earth equals that received from the sun—except for a small leak which has been stored in a reservoir over the millenia. These stored fossil fuels, resulting from a slight leak in an otherwise perfect food chain, are now being used at a rate comparable to the rate of solar energy.

A corollary of the conclusions about the energy-efficient mature ecosystem is that enforced simplification, for example the cultivation of single crops where hundreds would normally grow, decreases the sys-

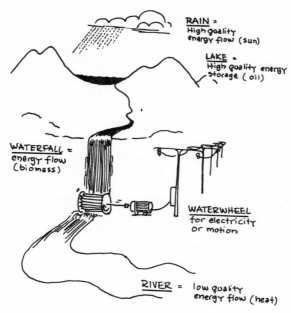

RAIN = High quality energy flow (sun)

LAKE = High quality energy storage (oil)

WATERFALL = energy flow (biomass)

WATERWHEEL for electricity or motion

RIVER = low quality energy flow (heat)

Figure I.4.4. Stored and flowing energy analogy: high-quality flowing energy (rain = sunlight) is stored (lake reservoir = plants, oil) and converted at will to a new energy flow (waterfall, waterwheel = mechanical energy, combustion) with ultimate loss as low-grade flowing waste (low-energy river = waste heat).

tem's energy efficiency, its stability in times of stress, and its susceptibility to epidemics.

Low human population densities engaged in hunting and food gathering may not disturb a mature ecosystem, but crop cultivation and animal herding are ecological simplifications which require increased energy inputs. Preindustrial energy inputs for agriculture simply required human and animal labor fed by the farm products it cultivated. The degree of simplification had a built-in natural limit—solar energy—now exceeded by modern Man. Modern agriculture, supporting urban development, represents a massive ecological simplification requiring equally massive nonsolar energy.

Ecology and energy efficiency, however, are not moral imperatives themselves. The energy costs of ecological simplifications result simply from the imposition of a specific goal on Nature: the preferential support of animal Man's existence. Like the high human population, Man's unique position in the world is an assumption basic to any discussion of development, but ecological principles must be respected to help sustain *all* organisms on which each specie depends. Energy-requiring "technological fixes" are possible within limits, but both energy and technology may soon be left wanting.

Modern agriculture should be compared with natural mature ecosystems only on the basis of long-term survival value. As modern energy inputs are not sustainable on a world-wide basis, ecology should serve as a guide to a new alternative.

RECYCLING

A discussion of recycling following upon a discussion of cycles is not as redundant as perhaps thought. Recycling focuses on the weak link in the biological cycle: by what process are the low-energy wastes/nutrients, the empty can of the food chain, made available to the crucial energy storage, plant growth step? The definition of "waste" vis-à-vis nutrient will be shown to depend principally on its location and density.

Components of this recycling problem are more easily understood by analogy: Consider the time when a single Coca-Cola factory manufactured and filled Coke bottles. Coke was distributed from this factory perhaps only to one city or local region, and empty bottles were returned directly to the bottler. But why were they returned? They had of course a "value" both to the bottler and to the returnee whose efforts were reimbursed by an amount considerably less than the cost of making a new bottle. As each consumer may be returning as few as one or two bottles, this initial concentration process is possible *only* because the trip to the store was to be made anyway. The money was an incentive to remember rather than compensation for work done. Consider however a Coke-hired door-to-door collector—a clearly untenable alternative.

Consider now an expansion to a region 100 km in radius. As before the first concentration stage began at the very low home Coke bottle density; each trip to the collection station (food shop) carried an average of perhaps two to four bottles. A local Coke distributor collected with his small truck perhaps a hundred bottles from each shop, and his accumulation was occasionally collected by one of a fleet of trucks sent out directly from the bottler. The extent of the distribution network could be extended by additional stages, but the total recycling costs must always lie below the bottle manufacturing cost.

Note now the problem of increased region size: not only does the distance from consumer to bottler increase, but the density, i.e., bottles per km², decreases as the distance *squared*. Distributorships required increase also approximately with the square of the distance. As both the incentive to the consumer and the manufacturing costs are fixed, the difference between the two represents the amount available for collection costs and hence limits the size of service region for each bottling factory. If the distance is exceeded, no money remains for the consumer and the recycling stops. Before this occurs, a new bottling center must be established.

This common-sense example becomes more interesting when one complements the cost analysis with energy considerations. Each economic cost has a corresponding energy cost:

- energy of Coke-bottle manufacture (fixed);
- energy of consumers' return (fixed);
- energy of collection (increases approximately as distance squared);
- energy of transport (increases approximately linearly with distance).

Peculiarities of oil economics virtually assure that energy analysis would require greater decentralization, i.e., smaller cycle dimensions, than cost analysis does. Note, however, that for either analysis, increased service-region size ultimately demands throwaway bottles or decentralized bottling facilities.

The obvious conclusion is that a good can become a waste when increased distance from the point of supply to need (transport costs) and/or decreased supply density (collection costs) make recycling costs (energetic or economic) greater than manufacturing costs.

The previous analogy is relevant to the overall question of decentralization, but has particular significance for food production. The most important building blocks for plant material (water and carbon dioxide) are so universal that recycling involves merely adding the "waste" to a general pool. In such cases recycling is no problem, and the energy for reorganizing the random particles comes from the sun.

Nutrients found in fertilizers (potassium, phosphorus and nitrogen) are, however, the basis of the Green Revolution and can potentially strangle any food self-reliance program. Nitrogen fertilizer is a special problem as plants make protein from nitrogen in an energy-*rich* or "fixed" form (i.e., as ammonia, nitrates, or urea). Nitrogen gas in the air is useless except to a few special organisms, while plant wastes or fertilizer run-off are an energy-rich resource.

The tightly interwoven energy and economic aspects of fertilizer production, pollution, and waste treatment appear repeatedly and point to the need for a radical change. Which can be deemed the most pressing: doubling of fertilizer costs with doubling of oil costs, the consequential strangling of the Green Revolution and loss of foreign currency, poisoning of ground waters due to fertilizer (nitrate) run-off, polluted (fertilizer run-off) waterways clogged with vegetation and later strangled of all life, or the increasing burden of urban waste treatment? Energy and capital technological fixes exist for the latter problems if present agricultural approaches must be continued, but economic and energy limitations will eventually put an end to the disposable Coke bottle. In Part I, Chapter 5 the magnitude of this problem is explored.

5

Hidden Energy Costs in Centralized Development

Primitive Man was one element of a mature ecological system in which nutrient cycles had dimensions of only a few kilometers and total daily per capita energy consumption could not exceed his daily food intake. He invested some of his own energy (labor) to gather food, whose energy content for him to survive had to be greater than his daily investment. This energy, or fraction of a day's labor, even for the Kung Bushmen in the harsh climate of the Kalahari desert in Africa, for example, is an astonishingly low two hours per person per day (Leach 1976). One can question the universal applicability of this culture where the per capita food-gathering area is greater than 10 km^2, but only such extremely low population densities permit (but do not assure) complex and efficient ecosystems. When population densities increase, mature ecosystems are no longer possible, and energy inputs must increase.

Agrarian Man perturbed his environment somewhat more by cultivating certain crops and animal herds. Such a self-sufficient "island" community is pictured in Figure I.5.1 together with the flow chart for the local nutrient cycle and energy flow. While this type of farm-community organization is normally limited to smaller villages, Paris during the second half of the nineteenth century supported its urban cropping system with the one-million tons of manure produced by the city transport system's horses; these 'marais'' of Paris produced 100,000 tons of high-quality crops and expanded the arable urban land by 6% per year (Stanhill 1977).

Considering again the smaller, self-sufficient and isolated community of the past, its nutrient cycle can be likened (see Part I, Chapter 4) to a wheel turned by the flow of energy from the sun to dissipated heat: dispersed (high entropy) nutrients are organized by this energy flow into ordered, high-quality energy forms (food and fiber) which in turn insure the survival of Man and other animals. Of particular importance in such a system, however, is that this stored high-quality energy is not just used

41

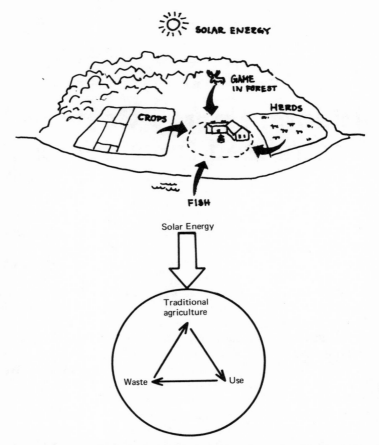

Figure I.5.1. Recycling traditional agricultural community (redrawn after Odum and Odum 1976).

for survival (against disorder in plants and animals), but a portion is also reinvested via a "positive feedback loop" of work—cultivating, watering, herding, harvesting, etc.—to reinforce the future energy flow. This investment as mentioned above can only be a fraction (less than one) of the flow, but the feedback loop can increase the fraction of the solar energy flow available to Man. Zero investment returns this fraction to the value characteristics of undisturbed, mature ecosystems which cannot support an excess of any specie, Man or otherwise. Draft animals are an important part of this invested capital (high-quality energy) whose recognized need for preservation is sometimes assured with religious sanctions,

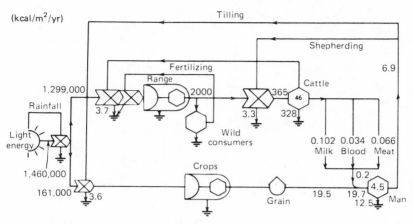

Figure I.5.2. Energy flow in an unsubsidized agricultural community (Dodo Tribe, Uganda; by permission of Odum 1971).

as in the Hindu tradition. The cow receives a portion of the energy flow (fodder) in return for its milk (to feed the population) and work; this feedback loop, tied to but separate from Man's, is explicitly drawn in the useful type diagram developed by Odum (Figure I.5.2).

MODERN MAN

Modern Man did not protect himself as well as his ancestors against his own greed and shortsightedness. His island farm is now as pictured in Figure I.5.3 where the solar flow is supplemented by a constant fossil-fuel input and food/nutrient cycles are extended the world over or more commonly broken.

The twentieth century, particularly since 1945, has witnessed a dramatic industrialization of agriculture. The major cause was an energy economy peculiar to this century which equated the cost of a barrel of oil to one hour of manual labor. As the barrel of oil has 4000 times as much energy as one man-hour, the replacement of Man by machine was a sound economic investment. This economically sound inefficiency was the shortsighted policy of an unsound "bank" whose capital (oil accumulated though the millenia) was expended rather than invested. Natural tendencies toward complex, efficient ecosystems were thus thwarted by energy inputs (oil) which subsidized urbanization and population growth. The subsidy may end, but the population imbalance remains.

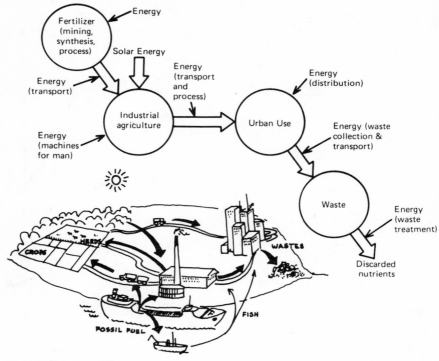

Figure I.5.3. Non-recycling industrial agriculture with energy (oil) subsidy and waste production (redrawn after Odum and Odum 1976).

ENERGY ANALYSIS IN AGRICULTURE

To facilitate comparisons among agricultural alternatives, the investment return of human labor is rephrased as the energy (food) output produced per man-hour labor. Energy-use efficiency is the unitless ratio of food-energy output to total (direct and indirect) energy input, inputs which include labor, fuels, and materials; incident solar energy is again not included for reasons discussed in Part I, Chapter 3.

Maize, the world's third most important cereal crop, is grown today at all levels of technology, from subsistence to almost industrial-type farms. Much of the energy analysis data (provided by Leach 1976, and Pimentel et al. 1973) are summarized in Table I.5.1 and Figure I.5.4 as an illustrative example. Note that among subsistence farmers the yields are invariably rather low and are unimproved by draft animals which only aid in doubling Man's hourly productivity. Major changes, however, result with

Table I.5.1 Energy Inputs and Outputs for Maize Production. (All unspecified units are GJ/ha-yr. Leach 1976)

	Guatemala	Mexico	Guatemala	Nigeria	Philippines	USA (1945)	USA (1970)
Labor:							
Human	1.13	.3	.56	.5	.24	.051	.018
Animal	—	1.6	2.5	.05	1.17	—	—
Machinery	.05	1.	1.	—	1.	1.9	4.34
Fuel	—	—	—	—	—	6.1	8.9
Fertilizer:							
Nitrogen	—	—	—	.9	.36	.6	10.
Phosphorus + Potassium	—	—	—	—	.02	.15	1.12
Irrigation	—	—	—	—	—	.4	.79
Insecticides	—	—	—	—	—	—	.1
Herbicides	—	—	—	—	—	—	.1
Drying	—	—	—	—	—	.04	1.24
Electricity	—	—	—	—	—	.3	3.2
Total Energy Inputs (GJ/hr-yr)	1.2	2.9	4.1	1.5	2.8	9.55	29.8
Energy Output (for crop energy content of 15.2 MJ/kg)	16.	14.2	16.	15.1	14.2	32.2	76.9
Energy Ratio (Output/Input, unitless)	13.6	4.9	4.	10.5	5.1	3.4	2.6
Labor Productivity (Output in MJ/man-hr)	11.	37.	23.	24.	48.	570.	3500.

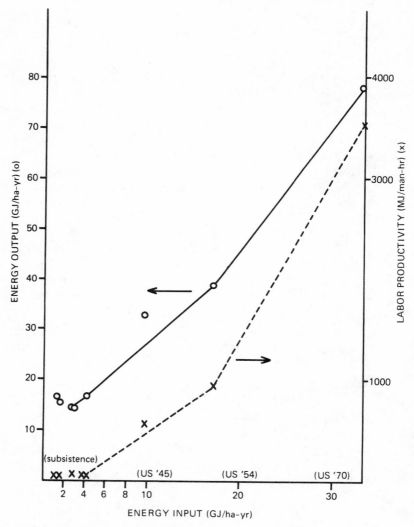

Figure I.5.4. The dependence of energy output and labor productivity on energy input for maize production (data from Leach 1976).

the employment of fossil fuels and modern machinery: a 30-fold increase in energy inputs and a fivefold increase in maize yield. Labor productivity (the energy in the maize produced for each hour of manual labor) is increased 200-to-300-fold with industrialization. As Man's metabolic requirements are fixed (except to increase waste, stored fat or relative meat consumption), this 200-fold change roughly translates to a 200-fold decrease in the time the population is involved in agriculture. The price paid for this labor "efficiency" in modern agriculture is of course the fossil-fuel subsidy expressed by the more than fivefold decrease in energy efficiency. A secondary subsidy is required, however: those individuals no longer occupied with agriculture migrate into urban centers and there incur energy costs for the collection, transport, and distribution of required foodstuffs. Now, outside of all natural food chains, urban Man's nutrient-rich wastes are no longer degraded and recycled, but rather require energy-intensive collection, transport, and disposal processes. A comparison of the two population distributions are shown in Figures I.5.1 and I. 5.3. The social effects of this population dislocation from the land cannot be dealt with here, but many of these effects also require an energy subsidy.

A closer look at the maize example in Table I.5.1 is even more revealing because it is based on an analysis which includes all energy costs including mining, production, handling, and transport, in addition to all direct fuel uses. With such data one can determine the magnitude and benefits of each energy subsidy. Oil subsidies to industrialized agriculture have:

1. increased energy inputs (fertilizer and irrigation) which expanded total agricultural production via increased areal yields and increased total arable land, i.e., use of marginal land;
2. replaced manual labor by fossil fuels;
3. increased fertilizer inputs to compensate for nutrients not recycled.

The left-hand scale of Figure I.5.4 (points marked "0") shows the linear relationship between energy inputs and areal energy yields without differentiating among the three points above. More careful analysis shows that one need not be restricted to this linear curve for increased agricultural output.

Point 1 above represents the pressure of population in relation to the availability of and competitive uses for arable land. As the land portion of this land-energy-population equation is relatively fixed, it is the best starting point. (Note: In fact arable land is threatened by urbanization, erosion, and encroaching desertification.) Present world land use is as follows:

11% (1.5×10^9 ha): arable and nearly all currently cultivated
22% (3×10^9 ha): pasture, range of meadows for livestock
30% (4.1×10^9 ha): forested
37% (5×10^9 ha): too dry, too cold or too steep for agriculture

With these 1.5-billion arable hectares, one can determine the energy costs and populations supportable by two agricultural alternatives: subsistence maize production and US maize production (see Table I.5.2).

The figures in Table I.5.2 are *farm yields* (losses and wastes past the farmgate are not included) and assume a vegetarian diet. While most of the world is basically vegetarian, 34% of the grain protein produced each year is fed to livestock with at best a 10% efficiency: 10-15 persons can live on the plant protein required to feed one person meat. In the US, producer of 20% of the world's grain, 77% is fed to animals.

The population potentially supportable by subsistence farming of the world's 1.5-billion hectares is not encouraging. The present 4-billion population will increase to 7 billion in the year 2000 and may reach as high as 16 billion in the year 2035, while with subsistence farming as presently practiced only 6 billion can be fed. World-wide agriculture on the US industrial scale could feed a 30-billion population, but even if technically possible, the *annual* consumption for *agriculture* alone would be 2% of the world's fossil fuel reserves—a vegetarian diet for 50 years if no competitive uses for oil existed. But is energy used in modern agriculture to increase yield?

Examination of points 2 and 3 (see page 47) reveals that only a mere fraction of the high-energy inputs are required for increased agricultural productivity; these inputs include only irrigation costs and less than a third of the fertilizer costs. The bulk of the energy inputs promote the replacement of manual labor by fossil fuels (plus machinery). The population displaced from the farm incurs energy costs due not only to broken nutrient cycles and the synthetic fertilizers required to replace them, but an additional food processing and distributing industry (the latter two are not included in the above calculations). A comparison of local cycling

Table I.5.2   Maize Production—A Summary of Two Alternatives.

	Energy Ratio (Energy Output/Input)	Yield (kg/ha-yr)	Supportable World Population*	% of World's Fossil Fuel Reserves Used per Year for Agriculture
Guatemala Maize (a good but typical subsistence farm)	13	1056	6.6 billion	0
US Maize	2.6	5060	30 billion	2

*(yield $\times$ 1.5 $\times$ 10^9 ha $\times$ 13 MJ/kg maize/{10 MJ/person-day $\times$ 365 days/yr})

and the appearance of extra energy inputs required in modern societies was shown in Figures I.5.1 and I.5.3.

Fertilizer is a critical example: For maize grown in the US (the figures will be nearly identical for all "industrially grown" cereal crops), an annual grain yield of five tons per hectare contains a half ton protein of which 16% is nitrogen. Therefore, each year about 80 kg of nitrogen in its fixed form or about 200 kg fertilizer (i.e., ammonia or nitrates) are removed from each hectare and shipped as plant protein to stockyards, large cities or are exported. At the point of use the human and animal wastes form a waste/nutrient *disposal* problem, i.e., the energy-rich fixed nitrogen wastes must be converted to inert nitrogen gas and waste heat. Careless disposal and the resulting prolific algal growth in waterways demonstrates the wastes' fertilizer value. Meanwhile, back on the farm, crops cannot grow on the fixed-nitrogen depleted soil unless synthetic nitrogenous fertilizer is supplied, half of which is lost to the ground water (only a few crops "fix" nitrogen biologically, legumes such as soya are particularly important). Today production of these synthetic nitrogen fertilizers requires petroleum and capital-intensive, high-temperature and pressure reactors. The fuel consumed directly represents about half the total energy cost of fertilizer production, the remainder is used for machinery production and process costs. Added together nitrogenous fertilizers cost 50 MJ/kg nitrogen and represent one third of the energy input for US maize production (Pimentel et al. 1975), not including the disposal costs. As complete recycling eliminates this energy input *and* the disposal costs, both are attributable to population dislocation, i.e., the replacement of manual labor by fossil fuels. The feasibility of fertilizer self-sufficiency through recycling is exemplified by the Chinese peasant who collects pig, buffalo, horse, and human excreta not only from his own farm but from roads, inns, and local drains. However, recent Chinese investments in chemical fertilizer plants indicate the need for a net increase in nitrogen cycled as some nutrient losses are inevitable, and the net amount of nutrients in the cycle must initially be brought up to an adequate level. Some synthetic fertilizers will probably always be needed, but the biological fixation of nitrogen from air has a potential which has yet to be fully exploited (see Part I, Chapter 6 and Part II, Chapters 1, 2, and 5).

The third and final energy element in the labor-fuel exchange is the direct cost of the 200-fold increase in the hourly productivity of labor. These costs (see Table I.5.1 and Figure I.5.4) represent 62% of the total agricultural energy budget and include fuel, machinery (energy of production, transportation, and maintenance), crop drying and electricity. This energy investment yields little increased productivity.

Energy inputs for the industrialized food chain do not end at harvest time, but constitute a mere 20% of the energy to place, for example, a plastic-wrapped loaf of bread on the store shelf; other contributions have been summarized by Leach (1976) in Figure I.5.5.

What are the net results from the industrialization of US maize production?

	Percentage of the Total Energy Inputs for Primary Agriculture
• 3–5-fold increase in yield due to	
• improved grain seeds	-0-
• improved agricultural husbandry	-0-
• irrigation	3
• net increase in available nutrients (*not* equal to the total fertilizer used, but only the 20% lost in a completely recycled system)	7
• 100–300-fold increase in labor efficiency due to fossil-fueled mechanization	60
• broken nutrient cycles accounting for approximately 80% of the total fertilizer input (energy for mining of potassium and phosphorus plus synthesis of ammonia)	30
• energy for the food processing and delivery industry	500
• home transport and preparation	125
• energy costs for disposal of the 2.5 kg daily waste generated per person	undetermined

As shown above, only 10% of the energy input for modern maize production increases the yield, 60% of the energy is used to "free" man from agricultural production, and 30% to industrially produce nutrients lost by the fuel-induced dislocation of Man from the farm. To this must be added later elements of the food cycle which increase fivefold the energy cost to put food in the shop. The fossil fuel to cause the primary event—replacement of Man by machinery—is thus a small portion of the energy required to support the secondary effects—the displacement of populations to urban centers.

The dilemma of modern industrialized agriculture can be inferred from Figures I.5.1 and I.5.3: urban Man left his ancestral farm and now participates in an oil-powered industrial society which is supported by oil-powered agriculture. As long as oil was abundant it could power technological fixes to strengthen the weak links in society's organization, which but increases populations and environmental hazards and in turn increases the oil bill. With the envisioned end to oil reserves, two alternatives remain:

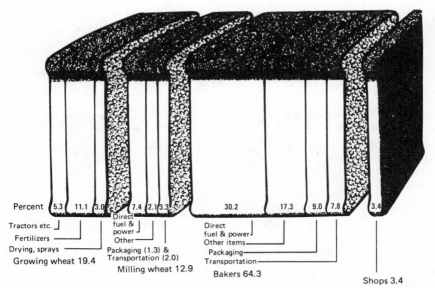

Figure I.5.5. Energy requirement for production and delivery of white bread loaf (reproduced from Gerald Leach *Energy and Food Production* (1976), courtesy of IPC Science and Technology Press, IPC House, 32 High Street, Guildford, Surrey GU1 3EW, England).

1. assume that the social organization spurred by oil *must* continue and therefore abundant energy *must* exist, or
2. invest existing reserves in solar energy systems which are then relatively self-sufficient and accept the reorganization required to realize long-term viability.

The former, "it must be, therefore it is," alternative may succeed for a time, but the next limitation—be it oil, coal, and uranium depletions, unsurmountable population densities and growth rates, or environmental catastrophes—may leave the earth with few or no options at all. The solar energy option requires an investment which cannot begin when the oil runs out. It requires careful planning, research, extensive development, and even social reorganization. It requires understanding of the previous chapters to optimize, to appreciate what can and cannot be expected, and to use optimally other complementary energy and material resources. Storage of solar energy by simple green plants has been optimized through eons of evolution, but agriculture as we know it today has only begun to tap its potential contribution.

6

Capture of Solar Energy Storage by Green Plants[1]

Green plants constitute the one and only gain in the earth's stored energy reserves and account for not only our daily food but also all the fossil fuels we consume today.

Green plants, including trees and algae, grow via a process called *photosynthesis; photo* because sunlight is the sole energy source and *synthesis* because this energy is used to construct complex, high-energy molecules from water and air. Farming, for example, can be called applied photosynthesis.

The major products of photosynthesis are sugar molecules (glucose, fructose, etc.) which are joined together in long chains (polymers or polysaccharides). Just as animals have a skeleton for structural strength and fat for energy storage, so plants have analogous materials, but both are made from sugar (see II.1, page 132).

- *cellulose* gives the plant its structural skeleton (stems, straw, branches, etc.) and constitutes about 50% of all plant material. Cellulose consists of long straight chains of up to 10,000 sugar molecules which are both sturdy and degradation resistant:
- *starch* (carbohydrate) is the plant's energy store and is, for example, found in all cereal grains. Sugar is also the sole constituent of starch, but the molecules are bonded together in a slightly different manner than in cellulose, and the chains are often branched.

Both cellulose and starch consist solely of sugar (glucose) linked together in chains, and both contain equal amounts of energy. A nearly trivial difference in the bond between molecules makes one digestible for Man and the other nondigestible. Plant matter that is both palatable and

---

[1]This chapter is a nontechnical summary of Part II's technical Chapters 1 and 2. More detailed discussions there are given in parenthesis footnote on first page of Chapter 6. Literature references are also given in Part II.

digestible is by tradition called food, that which is not (e.g., cellulose) is "waste." Cellulose is not indigestible by "misfortune" as structural integrity for plants has been achieved only by the evolution of these digestion-resistant materials. Furthermore, cellulose's label as a "waste" will be shown to be a misnomer.

Just as these complex sugar chains—starch and cellulose—contain only the single sugar, glucose, so glucose itself contains only carbon, hydrogen and oxygen, hence the unifying term *carbohydrate* and the chemical formula, $C_6H_{12}O_6$. (The term *hydrocarbon* for fossil fuels, also originating entirely from plant materials, is equally descriptive.) These elements, from freely recycled water and carbon dioxide (CO_2 in air), are reorganized by solar energy together with the release of oxygen to the air. When destroyed either by animal metabolism or combustion with oxygen, these molecules lose their high-energy organization and close the cycle by yielding water, carbon dioxide and energy (heat, muscular, chemical, or electrical energy). One can therefore summarize photosynthesis, and its loop closure via animal life, as shown in Figure I.6.1.

Figure I.6.1 emphasizes an important distinction between flowing and stored energy. Photosynthesis *stores* an uncontrollable energy *flow* (sunlight) as sugar, whose energy can be released at will to *flow* ultimately to low-grade heat; this store, while of limited size, is sufficient to maintain an alternative energy (food) flow to plant and animal life when solar energy does not flow, as at night and during the winter. The release of plant energy requires the oxygen evolved during the storage process, and, vice versa, the carbon dioxide and water released during metabolism or combustion is required for the next photosynthetic storage cycle. While these nutrients are commonly forgotten or taken for granted, controlling this cycle—particularly CO_2—is essential for full exploitation of the high efficiency systems which will be discussed later in this chapter. Increased carbon dioxide concentrations, for example, sharply increase plant productivity (see II.1, page 154), but while available in low concentration in

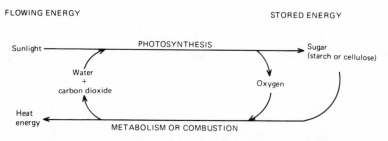

Figure I.6.1. Interconversion of flowing and stored solar energy via photosynthesis.

air (0.03%), carbon dioxide in higher concentrations is too expensive for general use. Smokestack gas represents a scandalous waste of this nutrient. While the gases are not lost of course, their dilution to atmospheric concentrations represents a needless loss of entropy and economic value. Accelerated algal growth will be seen to be one profitable reuse for concentrated exhaust carbon dioxide (see II.2, page 208).

While water and carbon dioxide constitute by far the bulk of the nutrients required for plant life, nutrient fertilizers provide relatively small amounts of other essential elements: potassium, phosphorus, and nitrogen (in its chemically combined or fixed form, ammonia, NH_3, see II.1, page 166). Nitrogen, discussed earlier in Chapters 3 and 5, is the nutrient required in greatest quantity after water and carbon dioxide and is about 1% of a plant's total dry weight. What really singles nitrogen out for special consideration is that, unlike air, water and all other soil nutrients, nitrogen in the form required for plants (ammonia, nitrates, or urea) requires considerable energy for its synthesis. All nutrients require energy for mining, purification and transport, but so-called fixed nitrogen is a type of fuel as its molecules store energy (primarily from fossil fuels) during their synthesis. This important problem will also be discussed later in this chapter.

LIMITS TO PLANT PRODUCTIVITY

World agricultural efficiency—the percentage of the annual solar energy stored in plant matter on cultivated land—is a seemingly unimpressive 0.1%. Comparisons with photoelectric cells, for example, whose peak efficiency is ca 14%, are often used in arguments against the exploitation of biological solar energy conversion. Earlier discussions in the chapter on energy analysis and the hidden energy costs in both agriculture and photocell construction showed the dangers of simple efficiency comparisons, but solar energy conversion efficiency is nonetheless important and makes necessary a short review of the losses and what role research and development might have to reduce these losses.

Light Energy.   All solar energy devices are plagued by the same characteristics of light radiation in general and sunlight in particular (see II.1, page 126):

- sunlight consists of several components—colors—which don't have the same amount of energy: red light for instance has about half the energy of violet light;
- light is an energy flow, not a stored energy reserve like gasoline, wood, sugar, or water (above a dam!);

- sunlight is a diffuse energy flow, i.e., the energy falling on a square meter is small, and hence the collecting area must be maximized;
- sunlight can easily be absorbed or blocked before reaching the collection device, e.g., by clouds;
- sunlight varies with time of day and season (same as in the preceding point except the earth blocks the sun).

The first characteristic, the broad color (energy) spectrum in solar radiation, is a problem for the so-called *quantum* absorbers, but not for heat absorbers, such as water heaters and distillers. The former type exploits the quantum or particular nature of light which means that light consists of indivisible packets of a certain quantity of energy, that energy being characteristic for each color.

Each quantum or packet of light, in both photosynthetic and photoelectric cells, can excite one and only one electron, but only if the quantum has a sufficient amount of energy, a so-called threshold value. This threshold is characteristic for each type of light-absorbing molecule (pigments, e.g., chlorophyll) and corresponds to that pigment's color. Figure I.6.2 summarizes this problem for three case (color) examples:

1. Absorbed light, e.g., infrared light, doesn't reach the energy threshold, and therefore it cannot excite any electron. Half of the solar radiation is below this threshold and is useless for photosynthesis or photoelectric cells.

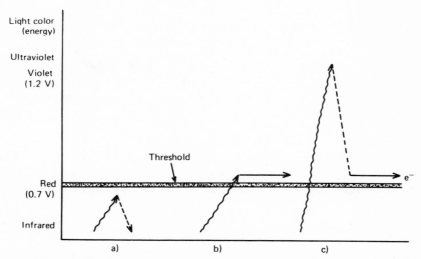

Figure I.6.2. Effect of light whose energy (color) is (a) below (e.g., infrared light), (b) equal to (red light), or (c) greater than (e.g., violet) the threshold energy required for photosynthesis.

2. The absorbed light, e.g., red light, just meets the threshold energy requirements. This energy is used with high efficiency (ca. 37%) and in the biological case is analogous to a photoelectric cell with a potential of 0.7 V.
3. The absorbed light quantum, e.g., violet, contains energy far exceeding the threshold. Only energy corresponding to the threshold value can be used, however, and energy exceeding this value is lost. Half of violet light's energy is lost (as heat), and hence its efficiency is only half that of case two or ca. 18%.

The low efficiency of case three could be doubled if the threshold value were raised to the level of violet light, but then the unusable portion of the energy spectrum would increase from 45% to perhaps 90%; conversely, that 45% lost energy could be reduced by lowering the threshold, but the use efficiency of green, blue and violet light would decrease together with the output voltage.

Nature has evolved a compromise threshold which can be shown to be near optimal (Porter and Archer 1976). An extra clever evolutionary development was required because this optimal threshold value gave a photocell voltage (O.7 V) below what was required to make available elements (e.g., hydrogen) react. Man solved the same problem with his first battery over a hundred years ago as Nature did eons before: two photocells are connected in a series to give a two-photocell battery with ca. 1.2 V (since the connecting "wire" isn't perfect, the battery doesn't have 1.4V!).

To summarize, the first several solar energy conversion losses are unavoidable for any quantum absorber:

- 35% of the solar radiation is reflected by the earth's outer atmosphere, and 17.5% is absorbed in the inner atmosphere, the latter contributing to the earth's heat cycle. (These losses are normally accounted for whenever solar radiation at the earth's surface is quoted);
- 40% is less energetic than the threshold requirement (Figure I.6.2a) and is lost as heat (largely detrimental to the leaves and requires compensatory evaporative cooling);
- a 20% loss accounts for that light which is more energetic than the required threshold.

Thus only 28% of the incoming radiation *can* be photosynthetically converted, a process itself which is ca. 37% efficient (for red light). The overall maximum conversion efficiency is ca. 13% of all solar radiation at the earth's surface (see II.1, page 152).

Rate Limitations. This 13% photosynthetic conversion efficiency applies for low-light levels. The *rate* at which plants can absorb light is also limited, however, and while this is not normally important for agriculture, it can severely limit maximum exploitation of algal growth or semi-synthetic bioconversion systems. The uppermost algal cells or plant leaves, i.e., those receiving direct sunlight, will have a noontime conversion efficiency well below 13%. These so-called photosaturated or photo-inhibited algal cells are not only inefficient, but wastefully filter out some light and diminish growth in lower-lying cells.

The culprit behind this saturation effect—which can lead to "photodestruction" if prolonged—is the connecting "wire" between the two photocells of the photosynthetic battery. There is, of course, no connecting wire, but rather protein molecules which carry electrons one at a time to fulfill precisely the same function. These molecules cannot carry electrons fast enough during high-light intensity, and just as excessive current in a normal wire causes heating and increased resistance, these enzymes form a rate bottleneck. While decreased efficiency can often result at noontime, natural conditions and built-in cellular repair mechanisms prevent any serious plant damage.

Future Improvements. Having listed both the efficiency and rate limitations in solar energy storage by plants, one might ask what the prospects are for improvements:

Atmospheric absorption. Little can be done except to minimize smog and other pollution-related cloud cover. A basic reflection and absorption loss must be accepted.

Pigment absorption. The threshold effect has been discussed and is basically unavoidable in any biological, physical, or photochemical device. A pigment whose second excitation threshold is low enough so as to emit *two* electrons at about the same potential from one photon of violet light and one electron from red light is desirable but unlikely. Calculations have shown that the evolved threshold level and antenna pigment system would be hard to improve for temperate regions.

Rate limits. The connecting link between the two photocells represents both a rate limitation and an efficiency loss: the two-cell "battery" has an effective potential about 25% less than the cell's sum. An unknown portion of this loss is converted to another energy form (ATP, see II.1, page 130). Rate variations do exist in Nature, such as between desert and temperate plants, and represent an important avenue for basic, genetic and breeding research. A less fruitfull approach has been to try to reduce the size of the light-absorbing system (i.e., number of pigment molecules). The idea is that by reducing the size, but hopefully increasing the number of these so-called antenna systems, more "wires" will exist with less of a load on each. This size has unfortunately proved to be relatively constant. An alternative perhaps available to synthetic systems, i.e., those using plant organelles in a nonliving system (see II.5, page 338), is to replace the connecting "wire" proteins with synthetic molecules. The basic photosystems require extensive research before they are sufficiently understood and before the can be fully exploited.

PRACTICAL AGRICULTURAL YIELD LIMITATIONS

Agriculture, even modern industrialized agriculture, demonstrates not 13% efficiency but rather less than a hundredth that value. The world average is about 0.1% and modern industrialized agriculture about 1%. Maximum solar-conversion efficiencies achieved, e.g., from algae and sugar cane under ideal conditions, are about 3.4%. What are the major limitations?

Leaf Area.   Obviously, water and nutrients are limiting in all but the most industrialized agriculture. These factors will be discussed next, but first two less obvious factors must be mentioned. First is the percentage of crop land actually covered with green plants or leaves, the so-called canopy cover. Modern agriculture begins with a barren field and planted seeds. Only during a small fraction of the growing season is the canopy cover "closed," i.e., the leaf-surface area is equal to or greater than land area; a closed crop canopy has an area four to five times greater than the land area and insures the absorption of not only direct sunlight but also the diffuse light reflected or transmitted through other higher leaves (see II.1, page 149). Such canopy cover is typical for natural ecosystems in which the proportions of various species vary during the growing season to assure a closed canopy. Exploitation of this same principle is practical mostly in "primitive" agricultural systems where a dominant tall-growing crop, e.g., maize, is planted and harvested first, and a second shorter-statured crop, e.g., mung or soya bean, is planted later in the rows between the first plants. In this intercropping manner, sometimes in conjunction with multiple cropping if the growing season permits, one can greatly increase the annual effective canopy cover. This annual average varies greatly with region, but typical calculated losses due to canopy coverage correspond well to the 30-60% productivity increases from intercropping (IRRI 1974).

Although the much touted advantage of the so-called "C_4" plants is to some extent real (see II.1, page 157), the major differences in annual productivities result mainly from the longer growing season and annual effective canopy coverage characteristic of their tropical regions; tropical forests' leaf areas can be twenty times the land areas. The ability to increase the annual effective canopy coverage through intercropping refers to all plants, but such practices will complicate mechanized agriculture and increase labor requirements. As discussed in Part I's Chapter 5, mechanization of agriculture was motivated by old oil economics which is less relevant now and will continue toward obsolescence. Energy analysis (together with present and future economic analyses) is a complex problem which must be solved for each region and crop to determine an optimum balance between the factors—machinery vs labor. While solar energy is by convention *not* included in energy analysis, the canopy coverage is one of many factors determining net crop productivity.

Carbon Dioxide. The second limiting factor not normally considered—and one which rarely can be affected—is the carbon dioxide (CO_2) concentration. Carbon dioxide in air diffuses into leaves through pores or stomates where it is enzymatically transformed to a more chemically useful form. Because of its low concentration in air (0.03%), diffusion and enzymatic uptake of carbon dioxide, the only source of carbon for all sugars and other plant products, is relatively ineffective (see II.1, page 154). A tenfold increase in carbon dioxide concentration has, for example, doubled soybean yield simply by increasing the carbon dioxide absorption rate. Soya is a relatively low-yielding plant, but if one considers the problem of a fast-growing species like maize or sugar cane, the gas requirements become enormous: a day's growth requires the equivalent of *all* the CO_2 in a layer over the field which is 50 meters high! A still day can result in a considerably decreased carbon dioxide concentration in the lower reaches of a plant canopy, and a lower concentration means even slower uptake rates.

Some evidence suggests that atmospheric carbon dioxide concentrations were greater during evolutionary times, and it is therefore that gas uptake rates now limit photosynthesis. The validity of this unproven theory is of little practical importance as little can be done in general due to economic limitations. Useful consideration of this problem should however be made in certain cases:

- carbon dioxide which results from all metabolic and combustion processes should be treated as a valuable by-product and not as a waste. Smokestack gases can be cleaned and used in certain special agricultural applications, the immediate development of which is possible and should be combined with energy analysis to determine its practicality;

- algae (see II.2, page 204, and further on in this Chapter) require a constant artificial carbon dioxide flow to grow optimally, and for this need "waste" CO_2 is the only economical source;
- as CO_2 diffuses in through the same pores (stomates) as water transpires out, increased carbon dioxide concentrations and/or increased uptake effectiveness are important for water conservation. C_4 plants have lower transpiration rates for this reason (see II.1, page 157), and
- carbon dioxide uptake poses several unsolved research problems which may in the long term have great practical significance. Among these are the differences between so-called "C_3" and "C_4" uptake systems and the relative carbon dioxide affinities of the enzyme involved (see II.1, page 157).

Water Conservation.   Water, which limits plant productivity when in either insufficient or excessive supply, represents an important energy factor:

- considerable energy is required for the irrigation process and construction of the materials used;
- energy is required to clean dirty water, e.g., via desalination or sewage treatment plants;
- the energy cost for water should be related to the increased crop energy yield.

While clean water can be related to energy costs, our ultimate dependence on this diminishing resource must be taken with increased seriousness as the contamination of open and ground waters, both from urban and rural pollution, threaten long-term survival in several areas. Excessive application of both fertilizer and irrigation water has caused soil salinization and increased ground water nitrate to dangerous levels, but the dangers are not restricted to the developed world: 40% of a 500,000-hectare Nile River irrigation project is threatened by salinization, while 50% losses have already been suffered in other Middle Eastern regions (de Bivort 1975).

Other Nutrients.   Volumes have been written on plant nutrition, and, as discussed earlier, fertilizers have become a double-edged sword in the Green Revolution (see Part I, Chapters 4 and 5). The need for nutrients is no less obvious for plants than it is for Man: a typical 5000-kg maize harvest constitutes a soil nutrient loss of 80 kg nitrogen (in its "fixed" ammonia-like form in protein) and considerably lesser amounts of potassium and phosphorus. Few harvests are required before the soil is barren. Consider, however, the difference between a harvest exported to an urban center and one consumed locally with the wastes (i.e., nutrients) returned to the same local soil.

Potassium, phosphorus and many trace minerals are not of less importance than nitrogenous fertilizers—the lack of any one can halt growth—but nitrogen is special in several ways:

- It is the element required in greatest quantity after carbon dioxide and water;
- In its biologically active complexed form (as ammonia, nitrate, or urea) it requires considerable energy for synthesis:
- Nitrogen fertilizer in some forms is volatile (ammonia and nitrous oxide) while in others (nitrates) it can be leached into ground or surface waters. Both result in a net nutrient (energy) loss, but in addition the former may ultimately detrimentally affect the atmospheric ozone level (Crutzen and Ehhalt 1977), while the latter both contaminates drinking water to levels poisonous for Man (Magee 1977) and stimulates aquatic weed and algal growth in waterways;
- Nitrogen fertilizer (i.e., ammonia) can be synthesized or fixed biologically from nitrogen gas in air as an alternative to industrial fossil fuel requiring synthesis.

This final and most significant point about nitrogen fertilizer is in contrast to all other nutrients and will now be discussed further.

Biological nitrogen "fixation" (see II.2, page 137) is an unfortunate use of a word whose everyday meaning provides little useful information in this context. Fixation of carbon dioxide for example implies a reaction in which it is combined with other elements (in this case hydrogen) in a more complex and energy-rich form, sugar. Similarly, nitrogen fixation means the energy-requiring process which leaves nitrogen, a low-energy inert gas, in a form containing some of this energy plus hydrogen. Chemically it is simply,

$$\text{nitrogen } (N_2) + \text{hydrogen } (H_2) \rightarrow \text{ammonia } (NH_3).$$

While no energy is written explicitly in this reaction, hydrogen (H_2) is a high-energy fuel. Normally methane (CH_4) from oil fields is the starting material for a high-capital, industrial catalytic process requiring high temperatures and pressures.

Biological nitrogen fixation is completely analogous to the industrial process except that: sunlight is the sole energy source (methane from fossil fuels also originated from solar energy but after a lapse of millions of years); the reaction proceeds at room temperature and atmospheric pressures, and the catalyst is an enzyme (nitrogenase) synthesized according to need (a rare earth metal, e.g., platinum, is used industrially). Furthermore, nitrogen is fixed when and where it is needed according to a finely tuned built-in regulatory system; from Man's point of view, this regulation is usually too effective.

This one-sided portrayal of biological nitrogen fixation evades the question of why nitrogenous fertilizers are synthesized industrially. The reason is simply that none of the higher plants, i.e., all plants of agricultural interest, themselves fix nitrogen. Only microorganisms, primarily bacteria plus some algae, fix nitrogen, and while these organisms can be very nutritious, culinary interest in them is limited. Fortunately, some of these organisms do cooperate with some plants of interest to Man, such as the soya bean.

Nitrogen-fixing bacteria, with rare exceptions, are nonphotosynthetic, that is they require organic "food" (typically plant wastes) just as higher animals. Because both energy and users of the fixed nitrogen are limited, their regulatory systems prevent these bacteria from fixing more than 1-15 kg nitrogen per hectare per year.

The roots of certain plants (legumes), however, are infected by nitrogen-fixing bacteria and thereby have a constant energy source. The reciprocal part of this so-called symbiotic relationship is the synthesis by the bacteria of fixed nitrogen (ammonia) both for their own and the host plant's needs. Ready access to energy and a constant consumption of the fixed nitrogen product increases net fixation in extraordinary cases to as much as 900 kg nitrogen per hectare per year (see II.1, page 157). As normal nitrogenous fertilizers are used with only 50% efficiency, i.e., 50% is leached to ground water or volatilized, fixation of 100 kg/ha-yr (typical for soya) corresponds to the application of 200 kg/ha-yr fertilizer, a figure typical for maize.

In order not to misrepresent bacterial nitrogen fixation, several disadvantages or limitations must also be mentioned together with possible improvements:

- regulatory mechanisms within the bacteria limit fixation rates below optimal levels;
- energy requirements for fixation may account in part for the relatively low legume crop yields;
- fixation is not always as efficient as it can be;
- legumes grow in rather special temperate climates;
- legumes are mostly beans (soy, peanut, chick peas, wing, and string beans) and have a characteristic taste which is not universally popular;

- the world's most significant crops (cereals such as maize, rice, wheat, and rye) are nonleguminous.

Regulatory mechanisms are analogous to the farmer deciding how much fertilizer to apply, except that these organisms have evolved principally via biological energy analysis rather than economics; the well-tuned symbiosis emolved to survive, not to maximize crop yields for Man. Research evidence suggests that several improvements can be made:

- effective infection and fixation rates are species selective and can be aided by inoculation with the proper bacterial strain;
- breeding has extended both the northern and southern limits of effective legume growth;
- growth under increased carbon dioxide concentration not only doubled the crop yield, but *quadrupled* the nitrogen-fixed fertilizer from 90 kg/ha-yr to 425 kg/ha-yr. This increase resulted partly from increased fixation rates, but even more important was a one-week extension to the fixation period. While increased CO_2 concentration is rarely practical (except, e.g., smokestack CO_2), extending the bacterial fixation period through the whole host growth period may prove possible by other means though as yet these periodic regulatory mechanisms are poorly understood;
- bacterial mutants without any regulatory mechanisms have been made in laboratory experiments, but it is not yet known whether it is possible or desirable to infect legumes with these species;
- nonleguminous associations of bacteria with important cereal crops (e.g., maize and wheat) have been studied recently, but their practical importance is not yet known. In these associations bacteria collect on or possibly within the hosts' roots, but do not form nodules as in legumes. The energy supply and rate of ammonia exchange with the host is slower and less effective, and hence 80 kg/ha-yr nitrogen is a typical maximal fixation rate.

The subject of nitrogen fixation will return in two contexts later in this Chapter (in the discussions on blue-green algae and long-range research), but one final relation between fertilizer and energy analysis should be made: It has been shown (see Figure II.1.22) that the energy and crop yields increase up to about 200 kg/ha-yr (for maize), while maximal energy efficiency occurs at ca. 120 kg/ha-yr. The energy ratio—output to input—is still greater than one at 200 kg/ha-yr, which means that increased fertilizer is a good energetic investment, but the *extra* return is less than, say, in a country where little fertilizer is used. Secondly, leaching rates will increase at increasing fertilizer application rates, but it is not yet known how much environmental costs would lower the energetic or economic maximum discussed above. Nor is it known what the long-term energetic returns would be on major one-time investments like trickle irrigation systems. These increase productivity, but more importantly they double fertilizer and water-use efficiency by supplying both when and exactly where necessary; pollution due to leaching is automatically eliminated.

A more complex consideration is where limited resources should be used. Synthetic fertilizers use a finite resource, fossil fuel, and hence constitute a flow for a limited time. An incremental increase in fertilizer use on an industrial farm increases yields but by less than the same increment would on an otherwise unfertilized field. As fossil fuels represent foreign exchange in virtually all countries and fertilizers are made from fossil fuels, they fall under the same special class of economic considerations, and the dilemma is obvious.

ALGAE

Algae (see also II.2) as a cultured crop are given special treatment here as a useful summary of all nutrient and rate limitations. Algae will also be seen to possess a unique productive potential role in recycling systems. An initial surge of algae's popularity as a potential food was inspired more than twenty years ago by algae's high protein content (42-60%) and yields as high as many of the best terrestrial plants (a maximum of ca 80 tons/ha-yr). Algae's more recent renaissance has been due even more to its important potential role in waste water treatment and nutrient recycling and somewhat secondarily to its direct use as food. What is clearly emerging from these trends is the development of an integrated role for algae in a system which will combine the reuse of water and nutrients with algae's potential for high productivity, nitrogen fixation, and genetic manipulation. Or, as summarized by one algae proponent: "algae grown on wastes will require 1/50 of the land area, 1/10 of the water, 2/3 of the energy, 1/5 of the capital, and 1/50 of the human sources for an equal amount of useful organic matter as terrestrial plants" (Oswald 1973).

Before proceeding further, this optimistic quote must be balanced by a few problematical realities. First of all, although nonarable land can be used and water-use efficiency is high, large amounts of water must be available and pumped. Although leaching can be prevented, evaporation is a problem as ponds are rarely covered. Secondly, it is very difficult to maintain sterility and species control, which, if required, will greatly complicate the cultivation techniques. The problem which dampened the original enthusiasm about algal growth is that low algal cell density and small cell size make harvesting, drying, and preparation both difficult and expensive; some progress has been made on this point which will be discussed below. Finally, mineral and gaseous nutrients are required in such vast quantities that wastes are the only economically viable sources. This is particularly true of carbon dioxide, but while sewage and other wastes are the economically preferable mineral sources, two other alternatives will now be discussed.

Growth.   Algae can be grown in any hole holding water if carbon dioxide, light, and basic nutrients are provided. Normally soils are sufficiently porous that to prevent water loss some material (e.g., sheet plastic) is needed to cover the pond bottom, but little more is required. Depth is not critical, but light intensity is attenuated so quickly that little growth occurs below 20-30 cm and hence optimal pond depths are 30-100 cm (see II.2, page 207).

Choice of nutrient source divides algal growth into three major categories, the last of which is the more interesting (see II.2, page 212).

Synthetic growth mediums.   A synthetic growth medium using clean water and chemical nutrients has the advantage of providing clean algae suitable for direct human consumption. This approach has been justified on the assumption that the product's high economic value—somewhat better than soya meal due to its higher protein content—can compensate for the added growth costs. Economic viability is not yet established, but further shifts in energy and food costs may yet alter the balance.

One profitable algal production is the semi-natural *Spirulina* ponds in Mexico. *Spirulina* is advantageous on several counts, and the natural alkaline lakes permit exploitation, but sales at present are for expensive health foods or chemical industrial use.

Natural mediums.   Natural algal growth, primarily sea farming, takes advantage of the vast nutrient and water resources of the sea and uses a portion of the earth's surface with few competitive uses. A major hurdle is the lack of nutrients in the surface waters where the light is; all nutrients are in the dark deep.

The most promising solution to this dilemma is a planned attempt to pump nutrients from the sea bottom to the surface waters where giant kelp, a water weed reaching 50 meters in length, will grow attached to sunken rafts. This seemingly farfetched idea is part of a well-integrated complex utilizing wind and/or wave power, local partial processing of the kelp, recycling of the nutrients, and shipment to land of dewatered kelp for food or fermentation to methane. Grazing fish are expected to be a by-product.

A second class of "natural" algae and aquatic weeds are those clogging polluted waterways in both developed and developing countries. Growth rates of 800 kg dry matter per day per hectare have been recorded for one common aquatic weed, water hyacinth (Wolverton and McDonald 1976) which in tropical or semi-tropical countries can form enormous rafts and flood onto and totally destroy rice paddies. Aquatic weeds also spread the water snail causing schistosomiasis and mosquitos causing malaria and encephalitus. Herbicides have been the traditional "technological fix" to control water weeds, but the cost is prohibitive, environmental risks are high, and the effectiveness is low. The basic problem remains of water rich in nutrients from urban and agricultural pollution. Harvesting of water weeds not only cleans the polluted waterways, it also provides a biomass rich in protein (10–26%) and nutrients for food and fodder. The nutrients eventually can be returned to the land via normal waste recycling systems. Nearsighted economics has generally overlooked this alternative as it clearly requires a coordinated effort: harvesting of relatively low density or dispersed water weeds is an underdeveloped technology, and the harvested material (normally 85–95% water) must be at least partially processed on site. A recent report on uses for aquatic weeds (NAS 1976a) marks an important step from standard technical responses to biological problems by turning it instead into a blessing: (aquatic weeds are) " . . . a free crop of great potential . . . for exploitation as animal feed, human food, soil additives, fuel production and waste water treatment."

Waste-nutrient mediums.   Aquatic weeds lead directly into the third category of algal cultivation, that grown on "waste" nutrients. Algae have been grown in up to 10^7-liter sewage treatment ponds for the purposes of sponging up "waste" nutrients, and providing oxygen (from photosynthesis) for bacteria which in turn break down complex organic molecules in the sewage. These ponds duplicate, accelerate, and contain in a closed system those events which normally lead to uncontrolled aquatic weed growth and eventual eutrophication, the biological death of a waterway. When the algal "sponge" is removed from the cycle together with nutrients, the released water is clean as shown in Figure I.6.3.

Algal ponds have traditionally been viewed in the context of waste treatment. Viewed instead as a method of producing algae for food and fodder, one sees the sewage as a valuable nutrient source and the bacteria as a provider of otherwise expensive carbon dioxide. Both are provided free or perhaps with a positive credit for waste disposal. A third point of view, always relevant in arid regions but increasingly relevant elsewhere,

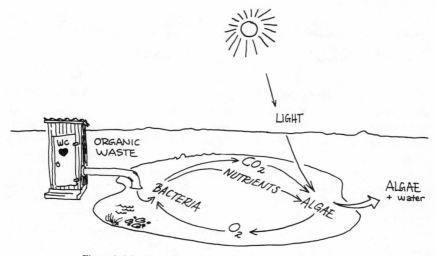

Figure I.6.3. Gas and nutrient cycle in an algal oxidation pond.

is that of water reclamation. Taken together, the system is extremely promising economically, ecologically and energetically.

Energetic considerations of this system are a bit premature, but one element should be pointed out. The organic wastes which are oxidized by the bacteria to basic nutrients and carbon dioxide are in fact energy-rich and in an optimal approach should not be wasted. This energy is lost to heat in the above scheme, but according to a discussion to follow, the wastes could be converted in the absence of oxygen to methane, the principal component of natural gas; the nutrients would still be freed, and CO_2 would be returned when the methane is burned (combusted with oxygen) in an energy-yield process. The nutrients are then free to be reorganized or "pumped up" to a high-energy form with sunlight.

Harvesting. (See also Part II, Chapter 2, page 224) Algae for sewage treatment have been grown on large scale for over 15-20 years during which the simple technology for this sometimes tricky process has been worked out. Removal of well over 90% of the limiting nutrient (e.g., nitrogen) with typical yields of ca. 40 tons/ha-yr is easily achieved. Cheap oil has previously dampened algae's product appeal and has to some extent prevented progress in one remaining step, harvesting. The algae which grow spontaneously in nonsterile ponds are the green unicellular microalgae (diameter of a few microns), and this small cell size together with low cell densities combine to exclude harvesting by simpler methods, such as filtration. The favored laboratory harvesting method is centrifugation, but both capital and operating costs are prohibitive even for industrialized countries.

Algae are presently harvested on large scale by so-called chemical flocculation: a chemical (e.g., aluminum sulfate or alum) when added to the algal culture causes the cells to flocculate or collect into clumps which can then be raised to the surface by air bubbles and skimmed off. This highly effective process removes nearly all of the algae (the water is recycled to the pond or released), but suffers from the disadvantage that the algae are contaminated with alum. A recent modification adds an acid wash which removes all but 4% of the alum which at least does not affect fish fed algal pellets. Fish, unlike Man, have a neutral stomach pH and cannot solubilize or digest the alum which is eventually eliminated in feces. Herbivorous fish and bivalves such as clams are of course a more direct and low-cost method for harvesting microalgae, while a bold proposal has been considered to use the manatee or sea cow, a herbivorous aquatic mammal, to harvest aquatic weeds (ICMR 1974). The former, or so-called aquaculture method, is highly developed in many regions and will be discussed further in Part I, Chapter 8, and in more detail in Part II, Chapter 2, page 230.

Fish are an important, but all too limited, algal user, and to be economically and energetically feasible, algae must be harvested by simple filtration or raking, as with aquatic weeds. Larger algae need to be cultivated. Some algal species are larger (e.g., *Spirulina*) and/or have a shape conducive to filtration (e.g., *Oscillatoria,* a spiral-shaped cell), but except for rare cases these species cannot compete with the fast-growing green microalgae. An important development is a technique which continuously filters off larger algal species and returns them to the growth pond; this artificially increased residence time for the chosen specie can eventually lead to its dominance (see Part II, Chapter 2, page 223). This approach has for instance recently successfully established a filterable algal population (*Micractinum*) and allowed harvesting of clean algae with the same simple equipment at a tenth the alum-harvesting cost. Hopefully this technique will provide the awaited breakthrough in algal culturing.

Species control is important for establishing an algal species which is easily harvested, but also because any algae destined for consumption must undergo years of toxicological testing. This testing would be wasted if a unique specie could not be maintained.

Algae species selected for sewage-water treatment, even when destined for food/fodder use, are functioning as nutrient sponges. An important class of Algae, the filamentous blue-green algae (*Cyanophycophyta*) are not only a large filterable sponge, but have an extra contributory function.

Blue-Green Algae.   (See also Part II, Chapter 2, page 204) Filamentous blue-green algae consist of long interconnected chains of cells which are identical except for every fifth or tenth one, called heterocysts. These cells are unique for their very thick cell wall (to exclude oxygen) and by their ability to fix nitrogen. A property of both bacterial and algal nitrogen fixation systems (i.e., the responsible enzyme, nitrogenase) is its oxygen sensitivity. Bacteria in legumes exclude oxygen from the host root nodules where they grow, and likewise heterocysts in blue-green algae exclude oxygen. Heterocysts also lack one of the two photosystems (see pages 129 and 140) and hence can neither release oxygen by splitting water nor can they synthesize their own sugars. Instead, in close analogy to bacteria in legumes, they have a symbiotic relationship to neighboring cells: heterocysts fix atmospheric nitrogen as ammonia for themselves and for their neighboring cells which in turn provide the necessary energy (sugar) for fixation and heterocyst metabolism.

Nitrogen-fixing, blue-green algae will thrive when and where they have a selective advantage: when all nutrients except the energy-rich ammonia are present. Fixation requires energy, and no organism will survive if it unnecessarily produces ammonia already available in the growth medium. For this reason fixation can turn on and off in blue-green algae according to need, but if ammonia isn't lacking periodically the faster-growing green algal species will reestablish dominance.

Blue-green algae have a unique potential in their ability to synthesize the one nutrient which is most easily lost (via leaching or volatilization) and requires energy for its synthesis. While urban sewage is not typically limited in nitrogen, the rural closed or semiclosed systems discussed later are.

Before modern science assumes too much credit for exploiting the unique potential of blue-green algae, their traditional uses should be mentioned. The nitrogen-fixing algae, *Anabaena,* has long grown in rice paddies and is now sometimes artificially implanted. In the paddy, algal growth precedes rice growth in a type of intercropping arrangement whereby fixation occurs before the rice canopy is completed at which time the breakdown of the algae begins to release ammonia for rice use. Even during growth, blue-green algae can excrete up to half of the nitrogen fixed, but the same regulation mechanisms discussed previously restrict fixation in blue-green algae to levels below those desired for rice growth. An interesting symbiosis exploited extensively in Viet-Nam and under present investigation at the International Rice Institute in The Philippines is seen in a higher water plant, *Azolla,* which has the blue-green algae growing and fixing nitrogen within it. Just as a close symbiosis in legumes increased net fixation rates over those observed in looser associations, so here too are fixation rates increased to about 160 kg/ha-yr. *Azolla* are grown in rice paddies, again during rice canopy development and thereby replace much or all need for synthetic nitrogenous fertilizers.

Upper limits to net ammonia production are not yet known, but estimates for natural ecosystems have gone so high as 300-500 kg/ha-yr and sometimes higher. While the highest figures may be a bit exaggerated, the effective regulatory mechanisms are known to be limiting in virtually all cases. Reasonable theoretical estimates would put the limit at several tons, but such rates have never been seen. Important avenues of research would be mutants without regulatory mechanisms and chemicals which inactivate regulation. Biologically fixed nitrogen in excess is just as polluting as synthetic fertilizer, and the dangers of large-scale genetic or chemical alteration must be appreciated. The high energy requirements of the process are nonetheless an ultimate safeguard.

AGRICULTURAL PRODUCTIVITY

(See also Part II, Chapter 1, page 185) How much of the sun's energy can be captured and stored as plant matter? The answer is of course "it depends," in part on those biophysical factors discussed earlier, but also on cultural techniques, species and chance. Table I.6.1 shows some of this variation, but these figures should not be taken as directly comparable: all other factors but specie are *not* identical. Hidden within this data are large variations in farming intensity, growth season, solar intensity, water and soil conditions and of course inherent species differences.

These reservations do not alter the nearly 200-fold difference between theoretical and world average agricultural productivities. Within 30° of the equator annual theoretical yields exceed 200 tons/ha, while the closest competitor, sugar cane, at best reaches half this value and is often somewhat lower. Algae are in the same productivity class, but with their 50% protein represent a very different product than sugar cane, a nearly pure carbohydrate.

Maize production, discussed earlier in Chapter 5, shows a typical yield dependence on energy inputs and a decreasing marginal return on energy investment. While maximizing grain *yield* is an understandable goal when energy is available, a maximal energy *ratio* (energy output to energy input) is optimal for an energy crop. As food crops are determined

Table I.6.1 Various Group Yields (Cooper 1975, Troughton and Cave 1975, and Schneider 1973).

Crop	Latitude	Total Yield (dry tons/ha-yr)
Cereals:		
Wheat	47°N	30 (12 grain)
Rice	7°S	22 (12 grain)
	38°N	22 (11 grain)
Forage:		
Napier Grass	14°N	85
Sugar Cane	21°N	64 (22 sugar)
Maize	45°N	40
Tubers:		
Cassava	3°N	38 (22 tubers)
Sugar Beet	21°N	31 (14 sugar)
Algae (and Aquatic Plants):		
Kelp	30°N	— (20 expected)
Water Hyacinth	—	30 (100 expected)
Green Algae	35°N	73
Trees:		
Rain Forest	—	44
Eucalyptus	—	49
Hybrid Poplars	42°N	21

by tradition and exhaustively dealt with in other sources, crop productivity here will concentrate on the ability to produce energy in the form of plant and tree biomass and fertilizer via nitrogen-fixing species, especially algae.

Energy Crops.   Crop wastes, if not used as fodder, are assumed available for collection and recycled for their energy and nutrient value. If crops are to be grown expressly for energy purposes, a number of factors must be considered:

- crop yield, i.e., solar-energy conversion efficiency;
- energy yield, i.e., ratio of energy in the crop yield to those energy inputs for plantation, cultivation and harvesting;
- water usage;
- ease and method of conversion.

All of these factors can be reduced to two for judging the overall energy crop performance: net fuel energy yield per hectare per year and the net energy ratio (fuel energy divided by all energy inputs, including conversion costs). The latter term, which must be greater than one for a net energy gain, is not even known for most modern or proposed nonbiological energy systems.

Table I.6.2 Energy Crops

	Crop Yield	Harvestability	Convertibility	Vulnerability
Trees	low	medium	low	low
Forage	high	high	medium	medium
Starch crops	high	medium	high	medium
Algae	high	low-medium	high	high
Wastes	variable	low	variable	low

Five plant categories considered for their energy value are characterized in Table I.6.2 according to their relative yields, the ease with which they are harvested and converted to the desired final form, and finally their vulnerability to changes in weather or other growth conditions.

Trees. Trees are among the easiest to cultivate energy crops, particularly if coppicing varieties are used to eliminate replanting. Harvesting every three to eight years lessens costs relative to alternative crops while allowing for a fuller leaf canopy and averaging of good and bad years. However, yields are low (ca 10-25 tons/ha-yr), and wood is as yet the most difficult cellulose product to convert biologically. Immediate applications concentrate on industrial, thermal conversions to methanol, but chemical or enzymatic breakdown to sugar for fermentation to methane or ethanol are conceivable in the future. Whatever the method of choice, wood use requires considerable mechanization, especially for pretreatment. Data adequate to choose among the alternatives is still lacking, although preliminary estimates have been made (see Part II, Chapter 1, page 192). The proper choice will also depend upon growth, harvesting and conversion conditions, level of industrialization and available capital.

Forage crops. Forage crops such as grasses share with trees the avoidance of replanting, but have the advantage of providing several harvests per year and among the highest yields (max ca. 85 tons/ha-yr). Several of these are the so-called C_4 type (see Part II, Chapter 1, page 157), which, in addition to having more effective carbon-dioxide uptake, less photoinhibition, and perhaps better water-use efficiency, are also the plant class which best forms (nonleguminous) associations with nitrogen-fixing bacteria (see Part II, Chapter 1, page 181). While containing considerable cellulose, the lack of lignin, an additional structural compound, makes them far more successful candidates for bioconversion than wood.

Starch crops.   Starch tuber crops, e.g., cassava or potatoes, contain very little cellulose and hence are ideal for biological conversions; most modern schemes envision yeast conversion to ethyl alcohol (see Part II, Chapter 3, page 249). Tuber crops are also traditional foods, and usage for fuel can build on cultivation experience. Yields are not as high (ca 30 tons/ha-yr) as other types, but the conversion ease is energy-saving which may more than compensate.

Algae.   Algae have high yields (ca 50 tons/ha-yr), can use low-quality land, are resistant to storm damage and can play a central role in nutrient reuse, water purification and fish culture (see Part II, Chapter 2, page 204). They are, however, susceptible to microorganism predators, and as discussed above, larger filterable varieties must be cultivated to avoid serious and costly harvesting problems. While water-use efficiency is high from the productivity point of view, water availability per unit area must also be high.

Crop wastes.   Wastes are, from the competitive use point of view, ideal, but their low density and variable quality make them most conducive to local use. As approximately half of each crop is "waste" cellulose, increased crop yields will automatically provide an increasing waste supply. Their periodicity with crop harvests requires compensation from some other energy crop, e.g., algae or wood. Water weeds are a special class of weeds whose harvest not only cleans polluted waterways, but yields energy and nutrients.

Special-purpose plants.   Nature's incredible adaptability has resulted in the evolution of special plants whose unique products present a largely unexploited and in many cases unknown potential (see Part II, Chapter 2, page 244):

- *Hevea brasiliensis* is the best known latex-bearing plant whose yields have been increased from 400 kg to 2 tons of rubber per hectare per year. Rubber is, of course, valuable as such, but it can also be viewed as a petrochemical like oil, i.e., both a fuel and chemical base.
- Guayule (wy-oó-lee, *Porthenium argentatum*) is one of the 3000 other latex-bearing plants which in contrast to the tropical *hevea* plant, grows on largely unproductive semi-arid land.
- Jojoba (ho-hó-ba, *Simmondsia chinensis*) is another bush of semi arid and even saline or alkaline soils whose seeds contain 60% of a light, odorless wax, a potential replacement for the important sperm whale oil.

- *Dunaliella,* a unicellular algae growing in saline waters (even salt-saturated!) typical of arid/semi-arid regions, survives only because it contains up to 85% glycerol, an alcohol useful as a chemical, lubricant, and energy resource.
- *Euphorbia lathyrus* and *tirucalli* also produce a petroleum-type hydrocarbon.

LONG-RANGE RESEARCH

Food is the fuel most constrained by tradition, and for that reason is least relevant in discussions of long-range research. Little reason exists for complicating food habits with preprocessing except in cases demanded by special health or malnutritional problems. If anything, Man's development has narrowed his choices of foods from the 3000 varieties he has historically used to only 150 which have been commercially grown. Now virtually all food is limited to the most common 20 or so species. Food research will profit most by retrospective studies exemplified by a recent symposium on primitive food fermentations (UNEP/Unesco/ICRO 1977) which have traditionally been used to preserve or increase the digestibility of a host of more recently neglected plants. Increased quantity is of course relevant, but has been dealt with earlier.

Fuel and fertilizer production are relevant to agricultural and other needs, but if solar-based biological development is followed, it is reasonable to ask what type of research and development needs to be stimulated and what results can be expected? Goals and possibilities will undoubtedly change with time, but some important avenues can be mentioned.

A look at the stepwise transformation of sunlight helps to reveal the potential for exploiting photosynthetic organisms. As sketched in Figure I.6.4, light strikes one of the thousands of light-converting systems in a leaf, an organelle called a chloroplast, which breaks a water molecule into oxygen (released to air), protons (H^+), and an energetic or excited electron. A chloroplast is nearly analogous to a photocell whose light energy is also converted to a current of electrons, but instead of having connecting wires, the energy is delivered to an enzyme in its surrounding membrane.

Conversion of light energy in a chloroplast leaves it in a form suitable for driving a host of biochemical reactions. The most common reaction is of course the synthesis of sugars and later starch, protein and fats, but alternatives do exist. Some organisms do fix nitrogen and others produce hydrogen, but these are relatively rare and carefully controlled occurrences, both for the same reason that they waste energy.

An organism needn't be alive for its organelles and enzymes to func-

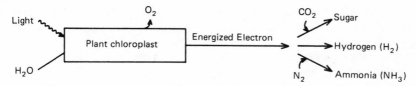

Figure I.6.4. *In vitro* synthesis of specified products.

tion, and in principle this allows one to remove the desired components from plant cells and artificially rearrange them to perform any of three functions:

1. biological photovoltaic cell to produce electricity;
2. biological system to produce hydrogen and oxygen from water (biophotolysis);
3. biological system to fix nitrogen.

The first alternative uses the chloroplast alone as a simple photocell to create an electric potential (ca one volt). As no wires exist, one must use electron carriers, enzymes as Nature does, or else substitute one of a class of synthetic compounds to carry the current to externally connected electrodes. While this has not yet been done, another biological light-absorbing pigment (rhodopsin) has been attached to a synthetic membrane and a light-produced potential has been measured. It is as yet too early to conclude what role photosynthetic systems will have for the direct production of electricity, but the required research is of critical and very widespread importance.

The other two alternatives, producing hydrogen and oxygen and the fixation of nitrogen to ammonia, both involve the same type of biological components:

a. chloroplasts to collect and transform light energy to an electrochemical potential (ca 1.2 volts);
b. an enzyme (ferredoxin) to carry electrons (electrical current) from the chloroplast to the catalyzing enzyme, and
c. enzymes to catalyze the reactions

protons + electrons → hydrogen gas (hydrogenase), or
hydrogen + nitrogen gases → ammonia (nitrogenase).

While the enzymes in items (b) and (c) above have nearly identical active centers (i.e., the functional portion of a much larger molecule), their sensitivities to oxygen are very different. Simple combinations of these elements in light have produced hydrogen or ammonia in the laboratory, but instabilities have severely limited functioning lifetimes. Immobilization and encapsulation have shown promising lifetime increases (see also page 330).

No organism normally evolves hydrogen because it is such an effective fuel. On the other hand, several algae and bacteria can use hydrogen as an energy source for which the same enzyme mentioned above (hydrogenase) is used. A method of electricity production from stored fuel (hydrogen/oxygen), the fuel cell, was discovered long ago, but recent development has mostly been for the space research programs. The fuel cell functions analogously to a car battery except that oxygen and hydrogen are used instead of lead and an electrolyte. Furthermore, by passing the gases over a metal catalyst, current can be produced as long as the gases last:

$$\text{hydrogen} + \text{oxygen} \rightarrow \text{water} + \text{energy}.$$

This same reaction applies if the energy produced is heat produced either explosively (e.g., the dirigible Hindenburg) or under catalytic control; metabolic energy as in certain microorganisms, or electrical energy as in a fuel cell with appropriate catalysts. In all cases, water is the only by-product which makes hydrogen/oxygen a uniquely appealing fuel.

Microorganisms which metabolize hydrogen constitute a variation of a fuel cell, and it is therefore not surprising that the enzyme involved (hydrogenase) has been considered as a substitute for the expensive platinum catalyst presently used. Fuel-cell research has entered a new phase, and while immediate results shouldn't be expected, they will undoubtedly play a critical role in decentralized low-capital electricity production.

Finally, one can consider the possibility of altering a plant's products, not just the normal balance of starch, cellulose, fats and proteins but even rarer products. A severely selective survival factor prevents plants and other organisms from any unnecessary or excessive functions. Thus, higher plants don't fix nitrogen because it is sufficiently available (for survival, but *not* for high productivities desired by Man), hydrogen gas isn't evolved because it is valuable energy, and the proportions of food vs waste (Man's terminology) is determined by survival needs, not according to Man's wishes. Alterations in plant products normally result in decreased survivability and correspondingly increased compensatory energy and protection requirements; lower energy-use efficiency, increased susceptibility to infection, and increased nutrient demands are a few examples. Various methods exist for altering a biological organism's characteristics:

- Mutation and breeding are the most common methods and in essence are two sides of the same coin: mutations are genetic changes occurring spontaneously or intentionally induced with, for example, chemicals or radiation. Breeding is the selection and/or propagation of desirable strains, those mutants which have characteristics of interest. Breeding is the oldest method for selecting desirable biological strains while induced mutation is somewhat newer. A few examples from an exhaustive list of important successful mutation/breeding experiments are:
    - long-stemmed rice plants which withstand high flood waters;
    - maize with improved protein quality;
    - grain crops responding to heavy fertilization (a mixed blessing);
    - microorganisms lacking (or with diminished) regulatory mechanisms for the production of hormones, enzymes, amino acids, or (on a laboratory scale) ammonia.
- Plasmid insertion is a more recent research development which allows the *addition* of genetic material and not simply the alteration of existing genes. Plasmids have thus far only been added to microorganisms, but for instance the ability to fix nitrogen has been given to some originally non-nitrogen fixing organisms. Extension of this ability to major crops seems at the moment doubtful (due to oxygen sensitivity) though not necessarily impossible. Improved and extended exploitation of microorganisms, particularly in the realm of leguminous and nonleguminous nitrogen-fixing bacterial association is a more readily attained goal.

Whatever the plant form used, traditional, bred, or mutated, the ecological principles of Part I, Chapter 2 must be remembered: plant survival requires a stable energy source, and if its own energy efficiency is lowered by manual perturbations, Man must also step in with extra energy, nutritional, or protective supports. When the plant cannot be found or altered to make the desired product balance, then the conversion methods discussed in Part I, Chapter 7 are the required alternatives.

7

Turning Plant Products into Useful Energy Forms

Crops, once grown, are generally separated into food (e.g., grains or tubers because of their easy digestibility) and wastes. The wastes, principally cellulose, are sometimes fed to ruminants such as cows, but are most commonly wasted, as the word implies. The distinction between food and wastes has been established by tradition and is in large part arbitrary. Both have the same energy (sugar) content and differ principally in their digestibility.

Man's consumption of plant crops may perhaps stand first in a list of biomass conversion methods, but little greater uniqueness can be claimed. Society has a particularly strong interest in Man's survival, but strictly speaking, digestion and metabolism by Man is not the dominant method for releasing the energy in plant materials.

Plant matter consists of little more than carbon, hydrogen and oxygen; the rather small but important quantities of nitrogen in protein can be temporarily ignored. The same elemental content characterizes virtually all fuels (with the except of mechanical, electrical and heat energy) which are of course all direct descendants from fossilized plant and animal matter (coal, oil, methane, gasoline, etc.). Thus, as pointed out earlier, food and fuel have more characteristics in common than differences as any food can be used directly as either food or fuel.

potato + oxygen → man → energy + CO_2 + freed nutrients,
potato + oxygen → combustion → energy + CO_2 + freed nutrients.

In both the conversions shown above, the potato is the fuel, but the existence of oil and coal suggests that natural conversions existed long before Man. One alternative, for example, is,

$$potato \rightarrow oil, coal \xrightarrow{\text{oxygen}} energy + CO_2 + \text{freed nutrients.}$$

Alcohol can, of course, be an energy source for either Man or machine, albeit with varying efficiencies.

In each example above, it is the oxygen-consuming step (oxidation) which releases energy, and hence one can say that increasing the oxygen content of a hydrocarbon decreases its energy content. Some processes decrease the energy content only marginally in the absence of oxygen and leave most of the energy release to occur during a later oxidation step. Such is the case in alcohol production in which the yeast cells, growing in the absence of oxygen, metabolize some energy in the plant matter, but cannot release all the energy in the alcohol product; that energy is available only in later reactions with oxygen, e.g., via combustion or animal metabolism.

The growth of these organisms in the absence of oxygen (anaerobic) constitutes a class of processes called fermentations of which alcohol production is but one. A most complete list is shown in Part II, Chapters 1 and 3, pages 144 and 249, respectively, but it is the bacterial fermentations to hydrogen (H_2) and methane gas (CH_4) which are of particular interest here. Hydrogen is a particularly useful fuel not only for its high energy content, but also because of the fact that only one combustion product can result: water. Its production by fermentation has been shown to be feasible, but theoretically this conversion method can never achieve the efficiency of the methane process (note, however, that the production of hydrogen and oxygen directly from water by blue-green algae or possible biosynthetic reactions is quite another matter to be mentioned below).

Conversion of organic wastes to methane is a natural phenomenon, the most well-known example being the production of marsh gas by bacteria at the bottom of still, standing water. Another natural digester is the cow (see Figure I.7.1) whose rumen bacteria digest cellulose and release methane. The cow represents an advanced, multistaged walking fermentor; it is heated, stirred, and even has a carefully maintained acid/base balance. Of no less importance is the pretreatment that the feed material—primarily cellulose—receives, the mashing and mixing from teeth and tongue. The chief problem of the process—slow digestion times—is even exemplified by the multiplicity of stomachs (four) in the cow. Various digester designs have been tried and used, each a balance between cost and effectiveness as related to the same factors dealt with in the cow: pretreatment, heating, stirring, and digestion (residence) time. Pretreatment depends upon the feed material, which in the case of cellulose represents a serious and expensive problem while manure and farm wastes are less problematical. Digesters are rarely heated and compensate for lowered digestion rates with longer residence times (see Part II, Chapter 3, page 261), but the consequently large digesters required have higher

Figure I.7.1. Advanced anaerobic digester with built-in pretreatment, stirring, heating and pH control.

capital costs and lower overall digestion efficiency. Timely effort is now being directed toward simple, yet effective, digesters which take advantage of higher digestion rates at higher temperatures (ca 35°C). The so-called thermophilic bacteria which grow at much higher temperatures (ca 60°C) are advantageous for their more complete waste conversion and the resulting increased gas production rates (50-75%). The speeding of the process considerably reduces digester volume and consequently the material cost. A side-benefit of the high temperature is sterilization of pathogenic bacteria. This final factor, that of improved hygiene, is in some areas the most important advantage, but is one which is difficult to calculate in economic or energetic terms. Human and animal wastes, just like plant wastes, have both energy and nutrient value and the lack of proper disposal has been and still is one of the primary causes of world ill health. Increasing food intake in those with chronic dysentery is attacking the problem at the wrong end.

While the optimization of the methane-producing digestion process is still being studied, each application will require slightly different technology. Not only is this often ignored, but not even for a single application have the full economic and energetic costs for the two operating temperatures yet been satisfactorily determined. Much of this disinterest in anaerobic digestion of organic wastes to methane results from the process' bad reputation and misuse in the developed world. Practical process problems result primarily from variations in feed material and particularly the presence of industrial poisons in urban wastes. Economically and energetically the misjudgment results from viewing the digestion process in isolation, usually in the role of sewage treatment, and thereby missing the advantages of integration with other processes on a suitable scale. Integration is of prime importance if methane production is to be economically or energetically attractive. Using a high-grade fuel like methane for digester heating, for example, is wrong when simple solar heat collectors are more appropriate. Furthermore, digester sludge—the watery mixture of free nutrients, simple organic compounds, and undigested wastes—poses a severe remaining "waste" disposal problem unless recycled as a valuable nutrient. Integration will be discussed in the next chapter together with the need for system evaluation, not short-ranged comparisons of isolated elements.

ENZYMATIC CONVERSION

(See also Part II, Chapters 3 and 5, pages 278 and 330, respectively) Fermentation of organic matter is a class of useful conversion processes because a microorganism's waste product is a more useful fuel than the starting material; the little energy consumed to keep the organism alive is justified by the increased energy density and convenience of the fuel. The conversions resulting from the many reaction sequences within the organism are, of course, also a part of its normal metabolism. These reactions are made possible by enzymes, biological catalysts, which passively, (i.e., without being consumed) promote reaction rates, but do not affect the final equilibrium between starting material and product. Enzymes exist in thousands of shapes, sizes and functions, but each is made specifically for each reaction, according to need and from little more than carbon, nitrogen, hydrogen and oxygen. Enzymes are in no sense living, and several have been synthesized in the laboratory, though not with the ease and efficiency of a biological synthesis.

Like fermentations, there is nothing new about using enzymes without the organism which produced them: curdling milk in a cow's stomach for cheese production, possible due to the enzyme, renin, is a several century old process, only slightly altered today. Awareness of enzymes' remarkable specificity and efficiency has in recent times spurred their extraction, purification and characterization so that now renin is but one example of enzyme engineering (Part II, Chapter 5).

FREE SOLUBLE ENZYMES

The enzymes of greatest interest here are those of more general applicability in the conversions of organic wastes to useful fuels and food. As mentioned, cellulose is the dominant agricultural waste, so labeled by its resistance to digestion. Ruminants such as cows can digest cellulose and hence constitute a ready conversion system, but of course it is only the bacteria in a cow's rumen which break down cellulose to a form which the cow can in turn digest. The cow, like Man, lacks the enzyme required to break the bonds in the long cellulose chains of sugar (glucose) molecules. As these chains are too large to enter cells, bacteria make and secrete the relevant enzyme (cellulase, see Part II, Chapter 3, page 278) which diffuses in water to the cellulose and there breaks off simple sugars which are then absorbed by the bacteria in competition with the cow, of course. But sugars, even impure and dilute solutions, can have widespread use for both food and fuel uses, and the example of the process in the cow rumen can be duplicated without the cow.

Extensive efforts have been made in the last few years to find organisms, or make mutants, which produce the best and most effective cellulase. Some highly effective (so-called white rot) fungi have been isolated and their cellulase—a complex of several enzymes—extracted for the large scale digestion of paper. However, enzyme production and purification is an expensive process. Further, there are biological regulation mechanisms which become inhibitory if the product glucose is allowed to accumulate. One can have the organism-producing enzyme in the same reactor with the paper or other cellulose waste, but as in the cow rumen, both cost and these inhibitory effects make a two-stage process more effective. Purified enzyme is added to the cellulose in a reactor from which the enzyme is later recovered for reuse, and thereby separated from the sugar product. This recovery process is still the most expensive step after enzyme production/purification itself and is one of a considerable list of basic and developmental research problems remaining. A pilot plant is in operation, but lowered costs and improved cellulase production and activity are required before further expansion is possible.

IMMOBILIZED ENZYMES

Cellulose is the most abundant waste, and it is also unique in that its insolubility and enormous chain length require transporting the relevant enzyme to it rather than vice versa. Starch, the carbohydrate digestible by Man, presents a much simpler problem. On the one hand the need for alternative conversion methods is perhaps questionable when all starch is presently required for human food (precisely because it is so easily digested). On the other hand, as energy is required to grow and produce more food, even starch has a conceivable future fuel and industrial role in addition to being a basic food. In this discussion it is an illustrative counter-example to cellulose.

Because starch is both soluble and more easily digested, the enzyme which breaks its chains of glucose molecules has been immobilized on a porous matrix to eliminate the enzyme recovery and product-separation problem. The matrix can have many forms, but the simplest is a column into which a starch solution is poured, where its glucose molecules are clipped off by the immobilized enzymes. A glucose solution comes out of the column's bottom, and the expensive enzyme recovery and make-up stages are thus eliminated in a simple room-temperature process.

Several sugar-related enzymes have been immobilized and have potential in the food industry, both in initial processes and some waste treatments. A few illustrative examples are included:

1. Cow's milk contains a complex sugar, lactose (5%) and therefore cannot be drunk by those individuals lacking the enzyme required for its digestion, lactase. Lactase breaks lactose into two simpler sugars (glucose and galactose), both of which are easily digested. This enzyme can be immobilized and may even allow individual pretreatment of milk.
2. Cheese whey contains 75% lactose, and similar use of the above enzyme allows the breakdown of the lactose to glucose and galactose to yield a valuable corn syrup substitute.
3. Cornstarch conversion to dextrose (via glucoamylase) is already an industrial enzyme process. Immobilization of the enzyme would reduce the cost and complexity of this process, and a second step— the conversion of dextrose to fructose (via glucose isomerase)— would produce a sugar substitute.

Many enzymatic conversions will be added to this list in future years as immobilization techniques improve, a few of which will be discussed later. One breakthrough should, however, be mentioned: enzymes are advantageous in that the complex enzyme factory is the easily grown bacterial cell, but the separation, purification and replacement costs (their lifetime is a critical variable) are nonetheless considerable. In some instances whole bacterial cells have been immobilized (for example, on glass beads) or encapsulated (in tiny membrane bubbles/vesicles) with the bacterial enzyme "immobilized" in its natural cellular environment. By this method enzyme separation and purification stages are eliminated, and in some cases the enzyme lifetimes are extended. The cell is kept intact, but is not strictly speaking alive.

THERMOCHEMICAL CONVERSIONS

(See also Part II, Chapter 3, page 301) The similarities of all hydrocarbons (compounds containing hydrogen, carbon, and oxygen), whether they are plant matter and trees or petroleum fuels (oil, gasoline, kerosene, coal, etc.), reveal the theoretical ease of all types of interconversions. Fossil fuels which we now exploit were converted naturally from plant matter through millions of years by slow, chemical rearrangements, but similar atomic rearrangements can be more quickly performed under high temperature and pressure by using varying amounts of oxygen and specific catalysts. Different balances of these factors will result in various combinations of fuel products; including gases (methane, hydrogen, carbon monoxide, and carbon dioxide and oxygen with no fuel value), liquids (methanol and various oil fractions), and solids (coke or carbon and ash with no fuel value). The desired fuel, and to some extent the fuel stock material, will determine what conversion method is to be used. These industrial processes, most commonly either pyrolysis or hydrogenation, have an enormous advantage over biological or enzymatic conversions in that the feedstock material need not be extremely uniform nor free from chemical toxins. Biological systems are extremely sensitive to a wide variety of chemicals appearing in urban waste, particularly if industrial wastes are included, but when used with organic wastes, they are far superior. Organic wastes typically contain large amounts of water and require energy for drying in preparation for the thermochemical process. Valuable nutrients, particularly ammonia, are lost in the mixed ash fraction from thermal processes (which in addition are capital- and energy-intensive and require large scale).

Wastes unsuitable for biological conversion will be difficult to avoid in any semi-industrialized society, and to the extent they exist, thermochemical conversion methods will prove useful. Long-range planning requires, however, that a minimum of organic wastes take this destructive conversion pathway and that wastes are re-evaluated for their potential value. Industrial poisons, in particular, should be treated on site to prevent the contamination of otherwise valuable wastes.

8

Integration for Decentralized Development

Integration for decentralized development based on biological solar energy conversion involves several different aspects of the same word:

1. *Technical integration* in the sense that all biotechnical processes are interlinked in a manner and on a scale which allows the closure of all nutrient cycles: waste becomes an obsolete concept, and the high efficiencies of ecological systems are sought.
2. *Conceptual integration* of a society about that life factor which is held most important, such as growth (economic, energetic, population), religious purity, security, material quality within fixed resource limits, efficiency (e.g., solar-energy use), plus many others.
3. *Cultural integration,* where culture includes virtually all of Man's social and technical responses to the demands of survival. It emphasizes the man whom the technical system serves, the need to unite food and fuel with mental and physical health, education and social life in a manner, and on a scale, appropriate to his potential and needs.

This imperfect subdivision of the term integration emphasizes that it is an organizational concept whose form is determined by one or more primal factors, or centers. Thus, the roles of a particular technical element or process in each of two systems integrated about different centers cannot be directly compared. Religious significances of animals, plants, rites or even attitudes toward life, progress, cheating, or death are among the obvious examples. Technical examples include, for example, handicrafts, which have been (and are) alternatively denigrated as either being too simple and common, or prized for their uniqueness, each with its "human error." In the West, composting has gone from being an integral part of a near ecologically sound life, to being banished as old-fashioned and unsanitary, to a partial return in some circles as a fashionable token from an old system and appended to the modern system where it doesn't really fit. The modern industrial, urban society has an organization which conceptually—and sometimes legally but more often economically—rejects that which doesn't fit. Flush toilets are required by law, tin-can recycling is not sound either economically or energetically (in the urban organization), "wastes" can't be reused because they are polluted with dyes, pesticides, industrial poisons, etc.

This discussion is relevant to the question of integrated biological solar energy systems in two ways: First, the conceptual integration of the existing modern industrial state has an inertial effect which selects against processes inconsistent with that system; solar energy, for example, is inconsistent in a centralized system due to its diffuseness, sunlight being the ultimate in decentralized resources. Secondly, the technical integration of the biological processes, according to the concept of maximum solar energy utilization, and optimized by-product re-use, produces a sustained energy flow with an efficiency far exceeding the sum of the individual components. This is the true result of integration over and above the value of the separate processes themselves.

These arguments should not be misread as defending all rejects from, say, society type A simply because they might suit society B. C or D. A particular process, biological or physical, primitive or modern, can be highly valuable or entirely worthless under even the best of circumstances, but a process or resource like solar energy cannot be rejected or judged alone in a system whose organization is foreign and almost contradictory to its use. Consider, for example, the pricing policy of the electrical power utilities in Israel, which for many years had special low rates for electric water heaters in order to make solar water heaters uneconomical. Solar heaters cannot be declared unsound in general when they've simply been made artificially uneconomical in one context. Similarly, waste and tin-can recycling are not themselves unsound simply because they are uneconomical in the urban, industrial context.

Development has and always will proceed via a complex dynamic growth process where marginal alterations are made continuously according to price mechanisms and their suitability to the established. Elimination of the obsolete is motivated by the competitive survival of the new alternative and again is continuous, not stepwise. *If* the decentralized solar-based society truly is as necessary and advantageous as presented here, then the developing countries have a headstart on the path which the industrial world must eventually join. Centralized industrial societies, and the concentrated oil reserves which fuel them, have a conceptual integration whose inertia will be difficult to brake.

Oil has not only unleashed a series of transport and industrial processes, but has driven a lock-step development of social organization—urbanization. Urbanization is a social order which requires a constant energy flow to support the influx of materials, food, and people to the urban center, followed by the removal of wastes and products. A closer look reveals a more technical specialization; factories are built to use oil or electricity as a fuel, millions of transport vehicles burn only oil, and industries are tooled to make only compatible products. Economical establishments further constrain flexibility to adapt to stepwise rather than small incremental changes: well over 10% of the industrial work force is employed for transport, a million gas stations are specialized for gasoline, a half-billion toilets are specialized to the waste concept, and previously self-sufficient farmers are now growing cash crops. Institutional establishments add more inertia, which is then ideologically described as conservatism: the government, courts, education, the military and even the church supports the industrial state. Highways force the surrender of farmland, small farmers in some countries are induced to yield to large farms while new poverties—education, power, transport, time— are born with public tax money (Illich, 1973).

TECHNICAL INTEGRATION

Against this lapse into integration's full cultural implications, the simpler question of technical integration can be more fully appreciated for its critical, but subsidiary role. The need is hopefully clear that solar energy, and the biological and physical means to exploit its potential, must be judged with an open mind and not confined to its potential place in the industrial world as we know it today.

The first and simplest stage toward the development of an integrated solar energy utilization system, sketched in Figure I.8.1, consists of normal agricultural lands, an anaerobic digester for methane fuel production, algae ponds, and the necessary pumps, burners and motors basic to the unit's operation. This system is integrated not only in the technical interlinking sense, but also in a manner which coordinates the satisfaction of all basic survival needs: fuel and recycled nutrients from the digester will improve food production, while the improved waste treatment and increased food will increase hygiene and health. Finally, the improved health will aid production of more food. In closed, flowing cycles there is no starting, nor more or less important, point. Likewise, it would be contradictory to speak of the center or heart of an integrated system, but the biogas or anaerobic digester in Figure I.8.1 is indeed special in that it is a traditional bacteriological process whose new role energizes most of this simple system and those later adaptations.

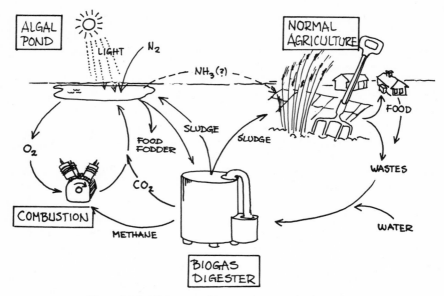

Figure I.8.1. Early stage of integrated system development.

Anaerobic digestion of plant and waste matter to methane gas, discussed earlier in Part I, Chapter 7, and later in Part II, Chapter 3, is one important example whose neglect and bad reputation have resulted from its use out of context. The process occurs naturally in marsh or swamp waters, but in modern times it has been used only for primary urban sewage treatment where the contamination of "good" wastes with "bad" (e.g., industrial chemicals) wastes regularly cause digester imbalance and/or failure. Furthermore, the nutrient liquid sludge from a digester is voluminous and unsuitable for disposal in natural waterways and hence poses a secondary waste problem unless recycling is possible. China, India, and several other Asian countries have employed these digesters in a more natural setting and with distinctly better results: plant and animal wastes are digested continously, methane is produced, and the nutritious sludge is returned to the land. In such a context, with even the simplest type digesters, the major technical problems are largely avoided and methane is produced for lighting, heating (including cooking and refrigeration) and internal or external combustion engines (e.g., irrigation or small industrial motors). An estimate for an "average" Indian village (500 persons and 250 cattle) shows that animal wastes alone can supply 10-15% more energy than is presently consumed from wood, cow dung, kerosene, and electricity (Prasad, Prasad, and Reddy 1974). In the same setting, a second calculation compared the energy costs to produce 230,000 tonnes nitrogen fertilizer per year in (a) a single coal-based high-technology fertilizer plant (in India), and (b) 26,150 biogas or anaerobic-digester plants. The former would employ less than 1% as many workers, and *consume* 10-million GJ (10^{10}MJ) energy compared with the biogas plants' *yield* of 23-million GJ (2.3×10^{10}MJ). Alternatively, one can ignore the fertilizer value and calculate only energy costs: methane cost (Indian estimate) is about half that of kerosene or electricity. Impressive as these results are, even they neglect the advantages of well-integrated but mechanically equally simple systems.

The high cost of construction of even simple digesters in capital- and materials- limited countries is a serious impediment to their use, but a long-range view reveals important advantages. Simple, unheated digesters begin the reorientation process necessary for successful recycling, and abundant solar energy justifies the construction of simple solar water heaters both for general use and, more specifically, for heating digesters to temperatures which maximize methane yields and help to sterilize the sludge. Addition of this single element, the anaerobic digester, can thus provide high-quality fuel, methane; low-quality heat from solar collectors and effluent sludge; recycled nutrients in the sludge for fertilizer, and purified waste water to decrease the spread of disease.

COMBINATION WITH ALGAE GROWTH

Recycling via an anaerobic digester for fertilizer and fuel is but a first stage of development, which still poses several problems and inefficiencies. Foremost among these are the periodic nature of crop waste availability (hence methane production) and fertilizer (sludge) requirements. Sludge will be produced even when not needed as fertilizer, and the digester will often be operating below capacity when crop wastes are not available. The "food chain" can be extended with the previously discussed algae pond (Part I, Chapter 6; see also Part II, Chapter 2), a simple earthen pond whose bottom is covered with, for example, a plastic sheet. As shown in Figure I.8.1, the sludge can provide nutrients required for intense algal growth when not needed for crop fertilizer (see Part II, Chapter 2, page 204). Thus, an extra closed loop is added to take advantage of the water-borne wastes and the high solar efficiency of algal growth. Algae in turn can alternatively be used for fodder (food?) or returned directly to the digester for conversion to (methane) energy. Methane can be used as much as possible for contained combustion, whose exhaust gases can in turn provide the carbon dioxide required for algal growth; some carbon dioxide will also be produced from oxidation of undigested organic material in the algae pond itself.

Algal growth leads to further developments, one of those deserving special mention is aquaculture. Aquaculture, in contrast to agriculture, is the cultivation of fish, especially in closed systems (see Part II, Chapter 2, page 230). Aquaculture is again nothing new as it has been practiced for centuries in both Asia and Polynesia where admirable production yields have been reached in closed recycling systems. Most table fish are however not herbivorous, i.e., most fish eat smaller organisms which in turn feed on algae and aquatic weeds. Whether aquaculture uses one of the relatively few herbivorous fish, bivalves like clams, or a longer food chain, is a technical detail; but more important is the natural algal-harvesting ability which fish life provides.

Fish, like all animals, are relatively inefficient converters of plant food to animal (fish) food; only about 10% of the food digested by fish is ever harvested as fish meat, while 90% is lost as metabolic energy or excreted as organic nutrients. In intense cattle feedlots, this same inefficiency results in prodigious quantities of manure, but in the closed system discussed here, the inefficiency is powered directly by the sun with no net nutrient loss. Even if fish are exported from the system, when they are produced in conjunction with the waste-recycling process, only a small fraction of the nutrients are ever lost.

Aquaculture can take several forms depending on the desired level of intensity, which to a large extent determines areal yields. The simplest form is, for instance, merely to combine aquaculture with irrigation ditches. A more intensive form bypasses a section of the algal growth pond through a separate fish pond (this separation is necessary as the fish pond must have an algal concentration lower than the algal pond, to which the fish excrement is later returned). A rather low-intensity pond a few meters in diameter is said to provide a five-member family's fish needs. As is true for digestion for methane production, the intensity of operation can be increased stepwise until production yields have been optimized for both fish and algal production.

The anaerobic digester assures the recycling of all nutrients, but not only must the cycled level be built up initially, losses must also be compensated for. This applies particularly to nitrogen (ammonia), which is lost via leaching to ground water or volatilized to air. Algal nitrogen fixation is particularly important not only because it compensates for lost nitrogen, but also because it can replace some or all of the need for synthetic nitrogen fertilizers (see Part II, Chapter 2, page 242). Other nutrients are not so readily lost from the closed system, but if productivity is to be maximized, their cycling level must initially be increased by imports. Nitrogen fixation and possible future research developments may increase fixation rates so that a fixed algae population will convert solar energy to ammonia with several percent efficiency, but for present applications the blue-green algae will contribute most if grown on low-nitrogen wastes. Wastes will, as the system becomes more complex, be used according to their relative values of nutrient and energy contents and ease of digestibility. Nitrogen content in wastes will be increased not only by the recycling, but also by increased use of nitrogen-fixing species in the agricultural crops.

Water resources are critical for both plant growth and hygiene. In all but the arid and semi-arid regions, water is available in sufficient quantity but often at the wrong time or place. Water considerations, therefore, include pumping (irrigation), storage, and conservation in all uses.

Water pumping, particularly for irrigation, is basic to even the simplest schemes and is thus one of the important uses for methane (see Part II, Chapter 4, page 324). Pumping from the digester to the algae pond can also use combustion engines, but the intermittency of need, relatively low volumes, and small height differences may make hand pumping more suitable. Irrigation pumping, whose energy demands vary with pumping height and distance, can be minimized via trickle irrigation to increase water-use efficiency from 20-40% to over 90%. In addition, nutrient uptake efficiency can be increased to nearly 100% with trickle feeding, and problems of salination, leaching and water-logging are simultaneously avoided.

Water storage is partially possible in conjunction with algal growth, particularly for the voluminous watery sludge from the anaerobic digester. Such considerations will, in some cases, justify a pond deeper than the 50-100 cm optimized for light absorption (see Part II, Chapter 2, page 207). In arid or semi-arid regions up to two meters of water can evaporate per year, which, in addition to being a loss of water resource, causes the loss of half the incident solar energy as evaporative cooling. Both the energy and water could be saved if the pond had one or two layers of transparent plastic covering. Algae can grow at temperatures of up to 30-35°C, while the excess heat can be removed for simple low-grade heat uses for domestic use or digester warming. In later stages of development, the same covered ponds can transfer their heat to a secondary fluid, which can in turn drive turbines for mechanical or electrical energy. The latter system has been estimated to convert solar energy to electricity with 3% efficiency (in addition to double use for algal growth). Covering also allows recovery of photosynthetic gases, oxygen for combination and later, perhaps, hydrogen gas for fuel cell and other uses (see Part II, Chapters, 2, 3, 4 and 5, pages 237, 294, 319 and 335, respectively).

ADDITIONAL ELEMENTS

The simple basic system described above, consisting of traditional agriculture with nitrogen-fixing organisms, an anaerobic digester and algae ponds, is but a first stage which itself can be developed (e.g., with nitrogen or hydrogen-fixing blue-green algae or covered ponds for heat and electricity production) or later combined with complementary processes. Most importantly, however, is the lack of planned obsolescence: the first-stage components would always be an integral part of future developments. Future additions would add to the system complexity and would begin to approach the efficiency of natural ecosystems by matching product or by-product quality to the end use: heating would use abundant low-grade and waste heat, while methane would be reserved for high-temperature use or combustion engines. Crop wastes would also be graded according to their quality, while crops grown for special end purposes or special local needs can motivate other processes:

- production of liquid fuels, e.g., ethyl alcohol, via fermentation of starch crops (see Part II, Chapter 3, page 249);
- enzymatic cellulose breakdown to simple sugars for a host of special fermentations (see Part II, Chapter 3, page 278);
- mushroom growth on digestion-resistant cellulose;
- rubber production from specialized plants.

This list could be extended to more futuristic developments from modern research level enzymology, microbiology and genetics (see Part II, Chapter 5, page 330).

Just as primitive biological processes (aqua- and agriculture, fermented foods) are combined with modern research level developments, the same component breadth applies to complementary nonbiological processes:

- windmills for mechanical energy: water pumping, stirring, small industry;
- solar energy devices for simple water heaters, water pumping, water distillation, photoelectric cells, and biophotolysis for production of hydrogen/oxygen from water;
- fuel cells for electricity production (e.g., from hydrogen/oxygen);
- wave power;
- geothermal power;
- microelectronic for optimization and control processes.

Biological elements have been emphasized here to establish a new balance and exploit their low-capital requirements, but clearly physical devices from past and future technology will raise the ultimate maximum level of sustainable energy use. As energy and other resources become limiting, both biological and nonbiological processes must be used on an investment basis for long-term viability. Many of these physical devices have been recently reviewed (NAS 1976a).

ENERGY YIELDS

The standard criterion for judging alternative energy systems is the per capita energy consumption in industrialized countries. These figures, shown in Figure I.8.2, cannot be used in simplistic comparisons, nor do they constitute justifiable goals or requirements. Firstly, as was pointed out in Part I, Chapter 5, a considerable fraction of energy costs for agriculture, transport, industrial, and personal use are attributable to centralized social organizations and not to the standard of living. Secondly, in each of these comparisons energy quality is ignored; 60-70% of consumed energy is used to produce low-grade heat which could instead be produced as a by-product or from other more appropriate sources. These two factors are corrections which are required *before* valid comparisons of energy vs. material standard of living can be made. Once known, several additional factors must be considered. For example, while energy is clearly required for life and comfort, a growing body of evidence indicates that energy use is not necessarily a measure of living standard or comfort, nor even directly related to gross national product. Beyond a certain point, low-energy services become increasingly important. Furthermore, pure waste and low energy-use efficiency cannot, in a world of limited resources, be assumed basic to a high standard of living. And finally, all goals are not achievable, and the future may show that the desired per capita rates of energy consumption are not sustainable. The existence of technological precedents cannot validate what may be a

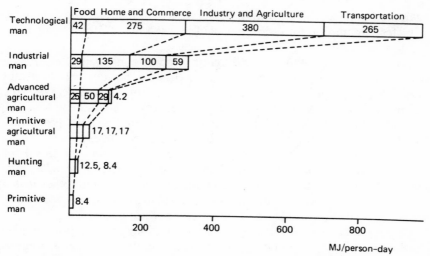

Figure I.8.2. Daily per capita energy consumption (data from Cook, 1971).

physically impossible assumption, and the assumed possibility of this wish should not distract Man from investing remaining resources in a viable alternative. Whatever goal is chosen, it cannot conceal built-in inequities nor retain those which exist today. Those inequities of present resource usage can be dismissed both on moral grounds and from the system stability point of view.

With these qualifications in mind, photosynthetic and other solar-energy alternatives can be analyzed. Reference to the world solar-energy distribution in Figure I.9.4 allows construction of Table I.8.1, which shows the annual energy capture per hectare and its variation with both latitude and conversion efficiency. While plant growth at the maximum theoretical efficiency (13%) would at 30°N or S produce over 400 tons/ha-yr (8000 GJ or 8 TJ/ha-yr), a more reasonable goal for the foreseeable future is a 4% conversion efficiency with yields of 100-150 tons/ha-yr (2-3 TJ/ha-yr); the latter yields are only somewhat better than the maximum yields presently observed for the most productive crops, such as sugar cane, water hyacinth, algae, or forage grasses (see Table I.6.1).

When speaking of crop yields in energy terms, however, these calculated gross yields must be reduced by the amount of energy invested in agriculture. Here begins a series of difficult assumptions. Reference to Part I, Chapter 5, shows an energy ratio for modern agriculture of between 2.5 and 3.0, but as shown there, up to 90% of the invested energy supports social organization rather than pure increased productivity. A recalculation for maize with 10% of modern energy investments increases the energy ratio to 25, i.e., 25 GJ produced in crop energy for each gigajoule invested. While future research will hopefully validate these estimates, a compromise ratio of 10 will be used here. Gross yields must therefore be reduced by 10%.

A harvested crop of sugar cane, grain, or wet algae can of course be used as food, but in a more general sense they are not universally useful fuels. The crop energy must be converted via imperfect processes (see Part II, Chapter 3), which reduce the gross energy value by between 10% (alcohol from starch) to 40% (methane conversion). These gross fuel values must in turn be reduced by the conversion-process costs. These figures are not known as the processes have not yet been optimized for energy efficiency, but are assumed for this calculation to be 67%. Net fuel yields for an assumed 4% solar-energy conversion (with other losses summarized in Figure I.8.3) are shown in Table I.8.1. Low-grade heat, not included here, will also be available at several stages.

Energy yields must also be related to population, or more importantly, population density per unit of arable land. Presently the world average is less than three persons per arable hectare, but by the year 2000 this figure

Table I.8.1 Biomass and Net Fuel Yields for Various Conversion Efficiencies and Latitudes.

Latitude	Seasonal Variation (highest/lowest month)	Solar Energy TJ/yr-ha		BIOMASS YIELD (ton/hr-yr.)*			NET FUEL YIELDS FOR % SOLAR-ENERGY CONVERSION	
				1%	4%	13%**	GJ/ha-yr	GJ/person-yr
0°	1.2	max.	87.4	43.7	175.	570.	1250	312
		min.	63.	31.5	126.	410.	880	220
10°	1.5	max.	85.	42.5	170.	550.	1200	300
		min.	59.4	29.7	119.	385.	830	207
20°	1.8	max.	77.	38.5	154.	500.	1080	270
		min.	53.	26.5	106.	345.	740	135
30°	2.2	max.	74.5	37.3	149.	485.	1040	260
		min.	44.2	22.1	88.	288.	620	130
40°	4.	max.	63.	31.5	126.	410.	880	220
		min.	34.	17.	68.	220.	380	95
50°	8.	max.	53.	26.5	106.	345.	750	187
		min.	26.	13.	52.	169.	360	90
60°	20–35	max.	30.6	15.3	61.	198.	430	108
		min.	15.3	7.7	31.	100.	210	52
70°	—	max.	10.2	5.1	20.	66.	140	35
		min.	5.1	2.6	10.	39.	70	18

*Biomass assumed 17.8 MJ/kg, but 23 MJ/kg for algae, 16. MJ/kg for sugar, 40 MJ/kg for fats, 24 MJ/kg for protein (Schneider 1973).
**Conversion efficiencies of total solar radiation at ground level.

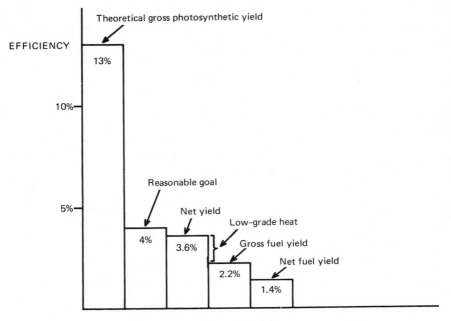

Figure I.8.3. Efficiency goals for energy crops.

will be up to just over four persons per hectare. Division by four then gives the net annual fuel yield per capita from arable land, which at the equator approaches the present US per capita energy-consumption rate. These data are summarized in the final column of Table I.8.1.

So far we have only included arable land in this calculation, but arable land is only 11% of the terrestrial surface, whereas pastures and range land is twice as large and forest land is three times as large. Plant or tree yields from these lands must also be added (Part II, Chapter 1, page 192) where appropriate in addition to heat energy obtained as waste by products or from physical solar collectors. The uncertainties in the crop yields are difficult enough, but these additional areas, while extremely important, have not yet been estimated. A careful examination of a given system in a specific region is required, but the potential is at least shown to be extremely promising.

SCALE

Energy yield of a biological solar-energy conversion system also depends on the relation between energy efficiency and scale. Whereas economy of scale is axiomatic today, is its validity specific to an industrial society fueled by cheap oil? What valid analogies can be made to the scale of organization in the biological world? Does one take guidance from cell specialization in complex organisms or the complex diversity of mature ecological systems, from the survivability of microorganisms or the extinction of dinosaurs? Apparent inconsistencies in the biological world are only superficial, as in every case adaptability at the most basic level is preserved. Each cell retains the information, and to varying extents the ability, to revive all functions for its survival. This ability, however, decreases in higher organisms and leaves them (read "us") more vulnerable to change. Retainment of capacities in spite of specialization of function, construction from simple recyclable, multipurpose building blocks, and network complexity with low transport requirements—these are among the evolutionary lessons.

Enthusiasm for biological analogies must, however, be tempered with the sobering realization that, much to Man's surprise, Nature (i.e., system evolution) maximizes total system stability, while Man wishes to maximize human population and system productivity. Exportable productivity in a complex ecosystem such as a rain forest is extremely low and increases only with compensatory energy inputs. This realization doesn't diminish the applicability of evolutionary principles, however, but instead brings long-term constraints more clearly into view.

But what are the practical consequences in questions of scale? Like other questions, the determination of an optimal size must be left unanswered. Intuition constrains us to experience, but this suggests that the individual and/or family level is too small a scale while an urban center is too large. An optimization presupposes only one or a few limiting factors and preferably ones with known effects. City planners accept large urban centers as a given fact, while others view cities as a means for optimization of social development (Doxiadis 1976).

While not proposing optimization on the basis of energy efficiency alone, population and support organization can at least be compared in such terms. Transport distances and associated energy costs vs. the energy contents of recycled wastes will set rather low-size limits: after a certain distance, a waste is indeed a waste and poses not only a disposal problem but one with an associated net *negative* energy value. Time and speed of transport has often been heralded as a triumph of modern technology, but estimates show that the time spent for the construction, use, repair and payment of all transport reduces the average rate of travel to 7 km/hr, i.e., walking speed (Illich 1974). No gain has been made in the average rate of travel, yet a tremendous amount of energy and resources has been spent to achieve variety and high speed for some.

Proponents of reduced industrial and population organization scale (such as Schumacher, *Small is Beautiful,* 1973, and Illich, *Tools for Conviviality,* 1973) argue from a techno-sociological point of view for man-sized processes and for tools whose cost and accessibility cannot be monopolized. High capital costs are shown to prevent equal accessibility; hence working places whose capital investment exceeds the worker's annual salary would not be considered "convivial" or appropriate.

Biological solar-energy systems fulfill most of the requirements for appropriateness even though economic and material investment costs have blocked their previous realization. Both solar energy and the basic biomass building blocks (air, water, microorganisms) are free, with the sole requirement being support equipment. But small is not necessarily beautiful, and process scale will vary from the highly diffuse small-scale (e.g., solar energy storage via plant growth), to the locally centralized (waste digestion and recycling), to the regional (e.g., liquid fuel production) and finally up to the national and international level for trade in specific items. Similarly trade flows in the reverse direction would increase sharply nearer to the local self-reliant unit where the bulk material flows are all internal. A corollary to this is the graduation of waste quality where low-quality wastes of necessity are recycled locally while higher quality wastes (e.g., starch, or low-grade sugars) *perhaps* can be collected and transported for specific functions.

Scale cannot be more specifically described without reference to a specific situation. Anaerobic digesters for methane production from wastes are probably not appropriate to each farm or family at this time in India, for example. Low crop (and waste) productivity and the unevenness of biomass availability due to land ownership and development are more conducive to village-level recycling where the landless can also partake of the gains. The modular nature of these processes allows—in fact will require—a proliferation and further decentralization as land productivity increases.

Scale, to summarize, cannot be defined absolutely, but rather represents a balance to decentralize those processes most conducive to local scale and, when justified, interdependency with centralized industry. The final balance is yet unpredicted, but the intermediate need can by no means be the same as the final one.

9

Steps to Development

It should be clear at this point that integrated decentralized production of food and fuel is required not only for maximum efficiency, but is also a minimum requirement for successful development as well. The failure of piecemeal solutions has been demonstrated repeatedly in developing countries: irrigation systems requiring fuel which isn't available or is too expensive, hybrid grains whose fertilizer-uptake ability can't be met, decreased child mortality for a life of starvation and sickness from lack of food and hygiene, cash cropping which leads to decreased local food supplies and subjugation to prices fixed by the overfed world markets, heroic feeding programs leading to increased populations with increased vulnerability, schools which educate for industrial life rather than self-survival, and so on.

This bleak and admittedly one-sided portrayal reinforces the inter-relatedness and interdependency of all survival factors. While it doesn't ignore that development is more a social problem than a technical one, health does require food, and food also requires healthy people to raise it. Both health and food production also require improved water quality and quantity. Cholera, typhoid, hepatitis, and leprosy are among the water-related diseases (Bradley 1974); acute diarrhea alone plagues 500-million Third-World children resulting in weakness, death, and inefficient use of scarce food supplies (Elliott 1976). Food, health, and water also require energy, but energy import in poor countries leads to economic bondage, while burning of cow dung or wood at only 5% efficiency wastes nutrients and leads to soil destruction (deforestation is contributing to a *loss* of productive soil at a rate of 1-3%/yr). Nothing will stop these trends until integrated development releases the need for wood burning, provides clean water, and offers a means to feed an upward spiral.

Integrated development is impossible without resource investment, and particularly that investment which catalyzes local self-reliance, not bondage to external imports and the profits of external markets. The economic burden of exporting all the products of the land to pay for fuel and fertilizer prevents long-term planning for the day when fossil fuels end. Investment must be a gift, an investment for our mutual future, and take both a capital and informational form. But first it must be accommodated to local needs and resources.

GEOGRAPHY OF NEED

The capital aid and specific development pattern has geographical variations, both according to need and resource balance. Population density, shown in Figure I.9.1, has in the past reflected Man's ability to survive in different climates and geographical conditions according to traditional farming and settlement methods. Local concentrations of economic and political power resulting from the fossil-fuel powered industrial revolution have changed this natural distribution. While the energy and economic costs for altering this balance are clearly excessive, population pressure has led to the settlement of marginal lands at, of course, higher cost: 20-25 times the expenditures are required to till each new hectare of marginal land compared to the fertile land tilled earlier (Kovda 1974). Most of these expenditures are for energy (direct or indirect) with the result that soil variations and population density together reveal those regions whose local food supplies are most threatened by decreased energy availability. Two groups are particularly relevant in this respect, those who live on the *expanding* edge of a desert—for example the northern Sahara Desert where 100,000 hectares are lost to desert each year (IFIAS 1976)—and those whose undernourished soil supports a considerable standing biomass but low annual productivity—such as tropical rain forests or savanna. Desertification has the advantage in that it is at least theoretically reversible in many areas where it occurs in East and South Africa, the Middle East, parts of Latin America and Asia. A climatological contribution is considered (Eckholm 1976a), but without question its principal causes are overgrazing, improper cultivation and firewood cutting (Eckholm 1976b). Soil destruction from certain pesticides, herbicides and industrial pollution (mercury, cadmium, lead, arsenic, fluorine, etc.) is mostly "an irreversible or almost irrevocable process" (Kovda 1974). Salinity and alkalinity, such occurred in the Middle East from improper irrigation, may in some cases be altered by water quality improvement, but water resources, shown in the second map (Figure I.9.2), are perhaps the greatest ultimate limitation.

As 400-500 kg of water are required for each kilogram of plant matter (McHale and McHale 1975), its availability has historically determined population patterns and even soil distribution. Desert conditions result principally from lack of water resources, but tropical rain forests have low

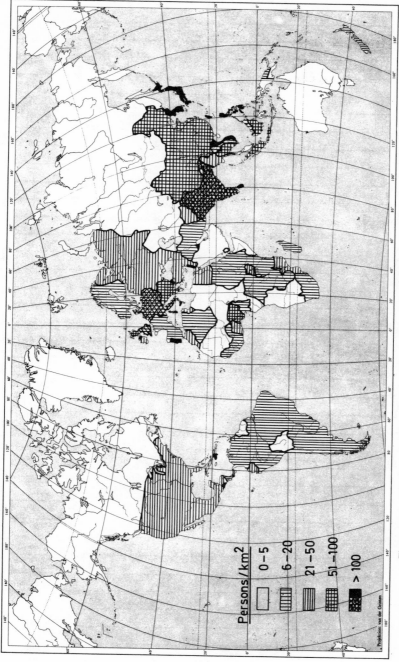

Figure I.9.1. Distribution of the world's population (redrawn from Ehrlich, Ehrlich, and Holdren 1973).

101

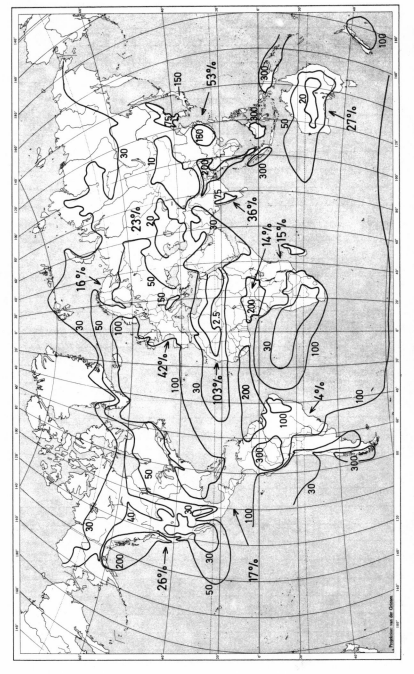

Figure I.9.2. Distribution of world water resources: contours of equal annual precipitation (cm) and regional water need's relation to resources (%) (data from Times Atlas 1973 and de Maré 1977).

102

productivity because an excess of rain leads to nutrient leaching. As over 100 cubic meters of water are required per person per year (de Maré 1977), rainfall plus other water resources can be combined to calculate a potential world water-carrying capacity of about 30 billion persons. On a regional basis only Northern Africa and the Middle East are at their maximums, while as shown in Figure I.9.2 no other region even exceeds 50% of its resources. Local variations exist of course, but the greater problems are periodicity (e.g., monsoon rains), irregular droughts, salination, and energy costs for water pumping. Water storage, flood control, and solar water pumping are among the greatest needs. Irrigation from these water reservoirs requires increased investment, but improper techniques have led to soil waterlogging, salination, alkalinization, nutrient leaching, soil erosion, and loss of 50-80% of the water. Trickle irrigation, when possible in conjunction with controlled fertilization, will increase water-use efficiencies up to 95% (Shoji 1977).

Solar-energy resources are abundant throughout the populated world and particularly in developing countries: over two times the annual solar energy is received in Central America, Northern Africa and India as in the extreme reaches of South America or Northern Europe. As shown in Figure I.9.3, the world maximum falls on the arid and desert regions of Northern Africa and the Middle East and calls for greater emphasis there on closed system farming, greenhouses and covered algae ponds, and a greater balance toward nonbiological solar devices. The greater capital costs involved in these physical solar energy devices is a happy coincidence with the petroleum reserves in the countries where they will be most appropriate. A unique opportunity for long-term planning would allow the development of a solar energy capacity based on petroleum investments which would provide not only long-term survival but technology applicable in part to less-fortunate regions. Oil and coal reserves are shown in Figure I.9.4 and reveal that none of the Third World regions with the exception of China are particularly well endowed with coal. The so-called "global coal option" in practice means that the USA, USSR and China own almost 90% of the world's coal (Grenon 1975). The fortuitousness of resource distribution and the limitations to its supply will hopefully begin to motivate more altruistic uses of what remains.

The geography of need is literally seen in the final map (Figure I.9.5) which shows the geographical distribution of low protein and/or low-calorie diets. While the future is normally portrayed as bleak, a review of the maps does promise hope: water, soil, and solar energy resources are adequate to support the populations of the future. To the extent that energy is required for food production, the world's future lies with its decisions about energy use and nutrient reuse. As nitrogenous fertilizers

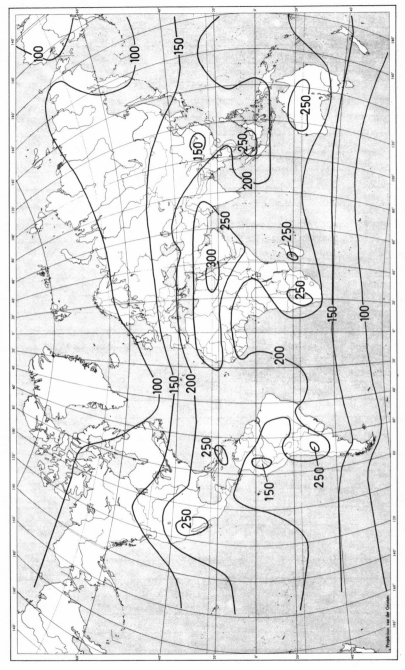

Figure I.9.3. Mean annual intensity of global solar radiation on a horizontal plane at the earth's surface (watts/meter^2 averaged over a 24-hour day; data from Porter and Archer 1976).

104

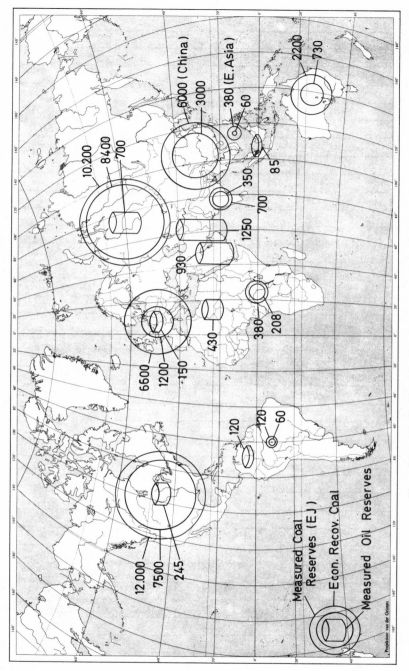

Figure I.9.4. Distribution of fossil fuel resources (data from WAES 1977).

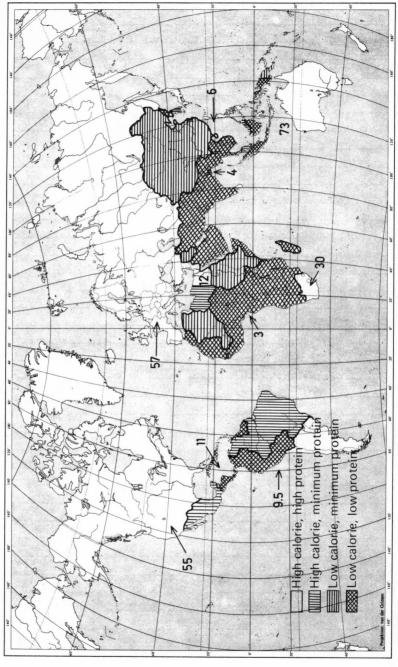

Figure I.9.5. Distribution of hunger and per capita fertilizer use (in kilograms, redrawn from Ehrlich, Enrlich, and Holden 1973; data from UN 1975).

106

combine the concepts of energy, nutrient and waste recycling in a particularly visible way, they will be used for a simplified example of this
choice: the magnitude of the economic and energetic burden for nitrogen
fertilizers with the modern waste concept will be compared to that for a
recycling society.

The Green Revolution is presently stymied by economic, not energy,
considerations. Oil is available, but as shown in Figure I.9.6 the seven-
and-a-half-fold increase in its cost in the last six years has increased
ammonia costs (a basic nitrogenous fertilizer) tenfold. These costs will
not decrease as the industrial synthesis process, already at a high-
efficiency maximum, depends on fossil fuels which constitute half of its
manufacturing cost.

World nitrogen fertilizer production deficit estimates for 1980/81 vary
from as low as one million tons to the World Bank's estimate of 12
million tons. These estimates are based on demand expectations resulting
from agricultural growth, but should not be confused with any ultimate or
optimal demand rate. *Real* world soil-nutrient needs can be calculated
from the arable land area (10^9 ha intensive plus 0.4×10^9 ha extensive)
and nitrogen fertilizer application rates typical of Europe and North
America; these rates are approximately 30 kg/person-yr or 70 kg/ha-yr.
Total world nitrogen-fertilizer need is therefore 120-million tons for the
present world's population; on a per hectare basis this fertilizer need is
100 million tons. Optimal areal rates are however 150 kg/ha-yr (50 kg for
extensive arable land) and increase the actual present need to 300-million
tons per year. (Breeding to increase optimal nitrogen uptake rates beyond
150 kg/ha-yr are assumed here to balance increases in use efficiency,
e.g., in conjunction with trickle irrigation.) The real fertilizer deficit is

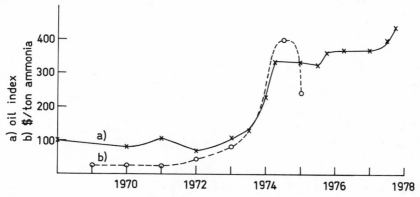

Figure I.9.6. Variations in fertilizer and oil costs (data from UN 1975).

thus 30-300 times what is normally quoted. Note that this astronomical sum is the logically concluded need of illogical modern agriculture.

The nitrogen need is 300 million tons per year while present production is 40-million tons. Man's industrial contribution is about 20% of the total, but future production can be double that of all natural fixation processes (150 million tons per year). Assuming industrial production can reach such rates, what would be the ecological effects? Can present pollution rates be increased eightfold? How does 300 million tons of nitrogenous fertilizer—double the annual natural production rate—look from the wrong end of our cities' sewers (see Figure I.5.3)? This view is but one reminder of the need for recycling.

Consider what India's capital and energy costs would be to produce nitrogen at both European (70 kg/ha-yr) and optimal levels (150 kg/ha-yr). Production of the optimal 18-million tons per year (which also corresponds to an equal per capita fertilizer use as in industrial nations, but only at the present Indian population) would require $7-billion capital for the 72 1000-ton per day fertilizer plants. In addition a constant supply of 1.4 billion GJ oil per year (200 million barrels) is required at an annual cost of $2-billion. This energy flow is half the present (1971) total Indian energy consumption of 2.8-billion GJ and three times its oil imports: 50% of the per capita energy consumption would be for nitrogen fertilizer compared with 0.4% for the US. With a GNP of $42.6 billion (1970), neither the investment nor continuous costs are even imaginable.

For the world to increase its nitrogenous fertilizer production to European or optimal rates would require $30 or $120 billion, respectively, plus an annual use of 5.6- or 24-billion GJ oil (0.8- or 3.5-billion barrels, respectively). Compare this to world oil consumption of $3000-million (130 billion GJ/yr and reserves of 88,000 million tons) 3800 billion GJ. Furthermore, recall that nitrogen is but one nutrient, and fertilizer is but a third of the agricultural energy costs. Can 60% of the world's energy budget go to agriculture?

Clearly the discussion has become absurd, but is it more absurd than defining a fertilizer deficit on the basis of an application rate 70 times less than the European rate of 1/300th the optimal rate? The absurdity doesn't end there. Consider world military expenditures of $334 billion per year.

Biological nitrogen fixation together with maximum nutrient recycling is an alternative which simultaneously eliminates the enormous production, transport, pollution, and energy burden associated with massive use of chemically synthesized nitrogen fertilizers. Some fertilizer production will always be required, particularly in those areas or for those crops where recycled nutrients or biologically fixed nitrogen doesn't suffice. Even with recycling and even with trickle application to approach 100%

uptake efficiency, some losses will occur. But present nitrogen fertilizer production plus biological fixation can more than compensate for losses. With this real alternative, the absurdity of the modern agricultural industry must be reviewed and an estimate made of the investment costs to replace the self-destructive goal with a viable one.

CAPITAL AID

The capital-aid requirement to launch self-reliant development would be specific to the region and technology available, but in general the capital equipment would be designed to realize the *first stage* in a development sequence without any planned obsolescence. The factors basic to this first stage are those referred to in Part I, Chapter 8, to support the simplified scheme in Figure I.8.1:

- Simple solar energy devices for distilling drinking water, solar (or methane) or wind-mill water pumps, solar crop driers, water heaters for digesters (below), etc;
- Anaerobic digester for producing high-quality fuel (methane), recycled nutrients (for fertilizer), and purified water. The superior sterilizing effect of thermophilic bacteria justifies in itself this process;
- Maximized use of microbiological nitrogen fixation via legumes, other soil bacteria, and blue-green algae;
- Non-nitrogenous chemical fertilizers to raise the soil-nutrient level;
- Simple machinery and burners utilizing methane gas.

This capital investment provides the means to increase food production, remove the two greatest sources of ill-health (impure water and malnutrition), produce fuel (methane) and at the same time recycle nutrients. A one-time investment can, from a capital materials' point of view, launch a self-reliant unit which can operate without further fuel or fertilizer imports. Closed units are not held as an ideal, but rather they are a model for a new balance, especially in relation to materials flow: crop matter and the incredible wastes involved can only be optimally used on a local scale. Exports, to minimize energy and nutrient loss, should be restricted to finished quality products from local small industries. Likewise, imports are mostly catalytic items which promote internal development.

EDUCATION AND INFORMATION EXCHANGE

While nutrient and energy flows in the future world of limited resources need to be minimized, education and information exchange is crucial to personal and community development. Education, instead of being a tool for industrial manipulation, would integrate the vast practical farming experience with scientific understanding of the whole biological world: nutrient and energy flows, the unity of plant, animal (including humans) and bacterial life, and their interrelations for health and energy. Such education is essential for effectively preserving hygiene, efficiency in energy and nutrient use, and controlling population.

Education and information exchange is not restricted to schools, but plays a central role in continued development. Particularly important for information exchange is the testing of soil and soil bacteria for optimal nutrients and the matching of plant and nitrogen-fixing bacterial species. The local experience of various plant and algae varieties would need to be shared as would the results of genetic research tested in practical applications. Suggested changes in system optimization could be tested with laboratory or computer modeling. These steps are but a beginning in an open-ended development whose optimum is yet unknown.

RESEARCH IMPERATIVES

The responsibility of the developed world is to invest resources for our mutual future and progress toward a mutually equitable goal. The positive technical results of the industrial world's period of excessive consumption should not, however, be ignored and can be further developed to repay a debt. While most of the scientific research and technical development cannot yet be done by the poorer countries themselves, elements for future stages can be developed through further research in accordance with their wishes. When possible, the research should be conducted in the interested developing country.

The attractive potential biological developments are most easily seen in the familiar schematic summary of green plant photosynthesis shown in Figure I.9.7. Sugars (and complexed sugars, carbohydrates) produced photosynthetically have efficiently fed and fueled the world for its whole biological history, but never before modern times has Man tried to replace evolutionary imperatives with His own. If the limitations, risks and costs are known, then a new look at the scheme in Figure I.9.7 reveals numerous critical areas for basic research and development.

Little can be done about the energy *source*: as pointed out by both nuclear energy promoters and detractors, we already have our first functioning nuclear fusion reactor. It is, however, at a safe distance and little can be done to alter it before its natural end.

The *process*, photosynthesis, evolved when conservatism was essential to survival. But today Man can intervene to aid those biological processes

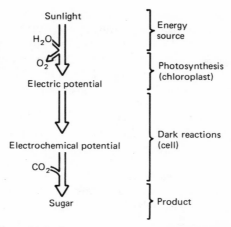

Figure I.9.7. Summary of green plant photosynthesis.

particularly useful to Him, but always at a price in energy and system stability. Plant growth can, however, be improved in several respects:

Rate Improvements.   Rate improvements have productivity as a goal.

Nutrient uptake.   Nutrient uptake can be limiting, and increased fertilizer utilization rates have successfully been bred into all of the major world crops, but whether universal use of these strains is justified with the resultant massive synthetic fertilizer dependence is uncertain as oil reserves dwindle. Carbon dioxide uptake rate often limits photosynthetic productivity in bright light, and most plants respond well to CO_2 "fertilizer," that is growth under higher gas concentrations. Some plants (see Part II, Chapter 1, page 157) have solved this problem, and their enzymatic uptake systems require research to be understood to determine if breeding or other means can transfer their properties to common crops.

Light absorption.   Solar-energy absorption and conversion is rate-limited in bright light by the (electron carrier) molecules connecting the two photosystems. The limiting processes, which can also result in plant damage, are poorly understood and require basic research to learn if natural variations in photosaturation levels (reaction rates and photosynthetic unit size) can lead to practical results. This problem is particularly serious for algae and synthetic choloroplast systems.

Nitrogen fixation.   Nitrogen fixation rates (ammonia production) are limited by nutrient supply and ammonia inhibition. The first factor in legumes requires extensive applied research to determine optimal bacteria-host species matching to minimize wasteful hydrogen evolution and to optimize fixation. Inhibition caused by the accumulation of ammonia (i.e., fixation product) is sound evolutionary policy but suboptimal for Man; regulatory mechanisms are only understood in part, and both genetic and chemical manipulation are feasible in the laboratory.

Efficiency Improvements.   Efficiency improvements may often also increase productivity but normally not to a maximum. Efficiency can have several definitions, however.

Fertilizer efficiency.   Nutrient or fertilizer-use efficiency is of immediate and critical importance. Breeding for maximal nutrient-uptake efficiency has been shown to be effective for potassium, but has been neglected in favor of increased uptake *rate* with excess fertilizer. Breeding together with carefully programmed trickle fertilizer application allows nearly 100% uptake efficiencies while present efficiencies of 10-60% result in energy and nutrient waste, plus water pollution.

Water efficiency.   Water-use efficiency is known to vary greatly (e.g., as for desert plants) in part due to reduced transpiration from closed stomates (pores). As transpiration and CO_2 absorption both occur via stomatal openings, the simultaneous occurrence of high productivity (requiring CO_2), high temperature growth optimum (i.e., low need for evaporative cooling), and high water efficiency are both perplexing and promising as an avenue of research. Control of stomate opening should be investigated as should the closely related question of growth temperature optimum.

Carbon dioxide uptake. CO_2 uptake difficulties limit overall growth rate and efficiency, and their origins require extensive basic research.

Photorespiration.   50% of crop productivity is needlessly wasted by photorespiration. The origins of this loss and methods to avoid it constitute a research imperative of highest priority.

Light efficiency.   Solar-energy use efficiency is probably unimprovable except for photosaturation effects (see ''Light absorption'' above, page 152) and husbandry practices which increase canopy cover and season length—i.e., intercropping and multiple-cropping, respectively.

Nitrogen-fixation inefficiencies.   Inefficiencies in nitrogen fixation (more than 30% of nitrogen fixation energy is lost as hydrogen gas) are in part controllable by bacteria-host species matching and breeding.

Product Balance.   Plants produce sugars first, but this energy is later converted to structural materials (sugar as cellulose, plus some proteins and fats), energy storage (sugar as starch, plus some fats or other hydrocarbons), and functional molecules (protein enzymes plus DNA, RNA, etc.). That the product balance varies with species is obvious from traditional food selection and generally is related to the plant specie's environmental niche via enzymatic and genetic growth control. Breeding with or without mutation is the traditional method for altering the product balance, and while its central role will continue, specific genetic manipulation, regulation blockage, plasmids and chemicals may play larger roles. All means should be tried for various product balance alterations.

Protein.   Protein content has been more than doubled in several grain species (e.g., maize, wheat, barley, rice, sorghum) along with dramatic increases in otherwise lacking essential amino acids (e.g., lysine). High protein content is essential for basically vegetarian diets.

Starch.   Increasing starch is not normally desirable to increase as tuber plants (potatoes, sweet potatoes and cassava) are already nearly pure starch. High starch plants are, however, ideal as energy crops.

Cellulose/lignin ratio.   Cellulose/lignin ratio in wood is important for energy crops: lignin greatly inhibits cellulose breakdown, and the additional pretreatment and treatment time requires substantial capital and energy investments. Lignin content is variable via species selection, but additional alterations in existing species would be advantageous.

Other hydrocarbons.   Hydrocarbons are produced in several plants whose importance will increase as fossil fuel hydrocarbons are depleted: rubber, waxes, and oils are among the most common forms. First, a survey and reevaluation of existing hydrocarbon-producing plant varieties are required. Secondly, breeding should be emphasized for increased yields of hydrocarbon product, especially in varieties tolerant of water stress, saline or alkaline soil. Later, manipulative techniques (e.g., genetics, specific metabolism inhibitors, etc.) may be developed for altering the product balance.

Semi-Synthetic Systems.   Elements from biological organisms can be extracted and used in isolation for performing their natural function in isolated nonliving systems. The two main subgroups in order of their probable development follow (see also Part II, Chapter 5).

Enzymatic product conversions.   Enzymatic product conversions to more desirable forms (e.g., cellulose to fuel or simple sugars) can exploit the specificity, mild operating conditions, and low production costs of these biological catalysts. Free, immobilized enzymes still within the producing microorganism are among the alternative forms; development has been rapid, but is still in its infancy.

Process alteration. Process alteration is accomplished by removing the solar energy *before* its storage as sugar:

- Ammonia (fertilizer) or, hydrogen/oxygen production from water, using plant chloroplasts and the final catalyzing enzyme, nitrogenase or hydrogenase, respectively.
- Electricity production using only the sunlight and the electrical potential converting capacity of plant chloroplasts or isolated pigment systems.

These are very long-range developments and will require considerable basic research.

SYSTEM DEVELOPMENT IMPERATIVES

Traditional research, including those items listed above, concentrates on the discovery, development and optimization of isolated processes. Normally, as discussed earlier, development strives for single variable optimization, economic return. The energy antecedents to this approach are not as valid as before, particularly if one compares the diffuse flowing nature of solar-energy resources to traditional centralized energy sources, process plants and product markets, each geographically isolated from the other.

System development and optimization requires the construction of models, both *in vitro* and *in situ*. Among the technically answerable questions through model building are:

- What energy (and economic) return is possible for a closed system of nitrogen-fixing organisms, traditional agriculture, algae-aquaculture ponds, and anaerobic digesters for energy (methane) production and nutrient recycling?
- How flexible is the above system to disturbance (climate, rainfall, infection) and to different environments?
- What physical devices and what type materials can catalytically increase the ability to exploit solar energy (electronic controls, solar stills, solar pumps)?
- With present technology, what is the *net* long-term energy optimum for a complex biophysical solar-energy closed system, and what is the energetic/economic investment cost?

- For a semi-closed system, can an initial capital gift provide the start for a self-generating or bootstrap development which trades for the additional capital requirements for consecutive stages of development?
- What research can alleviate the weak links?

System development, like basic research, requires a direct cooperative effort to prevent the imposition of inappropriate technology on developing countries. But risks are involved in any experimentation, risks which cannot be taken by the poor without some insurance against costs and crop failures. This too is the responsibility of the aid-giving developed nations.

AID IMPERATIVES

Investment aid required to begin an upward spiral of development must be given as a free gift to the countries in need and not for the donor's national security and trade interests. Trade agreements in return for aid, particularly of raw materials from the Third World, too often paralyze development toward self-reliance or along the lines desired locally. But the considerable capital aid required does mean sacrifice on the part of the developed world, and sacrifice is rarely chosen freely.

How can the necessary aid be motivated? Ideally, aid would result from *spontaneous altruism,* but barring this unprecedented phenomenon, *inspired altruism* motivated by moral arguments of suffering have on occasion proved successful. Slightly stronger are arguments of *debt repayment,* the realization that much of the modern world's development resulted from fortuitous local resources along with exploitation of weaker countries' resources. Much of the resulting progress deserves not pride at our success but shame for the price not paid. Combination of these arguments—with or without guilt for the past—has resulted at least in a UN resolution to pledge 0.7% of the developed countries' gross national products as "no strings attached" aid. As yet, Holland, Norway and Sweden are the only countries meeting this pledge. Total third world aid is only $25 thousand-million while its military costs are double this figure ($51 thousand-million); total world military expenditures are $335,000 thousand-million, or thirteen times its development aid (SIPRI 1977). Responsibility for this misplacement of scarce resources rests with the power elites of both developed and developing countries.

Self-interest, in both developed and developing regions, is the most effective motivation for development and change. Self-interest aid from industrial nations normally connotes aid with "mutual benefits", which supports the industrial base in the donor country. If one considers long-range self-interest, stability is prized above growth, and fortunately stability has predictable features discussible outside any ideological context. Aid to insure long-term stability may provide a motivation where morality fails. But what factors determine stability?

An enormous disparity in resource use and material standard of living has existed for generations, but the *relativeness* of poverty among the poor has not. Whereas poverty can be made bearable by necessity, the increase in *relative* poverty accompanying the awareness of another's wealth—particularly wealth gained at one's own expense—is not. This "shopping window" effect is a primal instability which, even if the modern world's moral motivations fail, will ultimately lead to an equalization, if not a catastrophic leveling.

Stability.   Today more than ever before in history, stability is necessary for our long-term survival. The uniqueness of this time is that development has pressed stability to its social rather than physical limits (Hirsch, 1976), and the potentially resulting physical catastrophe is of an unprecedented magnitude. While stability can be shown to be in the personal long-term interests of both the industrial and developing nations, the advantages of short-term stability are, however, less universal: the poor nations and peoples of the world, for example, cannot benefit by stability *if* stability precludes a redistribution of the world's wealth and resources. In addition, instability and social unrest, unfortunately, will always be profitable to certain individuals who are not poor.

The following discussion will probe this question of stability systematically, if in a simplified manner, to illustrate two important aspects related to development. Both are based upon a division of society's stabilizing factors into two qualitatively different groups, those which inherently or *intrinsically* stabilize, and those *extrinsic* stabilizers imposed externally upon the society. By beginning with the simple fact that massive wealth inequalities have created a destabilizing nonequilibrium, it will be shown that modern society's attempt to compensate for decreased inherent or intrinsic stability with extrinsic stabilizers (e.g., "law and order") has increased its vulnerability to small disruptive changes or *perturbations*. Secondly, decentralized development would reverse this trend by increasing society's intrinsic stability (e.g., by reliance on decentralized energy resources), decreasing the need for extrinsic stabilizers, and inherently reducing the potential for exploitive nonequilibria.

In order to illustrate and explore these factors further, Figure I.9.8 and an analogy to chemical reactivity will be used: low entropy (high order) is represented at A by, for example, reactive chemicals such as hydrogen and oxygen. In the societal model, low entropy at A can represent a

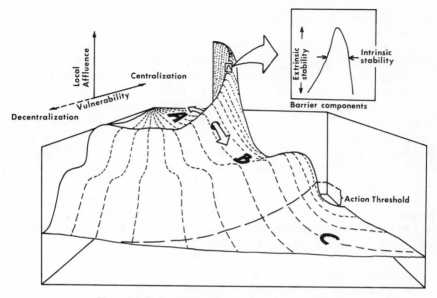

Figure I.9.8. Graphical representation of social stability.

greater than average per capita wealth which is prevented from the natural tendency toward random redistribution or high entropy (see Chapter I.2) only by the presence of a stabilizing *barrier*. The nature of these barriers is of particular interest and together with other terms affecting stability will now be illustrated more concretely.

Nonequilibrium.   While not characteristic of stability, nonequilibrium is a term which describes inequalities in the real world and places strains on the factors affecting its stability. In Figure I.9.8, the vertical difference between, for instance, groups A and B is a measure of the nonequilibrium, where ascending the vertical axis corresponds to decreasing entropy (increased order). Chemically it can correspond to energy or reactivity; while in the societal model, the most intuitive example is collected wealth, which is greater relative to the others at B or C. Wealth can equally well refer to that of an individual(s), group(s), region(s), or country(ies), while wealth itself can be replaced by, for example, power or status. For ease of discussion, however, group A in this figure will be assumed to represent those who have sequestered disproportionately larger quantities of material wealth in relation to other groups; this high-ordered, low-entropy state will, in the absence of a barrier, tend toward a new, more random equilibrium. Restated more simply, these isolated pockets of wealth will become more evenly distributed unless prevented by force.

Group B in Figure I.9.8 lies lower on the material wealth scale and to the right of the highest barrier. The vertical difference between A and B represents the material wealth which B might gain by disturbing this unstable state, while B's relation to C represents what he stands to lose. B's absolute position also measures the material, educational, and/or social forces which B could marshall in any attempted perturbation. Likewise, C's low wealth position has two opposing features: First, being positioned lowest gives C nothing to lose (except perhaps his life) and much to gain by perturbing the system. On the other hand, C's ability to act is opposed by lying beneath an ill-defined "minimum wealth" threshold. This threshold, paraphrased "the starving won't revolt," explains much of today's unequal distribution of wealth and resource use: a nominal diet, health and material wealth are prerequisities for the hope, energy and potential to effect a change.

Barriers.   Stabilizing barriers are required if a nonequilibrium state such as localized wealth is to be protected from the natural randomizing or distributional tendency. Whereas reaction barriers in chemistry can be clearly defined in terms of potential energy, social, economic or material inequality can be maintained by barriers of several distinct qualities. Information, for example, lacks a direct chemical analogue, but awareness of a socio-economic inequality is a prerequisite to motivate change. Unawareness in much of the Third World, together with sub-threshold living standards, have long preserved a stability favorable to a local elite and the developed countries in general. A virtual explosion of information has now left few regions unaware of the ever-increasing gap in material wealth. That the minimum action threshold may also be lowering with increased system vulnerability will be understood from the nature of the stabilizing barriers.

Barriers preserving an existing nonequilibrium (or equilibrium) in society (e.g., police, church, schools, labor unions, population distribution, communication, energy and legal systems) can be classified as being intrinsic or extrinsic. Intrinsic stability, depicted by the barrier width in Figure I.9.8, represents those cultural factors which inherently stabilize a society, factors which are spontaneously or automatically cohesive. Traditions, whether religious or social, here play the dominant role together with an additional important factor—social or population organization. This latter point is best illustrated by caricature: Compare two alternatives, one a delocalized population fueled predominantly by solar energy (whether primitive or modern) while the other is highly centralized, characterized by highly centralized urban populations, high occupational/industrial specialization, and a centralized energy source. The latter, a simplified approximation of our present society, is highly tuned for maximum efficiency, but like a monoculture hybrid grain crop, it is extremely sensitive to perturbations, natural or otherwise. This example of low intrinsic system stability demonstrates the attempt to compensate by consciously imposing externally-applied (extrinsic) stabilizers and controls. In Figure I.9.8 the third dimension, i.e., into the page, is an attempt to represent this change from a near equilibrium system to a nonequilibrium one with high extrinsic stability but low intrinsic stability.

Extrinsic stabilizers, depicted by the barrier height, have been the traditional responsibility of the police and military. Police would undoubtedly be necessary in even the most egalitarian utopia, but normally they function as neutral, extrinsic stabilizers of an existing nonequilibrium, irrespective of the nonequilibrium's moral merit. Of course, both police and military forces can also misuse their role or themselves be misused to establish a new nonequilibrium.

Extrinsic stability should not only be equated with military or police force. Other examples include back-up systems to central electric power failure, nuclear waste storage, or massive transport facilities to keep cities fed and free from wastes. Monoculture agriculture is a similar example of an instability discussed throughout this book: lack of diversity in modern agriculture trades efficiency (productivity) for stability; herbicides, pesticides, fertilizer and irrigation are the extrinsic stabilizers required when the intrinsic stability resulting from ecological diversity is removed.

Barriers to the equalization of wealth (or any other societal collection of low entropy) have been represented in the figure on two separate axes, implying stability is purely intrinsic or purely extrinsic. The examples chosen have been rather pure, especially those of wealth protected by force and a crop monoculture artificially maintained. Other examples are less clear, such as public education, religious organizations, the legal or economic systems. Each of these is based on culture and tradition (intrinsic stabilizers) which in response to centralization and specialization have been formalized with an operative arm, an extrinsic stabilizer. The police, schools, and churches are the formal replacements for the intrinsically stabilizing social traditions. The result is predominantly extrinsic, but arising from an intrinsically-stabilizing tradition.

Economics is one of the more pervasive stabilizers whose complexity is beyond the sketchy analysis presented here. The need for basic goods and the interdependency it breeds is obviously a powerful intrinsic stabilizer, such as in a barter society. As wealth becomes unequally distributed and economic power more centralized, economics begins to operate on several levels: the wealth component becomes one measure of the nonequilibrium while the business component becomes increasingly an extrinsic stabilizer. As most analogies, the one presented here is strained when pressed for detail.

One can ask if intrinsic and extrinsic stabilizers are equivalent or interchangeable. To a degree they are, but only under normal circumstances. Compare again the protected monoculture with diversified agriculture, centralized and decentralized populations (or energy supplies), or a democratic and totalitarian state. Each example appears to be equally stable under normal conditions, but as the intrinsic stability becomes minimal, the required increase in extrinsic stabilizers rapidly increases. The police/army can be increased, but their success in assuring stability—as for the monoculture crop—depends very much on the nature and magnitude of the perturbing or disrupting factor.

Perturbations.   An unstable chemical equilibrium doesn't change unless there exists a perturbation larger than the potential barrier. In such a case high energy compounds (A) proceed to the right in the figure to a more stable, lower energy stage, B. In the societal stability model, *A* represents order, or the affluent population, and a perturbation sufficiently larger than the barrier height (e.g., a larger army) causes a loss of that local order. Localized wealth can then flow "over the barrier" to an equilibrium, distributed state.

Just as stabilizing factors can be intrinsic or extrinsic, so can the nature of the perturbation take qualitatively different forms. Those perturbations mentioned above must compete with the extrinsic stabilizers: armed force must be greater than the stabilizing armed force, drought greater than the irrigation resources, famine greater than food reserves and/or transport facilities, etc.

A second type of perturbation can be called catalytic due to its immunity to extrinsic stabilizers and its ability to penetrate *through* a barrier narrowed by low intrinsic stability. (The analogy to quantum mechanical tunneling through a forbidden region would be more accurate but unnecessarily complicating). This situation is represented deep into the third dimension of Figure I.9.8.

Modern, centralized and specialized industrial societies have compensated for lost intrinsic stability with extrinsic stabilizers to a point where catalytic perturbations are the more probable threat. The obvious examples are purposeful efforts (guerilla or terrorist activities, the label in each case depending on one's ideology) designed to paralyze, or threaten to paralyze society via a host of simple means (bombs, kidnappings, hijackings), all real or threatened. A complex society can protect itself against threats to conquer it, but not against blackmail or disruption. Not all nuclear power plants, oil pipelines or industrial leaders can be protected indefinitely from isolated attack, or threatened attack. Numerous scientific developments increase the potential scope of the catastrophe while simultaneously decreasing the effectiveness of extrinsic stabilizers: centralized food and water supplies are susceptible to intentional contamination by toxins, virus, or bacteria; political and military leaders are vulnerable to chemical and biological agents, some with delayed or slow action. As sophisticated weapons (handheld missiles, home-made nuclear weapons) become generally available, as increasingly sophisticated scientific information reaches groups "B and C" and as intrinsic stability continues to decrease through centralization, all extrinsic protection effects will have decreasing, if any marginal, value.

Belaboring this point of world stability factors is not to generate a threat, but to emphasize two logical conclusions which are independent of any political or economic ideology:

1. Decentralization of energy, food, and water dependencies, i.e., development for self-reliance, is insurance against not only political but also environmental and accidental catastrophes. Decentralization has, in addition, the ideological component of intrinsically decreasing the destabilizing wealth or power gaps.

2. Aid-supporting the ability of poor nations to achieve self-reliance through maximum exploitation of solar energy is not only in their interest but in the self-interest of the developed world as well. It is in the interest of all, in fact, on this earth we share!

PART II

TECHNICAL BASIS
OF BIOLOGICAL
SOLAR ENERGY
CONVERSIONS

1

Photosynthesis

LIGHT PROCESSES—PHOTOSYNTHETIC MECHANISM

Photosynthesis, the conversion of solar energy to chemical energy (sugars, $[CH_2O]_n$) by biological organisms, is summarized simply as the transfer of electrons from water to carbon dioxide (CO_2), an endothermic reaction requiring light energy. More generally stated, where the electron donor is designated H_2A, (A generally being oxygen but can be, for example, sulfur), the net reaction is

$$\text{light} + CO_2 + 2H_2A \rightarrow \{CH_2O\} + 2A + \text{waste heat.}$$

Behind this simple equation lies a complex conversion system accommodating the various intermediate steps required:

- Initial light absorption by the photosynthetic antenna pigment systems;
- Transfer of energy from the antenna pigments to the reaction center pigment;
- Photo-oxidation of the reaction center pigment;
- Chemical transfer of the photo-excited electron *via* electron-transferring enzymes to a second photosystem (see below) or to the soluble form required in all dark reactions.

In eucaryotic cells this entire mechanism is confined in membranous organelles called chloroplasts, the number of which varies from one to a thousand per cell depending on plant type. Chloroplasts, in turn, contain complex stacks (grana) of folded lamellar structures (thylakoids) on which the photosynthetic pigments are located. Each photosynthetic unit consists of a variety of pigments, i.e., light-absorbing molecules, several of which belong to a class called chlorophylls. One end of the chlorophyll molecule, the so-called porphyrin ring with a chelated magnesium, is hydrophilic and may well project out of the thylakoid membrane to act as the actual light absorber as shown in Figure II.1.1. The phytal end of the

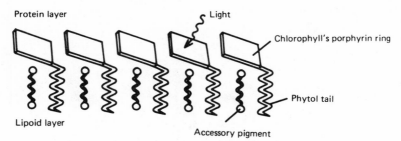

Figure II.1.1. Orientation of chloroplast pigments (redrawn from Govindjee and Govindjee 1974).

chlorophyll molecule, a chain of single-bonded carbon atoms, is hydrophobic and probably is embedded in the lipid membrane to hold the porphyrin ring in a position advantageous for light absorption and electron transfer.

Light Absorption

Photosynthetic pigments have the task of absorbing solar energy for later conversion to high-energy chemical forms, but only the visible portion is even potentially absorbed. Visible light occupies only a small segment of the total radiation spectrum shown in Figure II.1.2. It shares with all radiation forms a quantized nature which complicates the collection process: radiation is available only in discrete packages or quanta, each quantum being equally well characterized by a frequency, wavelength or energy. In free space these quantities are related as,

$$E = h\nu$$
$$= \frac{hc}{\lambda}$$

where

E = energy (joules)
h = Planck's constant = 6.6255×10^{-34} joule-seconds
ν = frequency
λ = wavelength (meters)
c = velocity of light (2.9979×10^{8} msec^{-1})

For each frequency (wavelength), the total energy available increases in discrete integral steps, n, giving

$$E = nh\nu$$

Just as emitted light arrives in integral numbers of quanta, so is energy absorbed in discrete quanta *and* of a fixed frequency (wavelength, or

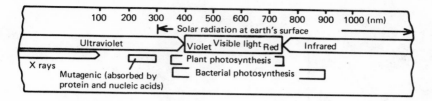

Figure II.1.2. Solar radiation spectrum.

energy); that is, $n = 1$ in the above equation for each quantum whereas light in general is characterized by n_1 quanta of several different frequencies, v_i,

$$E \text{ (total)} = n_1 h v_1 + n_2 h v_2 + \ldots n_i h v_i$$

Each atom or system of atoms has a set of quantum mechanically allowed electronic states, the lowest energy ones being the "ground states" and all others "excited states." At zero degrees Kelvin, absolute zero, only the ground state is populated, but in general the temperature of the system determines the equilibrium excited state population. The energy required to excite an electron is also discrete, that is each quantum mechanically allowed excited energy level is separated from the ground state by an amount of energy determined by the molecule's electronic structure. Generally a single quantum of the required energy—neither greater nor lesser—is required to excite one electron. This highly rigid criterion is relaxed without violating the laws of quantum mechanics by the presence of numerous smaller energy perturbations, e.g., atomic vibrations and rotations. These result in the generally observed individual but broadened absorption bands shown in Figure II.1.3.

Even with the vibrationally broadened absorption bands, a single type of organic light absorber can use only a very small portion of the solar spectrum and most commonly the ultraviolet. Pigments absorbing longer wavelength light have evolved in photosynthetic systems which employ various combinations of conjugated bonds (alternating double and single carbon-carbon bonds) in chains and rings. When these various pigments are combined as in most chloroplasts, the individual absorptions sum with a net light response or action spectrum as shown in the upper curve of Figure II.1.3.

Energy Transfer

Not only must the chloroplast absorb light over the wide frequency (energy) range of visible light, it must also absorb light over the largest possible surface area. The first problems were solved, as mentioned

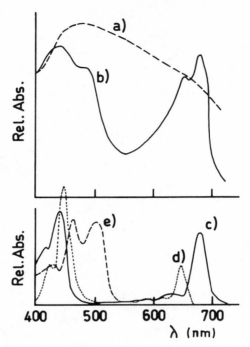

Figure II.1.3. Solar spectrum (a), chloroplast action spectrum (b), and the relative absorption spectra for plant pigments chlorophyll a (c), chlorophyll b (d), and carotenoids (e) (redrawn from Govindjee and Govindjee 1974 and Lehninger 1975).

above, by having a variety of pigments which absorb different colors of light. The second part of the problem should be solvable simply by spreading these pigments all over the plant surface, but absorption is only the first stage of a series which must also assure the effective transfer of this collected energy to the reaction center and later reactions. The alternatives depicted in Figure II.1.4 are basically of scale, whether a single antenna system should feed a central reaction center or whether there should be several smaller systems. Plant evolution has solved both problems by subdividing the chloroplast into photosynthetic units each containing about 300 pigment molecules and a single reaction center. These pigment molecules (carotenoids, phycoerythrin, phycocyanin and various chlorophylls) have complementary absorption energies and can act as an antenna or energy funnel for the reaction center. This antenna system's effectiveness depends, in turn, on the effectiveness of its transfer mechanism. Normally if a molecule is excited, the absorbed energy will be re-emitted (with a slight energy loss, i.e., longer wavelength) within

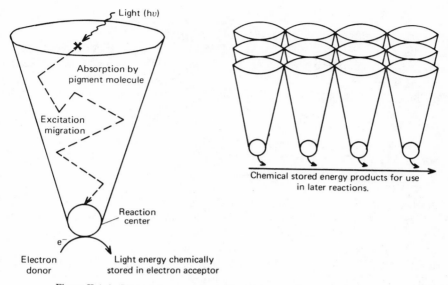

Figure II.1.4. Photo-antenna systems (a) and their spatial arrangements (b).

about 10^{-9} seconds, and such a phenomenon, called fluorescence, can be readily observed in a chlorophyll solution (Govindjee and Govindjee 1974). Fluorescence is not normally visible in green plants, but a reemission of about one of twenty photons can be detected instrumentally and is invariably red, even if the exciting light is blue (i.e., higher energy). The red fluorescence and its low intensity in plants both indicate that the transfer of energy (quenching of fluorescence) is extremely fast and efficient among pigment molecules and finally from the final reaction center pigment to the first stable reduced product. This photosynthetic unit organization is thus an effective compromise which in addition to the above factors is also determined by use rates of the photo-oxidation product.

Photo-oxidation

The absorption of visible light quanta by pigment molecules in the photosynthetic antenna system and the resonance transfer of this energy to the reaction center is followed by the first chemically significant step: the photo-oxidation (light-induced ejection of an electron) of the reaction center pigment, a chlorophyll a molecule (labeled P700 for its absorption peak at 700 nm; Lehninger, 1975). This photo-oxidation, analogous to the more familiar photocell, expels an excited or strongly reducing electron from chlorophyll P700 which is ultimately provided by water. First, however, an additional photosystem is required to raise the electron's energy (potential) to a level where it can perform its biological tasks.

The potential required to drive all of the necessary biochemical reactions is about 1.4 V, and just as multiple cells are added to make a stronger battery, so have photosynthetic systems evolved with the required two-photocell "battery." Each photon, depending upon its energy or the color of the absorbed light it represents, can produce a potential of only 0.8 − 1.0 V. Hence, the two series coupled cells shown in Figure II.1.5 are required. Note that the "wire" connecting the two cells, a series of so-called electron transport enzymes is not loss-less, and hence the resulting sum is only about 1.4 V.

Figure II.1.5 also shows the electron's path beginning in the low-energy state in a water molecule, being raised 0.8 V by photosystem II, and after losing some energy on its way to photosystem I, being reexcited by a second photon to the required potential.

The total potential difference between the final reduced product and the water molecule is 1.4 V, but electrochemical potentials are commonly referred to the hydrogen electrode. According to this rather arbitrary convention, the photoprocesses in photosynthesis end with the production of an electron at a potential of −0.6 V relative to hydrogen. This electron is actually an excited state of some unknown compound X which is

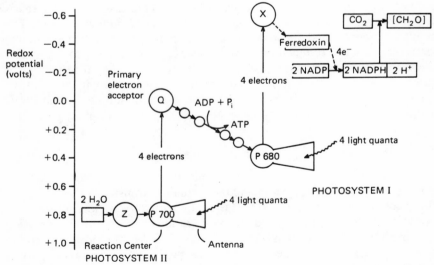

Figure II.1.5. Electron coupling in chloroplast photosystems I and II (redrawn from Lehninger 1975).

immediately transferred to other molecules and ultimately to a long series of dark reactions.

Two additional points are worth observing about the photosystems: first, the electron isn't "pushed" from water to the photosystems, but rather the photo-oxidized compounds are so reactive that a water molecule, a very stable compound, is forced to split. Each water molecule gives up two electrons, one oxygen atom and two protons. The enzyme system involved is only poorly understood, but it is known to remove sequentially four electrons from two water molecules before it evolves an oxygen molecule (Kok, Forbush, and McGloin 1970). Eight light quanta are therefore required to evolve one oxygen molecule.

The second point worth noting is that the connecting link between the two photosystems converts some of the lost energy into an energetic molecule, ATP, which is the universal energy "packet" consumed in endothermic enzyme-catalyzed reactions common to all biological organisms (Lehninger 1975). Photosynthetic organisms can also produce ATP in a shunted cyclic manner in photosystem I, while all other organisms synthesize it by metabolizing sugars. ATP and reducing potential, the net products of photosynthesis, provide the driving force for all biochemical reactions of which carbon dioxide fixation in the form of sugar is the major plant reaction.

Photosynthetic Products

As described above, the bound reduced compound X (in Figure II.1.5) at reduction potential -0.6 V is the ultimate photoproduct and the driving force for all dark reactions. Photoreduced X normally reduces one of two soluble compounds:

1. Via the cyclic photophosphorylation process alluded to above, ATP (adenosine triphosphate) is formed by the endothermic addition of a pyrophosphate group (phosphorylation) to ADP (adenosine diphosphate). ATP is the standard working chemical package of energy which drives later endothermic reactions with the chemical energy released together with the removal of each phosphate group:

$$ATP + H_2O \rightarrow ADP + P_i + 7.3 \text{ kcal}$$
$$ADP + H_2O \rightarrow AMP + P_i + 7.3 \text{ kcal}$$
$$AMP + H_2O \rightarrow \text{adenosine} + P_i + 3.4 \text{ kcal}$$

Cyclic photophosphorylation is believed to be relatively unimportant as most ATP is formed in the link between the two photosystems (noncyclically) while X normally reduces ferredoxin (Lehninger 1975).

2. Ferredoxin is a small soluble electron-transport enzyme whose active center contains either two or four iron atoms plus an equal amount of inorganic sulfur. Among their other important properties are (Sands and Dunham 1974),

	Plant Type (e.g., Spinach)	Bacterial Type (e.g., *Clostridium*)
Molecular weight	10,600	6,000
Redox potential	-420 mV	-395 mV
Electrons transferred	1	2
Iron atoms per active center	2	4

In each of the above ferredoxins, the iron atoms of the active site are bound to the protein's amino acid chain via cysteine sulfurs with two (or four) inorganic sulfur atoms completing the bonding scheme. This active site, with the protein folded around it, receives the photoexcited electron from reduced X (Figure II.1.5) via an intermediate enzyme as yet unidentified. Ferredoxin's iron-sulfur center is thus reduced, and this protein becomes the first soluble, stable photosynthetic product, essential to all reactions to follow. Ferredoxin is also included in detail here because the 2- and 4-Fe clusters, with nearly identical structures and electronic properties, are found in enzymes of every living organism, but with various functional roles: photosynthesis as above, mitochondrial respiration, hydroxylation reactions, or the fixation of nitrogen. These iron-sulfur clusters will also play a central role in discussions of synthetic nitrogen fixation or hydrogen production.

The normal fate of photoreduced ferredoxin is that shown in Figure II.1.5: the enzymatic reduction of CO_2 to glucose. The first step is the reduction of the cofactor NADP which then becomes the reducing power required for many later reactions. NADPH and ATP represent the reducing and energy sources, respectively, for all varieties of biochemical reactions. Whereas ATP was formed in the chloroplast, cofactors like ferredoxin are soluble and react as shown:

$$2 \text{ ferredoxin (reduced)} + H^+ + NADP \rightarrow$$
$$2 \text{ ferredoxin (oxidized)} + NADPH$$

DARK FIXATION REACTIONS

CO_2 Fixation

The principal product of photosynthesis is sugar as synthesized from water and carbon dioxide via the overall process shown below:

$$\text{light} \xrightarrow[H_2O]{\nearrow O_2} \left\{ \begin{array}{l} \text{strong reductant (ferredoxin)} \to \text{NADPH} \\ \text{energy (ATP)} \longrightarrow \end{array} \right\} \xrightarrow[\searrow]{CO_2} \begin{array}{l} C_6H_{12}O_6 \\ \text{(glucose)} \end{array}$$

While two alternative mechanisms do exist for CO_2 absorption from the atmosphere (see page 154), their important differences will not be discussed here, but rather the basic similarities of carbon fixation will be emphasized.

Glucose production or carbon dioxide fixation is a well-studied process which only requires:

a carbon source - CO_2;
reducing power - NADPH;
energy - ATP;
catalyzing enzymes—Calvin-Benson cycle enzymes.

Just as the major processes occurring in the light in photosynthetic organisms are the production of reducing power and a convenient energy packet, so the major product in the dark is the fixation of carbon dioxide, i.e., the synthesis of sugar from carbon dioxide and light reaction products according to the following stoichiometry:

$$6\ CO_2 + 18\ ATP + 12\ NADPH + 12\ H^+ \to$$
$$C_6H_{12}O_6 + 18\ ADP + 12\ NADP + 18\ P_i$$

As single carbon units (CO_2) must be added piecewise to ultimately form the desired six-carbon sugar molecule (glucose, $C_6H_{12}O_6$), the overall process requires a vast series of linking reactions. Each step requires carbon dioxide from the atmosphere, energy, and reducing power in the process now called the Calvin-Benson cycle. A five-carbon molecule, ribulose diphosphate, never leaves the cycle, but forms a molecular workbench on which the glucose is synthesized and eliminates the need to assemble the small CO_2 molecules directly to each other.

Photosynthetically produced glucose is rarely left in free form, but rather is normally linked into longer chains or polymerized into a number of forms serving either metabolic or structural functions. Of these, the two most important products are starch and cellulose, both homopolysaccharides with D-glucose as the exclusive monomer. The structural differences between these two polymers and their relative digestibilities are fundamental to what is traditionally called food and the limitations to exploiting nonfood, cellulosic plant products.

Starch, a storage saccharide formed under periods of excess glucose production, is the easily digested, commonly branched glucose polymer [$\alpha(1\to4)$ glucan] shown in Figure II.1.6; its molecular weights can ap-

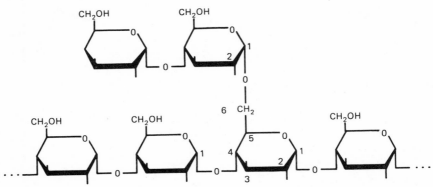

Figure II.1.6. Chemical structure of starch with $\alpha(1\to4)$ chains and $\beta(1\to6)$ branches.

proach a million. Cereals have long been an important part of Man's diet because starch, the principal constituent of most cereals, is so easily digested and energy-rich. Starch exists not because of its digestibility by other organisms but rather for long-term energy storage for the plant itself.

Cellulose, a structural molecule, is, however, the most prevalent photosynthetic product, it constituting more than 50% of biosphere carbon (Lehninger 1975). Wood, for example, is 50% cellulose, cotton nearly pure cellulose, and common plants 20-60% cellulose. Cellulose differs from starch only in the type of the glucose-glucose bonds, while starch is an $\alpha(1\to6)$ glucan, cellulose is a $\beta(1\to6)$ glucan; is never branched, and is formed by the stepwise addition of D-glucose monomers, first forming a disaccharide (cellobiose), and then chains with up to 10,000 residues (Figure II.1.7).

The seemingly small difference between chemical bonds in starch and cellulose allows long chains of the latter to lie anti-parallel and closely packed together. It is precisely this ability of cellulose to aggregate that gives the necessary strength to the fibers which are formed. Aggregation occurs at various levels and to varying extents to give both strength and

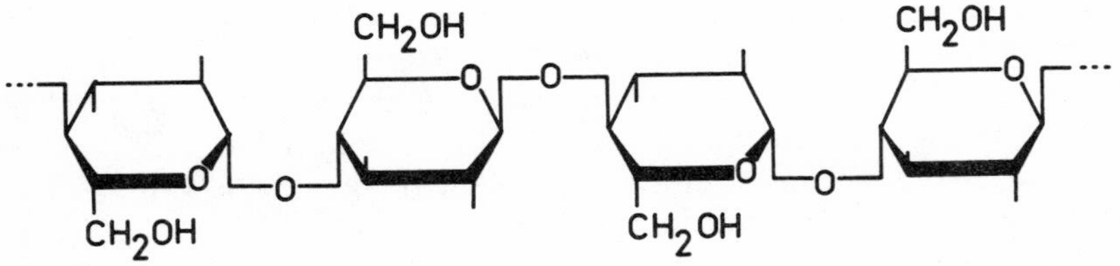

Figure II.1.7. Chemical structure of cellulose with $\beta(1\to4)$ chains.

flexibility as demanded by the particular function. The basis for this aggregation, however, begins at the molecular level where intermolecular hydrogen bonds between parallel glucose molecules from neighboring cellulose chains serve to form a microfibril of high crystalline order with some ''paracrystalline'' and amorphous regions of less order and hence less strength (Cowling and Kirk 1976).

Microfibrils of typically 50 Å × 100 Å cross section in turn cluster together as integral units (apparently no cellulose molecules pass from one microfibril to another) with microfibrils interfaced by a rather disordered paracrystalline sheath (Figure II.1.8a and b). Additional paracrystalline regions can also occur along the fibral axis (Figure II.1.8c) for increased flexibility while strength is achieved when the aggregated cellulose structure first folds to form a ribbon and then winds into a right-handed helix.

The balance between the need for strength and flexibility is reflected in both the amount of cellulose in a plant structure and its relative crystallinity. Cotton (90% cellulose) and wood (45% cellulose) are two extremes of cellulose aggregation although the basic cellulose fibers have similar structures. Differences can also occur in the manner or degree of aggregation when microfibrils form elementary fibrils: clustered microfibrils form two helically wound sheaths which are separated by a sheath of microfibrils lying parallel to the fiber axis. This middle layer, forming the

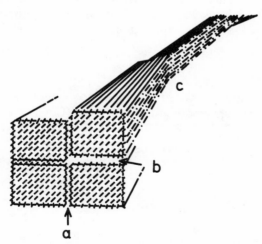

Figure II.1.8. Basic crystalline cellulose (cotton) fibril showing (a) coalesced surfaces of high order, (b) slightly disordered surfaces and (c) strain-distorted surfaces due to tilt and twist (with permission Cowling and Kirk 1976).

bulk of the cell wall, consists of concentric zones formed during consecutive days of fiber growth. These elementary fibrils (ca 4μ diam) are then covered with a thin primary wall (Cowling and Kirk 1976).

While all cellulosic compounds are similar at the molecular level, differences between cotton and wood for example do become apparent at the level of aggregated fibrils: wood, in contrast to cotton fibrils, contains large amounts of intercellular substances which bind fibers into a complex, cohesive three-dimensioned structure. These intercellular substances even penetrate into the primary fibril sheath and between cellulose molecules in amorphous regions, forming as much as *half* the polysaccharide content even in inner layers. As shown in Table II.1.1, the two prime constituents of this intercellular material are hemicelluloses and lignin (Cowling and Kirk 1976). Hemicelluloses are polysaccharides not related to cellulose, but rather consist of up to 200 molecules of D-xylose, a five-carbon sugar; this polymer is highly branched with side chains of arabinose and other sugars. Lignin is an extremely complex three-dimensional polymer of aromatic alcohols (e.g., coniferyl and syringyl) which with its multifarious binding possibilities can together with hemicelluloses form a cellulose package highly resistant to chemical and biological degradation. While this resistance accounts for trees' structural strength and integrity, it limits the usefulness of wood for nonstructural functions. These difficulties will be discussed in Part II, Chapter 3. Various extraneous substances also occur in wood including waxes, fats, oils, tannins, resin and fatty acids, terpenes, alkaloids, starch, soluble saccharides, and numerous inorganic elements. While these are valuable compounds, however, their exploitation is prevented by extraction difficulties.

Table II.1.1 Composition of Cotton and Typical Angiospermous (birch) and Gymnospermous (spruce) Fibers (Cowling and Kirk, 1976).

Constituent	Cotton	Birch (% by wt)	Spruce
Holocellulose	94.0	77.6	70.7
cellulose	89.0	44.9	46.1
noncellulosic polysaccharides (hemicellulose)	5.0	32.7	24.6
Lignin	0.0	19.3	26.3
Protein (N × 6.25)	1.3	0.5	0.2
Extractable extraneous materials	2.5	2.3	2.5
Ash	1.2	0.3	0.3

Nitrogen Fixation

Nitrogen is required for protein synthesis but not in the inert diatomic gaseou norm (N_2) which is the principal component of air. Nitrogen must first be "fixed" or chemically reduced to ammonia (NH_3) or to the ammonium ion (NH_4^+), a process which only certain microorganisms and the chemical industry can perform (Hardy and Havelka 1975, Brill 1977). As no animals or higher plants can fix nitrogen, all protein production is ultimately dependent on one of these two processes.

Fixation Reaction. Biochemical reduction of a nitrogen molecule to two ammonia molecules is a dark reaction which requires six electrons (reducing power, e.g., reduced ferredoxin), 20 ATP molecules (energy), plus water. (The energy requirement may be as high as 27 ATP under normal conditions; Andersen and Shanmugam 1977.) The stoichiometry, according to present theories, is

$$N_2 + 6H^+ + 6e^- + 20ATP + 12H_2O \rightarrow 2NH_3 + 20ADP + 20P_i.$$

After fixation, ammonia is immediately combined with glutamate, for example, to form one of the amino acids, glutamine, and to hold the ammonia concentration low. In higher concentrations ammonia is both toxic and causes regulation to stop the fixation reactions. This particular amino-acid synthesis, catalyzed by the enzyme glutamine synthetase, is particularly relevant for the regulation of the nitrogen-fixation process (discussed later) and can be altered to form several other amino acids. These amino acids (glutamine is one of the twenty amino acids used in proteins) are linked together to form protein using ATP for energy and messenger RNA as a synthesis template. The overall process,

$$N_2 \xrightarrow{ATP} NH_3 \xrightarrow{ATP} \text{amino acids} \xrightarrow{ATP} \text{protein},$$

demonstrates the central role of nitrogen fixation for plant (and, therefore, animal) protein production. It is the first nitrogen-fixing step which requires the most energy and which will now be discussed.

Catalyzing Enzyme. The nitrogen-fixing or ammonia-production reaction is catalyzed by an enzyme complex called *nitrogenase*. Nitrogenase is an extremely oxygen-labile enzyme complex whose ability to bind tightly and reduce inert nitrogen is poorly understood. From the overall process one knows that six electrons from six ferredoxin molecules must be transferred to a diatomic nitrogen molecule to produce two ammonia molecules, but the multiplicity of electrons transferred and the binding of inert nitrogen are difficult mechanisms. Only recently have model compounds been synthesized which can bind but not reduce nitrogen (Holm 1976 and Leigh 1976).

Nitrogenase has invariably been shown to consist of two subunits, each alone having no activity. The smaller component (see Table II.1.2) or "iron (Fe) protein" is a protein dimer whose active site is similar to a bacterial ferredoxin, i.e., a single cluster of four iron atoms and four sulfur atoms. Like bacterial ferredoxin, the Fe-protein transfers a single electron, but has in addition two Mg-ATP binding sites, is strongly inhibited by carbon monoxide, and is ten times larger (MW = 65,000; Smith et al. 1976a).

The reduction potential of the Fe-protein decreases from -290 mV to -400 mV when the Mg-ATP complex binds to it due to or accompanied by a conformational change in the active site (Fe-S cluster). This conformational change is indicated by the electron spin resonance spectrum change from a rhombic ferredoxin-like spectrum ($g_1 = 2.053$, $g_2 = 1.942$, $g_3 = 1.865$; $S = \frac{1}{2}$) to an axial spectrum ($g_{11} = 2.036$, $g_1 = 1.929$; Burris and Orme-Johnson 1976).

The two ATP-binding sites on the Fe-protein also provide a fast regulatory mechanism with Mg-ADP as an effective competitive inhibitor of nitrogenase activity; if one site contains ADP, the low-energy product of ATP metabolism, the second site with ATP has an increased binding

Table II.1.2   Properties of Nitrogenase Proteins from *Klebsiella pneumoniae* and *Azotobacter chroococcum* (Smith et al. 1976a).

The maximum specific activity is expressed as nmol of the ethylene formed/min per mg of protein (complementary protein in excess).

	K. pneumoniae		A. chroococcum	
	Mo-Fe Protein (Kp1)	Fe Protein (Kp2)	Mo-Fe Protein (Ac1)	Fe Protein (Ac2)
Mol wt	218000	66700	227000	64000
Subunits	51300 59600	34000	Tetrameric	31000
Molybdenum (g-atom/mol)	1.85–2.15	0	1.9	0
Iron (g-atom/mol)	29–35	4	21–25	3.9
S^{2-} (g-atom/mol)	(>18)	3.85	18–22	3.9
O_2-sensitivity ($t_{1/2}$, min)	10	0.75	10	0.50
Maximum specific activity	2200	1200	2300	2350

constant with the result an overall decrease in activity (Orme-Johnson 1976). Equimolar concentrations of Mg-ADP and Mg-ATP result in 25% activity while 50% activity occurs at a 0.7 ratio. Any energy crisis quickly results—both *in vitro* and *in vivo*—in shutting off the energy-intensive nitrogen-fixation process (long-term control, discussed elsewhere, is controlled by nitrogenase and/or heterocyst [blue-green algae] synthesis). The conformational change induced by ATP binding also increases the enzymes O_2 sensitivity which, when damaged gives a Mössbauer spectrum reminiscent of *Chromation* high potential iron protein (HiPIP), another electron transport, iron-sulfur protein (Smith et al. 1976a).

The larger nitrogenase component (MW = 220,000) contains molybdenum plus many iron-sulfur centers and is called the "Mo-Fe protein" in Table II.1.2. The Mo-Fe protein subunit is itself a heteromeric tetramer (containing two pairs of subunits) which presumably becomes reduced by the Fe-protein-Mg-ATP complex. Nitrogen is also bound to the Mo-Fe protein, but its reduction can be prevented by the binding of carbon monoxide to one of two sites. While the separate nitrogenase subunits catalyze no partial reactions, the intact complex (with Mg^{+2} and ATP) reduces nitrogen gas to ammonia with no detectable intermediates. A number of triple-bounded nitrogen analogues (C_2H_2, HCN, N_2O, CH_3NC) can also be reduced. If no other substrate is available, protons can be reduced to hydrogen gas (but not the reverse reaction) but apparently not at the N_2 binding site; reduction of protons, unlike nitrogen, is *not* inhibited by carbon monoxide.

Reduction of nitrogen proceeds according to Figure II.1.9 (Smith et al. 1976a). Reduced ferredoxin or flavodoxin first reduces the Mg-ADP-Fe protein complex following which the addition of Mg-ATP allows the Mg-ATP-Fe protein to form a tight one-to-one complex with the Mo-Fe protein $-$ N_2 complex. The latter complex is reduced by the Fe-protein, and the reduced product (e.g., ammonia) is released together with ADP + P_1 (hydrolyzed Mg-ATP). At least one and probably two ATP are required for each electron transferred, and six electrons are required to reduce N_2 to two NH_3. The exact reduction process is not understood, but the nitrogen presumably remains bound to the nitrogenase complex until all six electrons have been accumulated. Complexes binding nitrogen reversibly are very uncommon.

Energy and Reductant Source. The electrons required for nitrogen fixation originate ultimately in water and are released together with oxygen by the two photosystems. Evolution has not, however, selected for the simple system sketched in Figure II.1.10, although in principle it

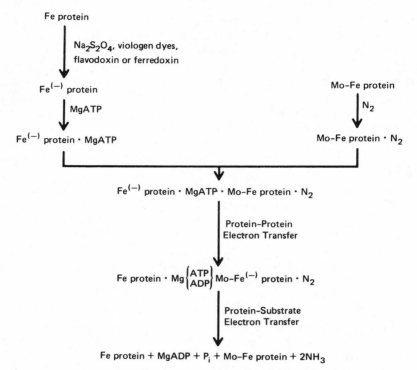

Figure II.1.9. Steps in enzymatic nitrogen fixation (from Smith et al. 1976a).

would be the most efficient and has functioned *in vitro*. This scheme has not evolved because nitrogenase is extremely oxygen-labile, and oxygen is a normal by-product of biophotolysis: the process which provides the necessary electrons releases oxygen which can destroy the electron-accepting enzyme system nitrogenase. Such oxygen lability will handicap all attempts to exploit biochemical nitrogen fixation *in vitro,* but some lessons can be taken from Nature's compartmentalization schemes.

The system most nearly approximating Figure II.1.10 exists in certain blue-green algal (see Part II, Chapter 1, page 175) cells which possess only

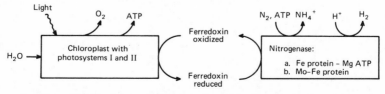

Figure II.1.10. Schematic representation of *in vitro* nitrogen fixation.

photosystem I (Stewart 1973; Stewart, Rowell, and Tel-Or 1975). Ferredoxin in these specialized blue-green algal cells, called heterocysts, is reduced directly from the photosystem, but as photosystem II is lacking, oxygen is not evolved. The electron donor for photosystem I in these cells is not yet known, but it is most certainly some organic compound produced by the normal vegetative cells neighboring the heterocyst. Within the heterocyst the fixation process is,

$$\text{Organic electron donors} \xrightarrow[\text{Photosynthesis}]{\text{light}} \text{ferredoxin} \xrightarrow[\text{reduced}]{\text{ATP}} \text{nitrogenase} \longrightarrow NH_3$$

The reducing electrons and energy in most nitrogen-fixing organisms originate from sugars photosynthetically produced in the same cell. For such organisms, the scheme is summarized as,

$$H_2O + CO_2 \xrightarrow[\text{Photosynthesis}]{\text{light} \quad O_2} \text{Sugars (glucose)}$$

$$\text{Sugars} \xrightarrow[\text{glycolysis}]{\text{ADP} + P_i \quad \text{ATP}} \text{Pyruvate} + NADH$$

$$\text{Pyruvate} \longrightarrow NADH + \text{isocitrate, } \alpha\text{-keto-}$$
gluterate, succinate,
malate, (tricarboxylic acid
intermediates)

$$NADH + \text{ferredoxin (oxidized)} \xrightarrow{\text{NADH-ferredoxin reductase}} \text{ferredoxin (reduced)} + NAD^+ + H^+$$

$$\text{ferredoxin} \xrightarrow{}_{\text{reduced}} \text{reduced nitrogenase}$$

$$\text{reduced nitrogenase} \xrightarrow{N_2} NH_3$$

Or, summing the above reactions one obtains the same overall net reaction,

$$H_2O + CO_2 \xrightarrow[\text{Photosynthesis}]{\text{light}} \text{(Sugar)} \xrightarrow{N_2} NH_3$$

The latter scheme involves the separate photosynthetic synthesis of sugars followed by the metabolic release of energy-reducing power. Such a separation prevents direct oxygen problems, but lowers the efficiency. Specific morphological arrangements will be dealt with later.

Regulation and Genetics of Nitrogen Fixation.   Of the relatively few organisms which fix nitrogen, all are microorganisms, and many of these are primitive, indicating an early evolution of the nitrogen-fixing function. Blue-green algae is the highest nitrogen-fixing organism while the absence of this ability in the higher plants evolving later indicates both the availability of ammonia (from microorganisms) and the high energy costs involved. Earlier it was believed that 15 ATP plus six reduced ferredoxin molecules were required per N_2 fixed, but more recent experiments have shown that this requirement can be as high as 27 ATP (Andersen and Shanmugam 1977). In some symbiotic associations this energy requirement can increase even more due to inefficiencies and can constitute 32% of the total photosynthetic stored energy product (Minchin and Pate 1973).

The biological requirements for nitrogen fixation are summarized in Table II.1.3. Note the appearance of some of these requirements in higher plants and animals; the possibility of genetic manipulation to extend the nitrogen-fixing ability to these organisms will be discussed later.

Production of nitrogenase requires first of all the genetic material coding for the proper assembly of the enzyme's amino acids. The gene involved, labeled *nif,* has been studied with modern genetic techniques, and its position has been "mapped" with respect to other genes; synthesis of the amino acid histidine, for example, is controlled by the gene next to *nif* and serves as a useful guide in genetic "engineering" experiments. One such experiment showed that the *nif* gene can be transferred to a second non-nitrogen-fixing organism (bacterium *E. coli*; Shanmugan and

Table II.1.3   Biochemical Requirements for Nitrogen Fixation in Different Potential Recipients of *NIF* Genes (+, observed; −, not observed) (from Shanmugam and Valentine 1975).

	POTENTIAL RECIPIENT		
Requirements	E. coli (microbe)	Chloroplast (plant)	Mitochondrion (animal)
Nitrogenase	−	−	−
Reduced ferredoxin or equivalent	+	+	−
ATP	+	+	+
Glutamate synthase pathway of NH_4^+ assimilation	+	−	−
Genetic activator	+	−	−
Oxygen protection	+ (anaerobic only)	−	−
Molybdenum uptake	+	+	+

Valentine 1975, Postgate 1977). In these experiments the *nif* gene, plus a few others, forms a small separate or extrachromosomal DNA called a plasmid which does not combine with the recipient's chromosome (its principal genetic material). Plasmids are circular and replicate simultaneously with the host organism. In such a manner, the non-nitrogen-fixing bacterium *E. coli* was provided with *nif* genes and began to fix nitrogen. *E. coli* was a particularly fortuitous choice as all other functions required for nitrogen fixation (Table II.1.3) were present. It is not surprising that naturally fixing *E. coli* have since been isolated; common species seem to have lost only the *nif* gene, but have retained all accessory functions.

Transfer of the *nif* gene is easy relative to most of the accessory functions. Particularly important among these accessory functions is the regulation of the energy-intensive nitrogen-fixation process. Short-term regulation, as if a brief energy "crisis" should occur, is provided by ADP inhibition of fixation: ADP, the low-energy form resulting from expended ATP, competes for the ATP-binding sites on the nitrogenase enzymes. Long-term regulation assures that nitrogen fixation occurs only in the absence of ammonia, the product of nitrogen fixation: ammonia inhibits the *synthesis* but not the function or activity of nitrogenase. Accumulation of ammonia is normally prevented by its rapid reaction with glutamate to form the amino acid glutamine. Glutamine synthetase, the enzyme catalyzing this reaction, has a high affinity low K_m for glutamate and ammonia and assures only a low ammonia concentration. Evidence exists, however, suggesting a more complex role for glutamine synthetase. According to these schemes, ammonia promotes the biochemical modification of glutamine synthetase to a form which cannot bind to the *nif* promoter. This more complex proposal is perhaps still oversimplified as several regulatory functions may interact. Ammonia is, however, not a simple inhibitor and only results, whatever the detailed mechanism, in the inhibition of nitrogenase synthesis, not its function (Shanmugam and Valentine 1975).

Nitrogenase activity is severely inhibited by oxygen, and some protection mechanism is required. Incorporation of *nif* genes and some other functions in Table II.1.3 are genetically possible for certain bacteria and even blue-green algae, but oxygen sensitivity is the most severe limitation. The natural oxygen protection mechanism is cellular compartmentalization to separate physically the photosynthetically evolved oxygen from the site of nitrogen fixation. The unlikeliness of breeding such a morphological change precludes the functional effectiveness (although the transfer is genetically feasible) of nitrogen-fixing plasmids in photosynthetic leaves (Postgate 1977). The many aspects of this complex problem were the subject of a recent symposium (Hollaender 1977).

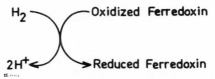

Figure II.1.11. Enzymatic reduction of ferredoxin by molecular hydrogen.

Hydrogen Production/Utilization via Hydrogenase

Hydrogen gas, a high-energy and extremely useful fuel, can be produced biochemically from photosynthetically (or otherwise) reduced ferredoxin and hydrogen ions in solution. This reaction is completely analogous to biochemical nitrogen fixation except that the catalyzing reaction, hydrogenase, does not require ATP, and the product is a fuel, not a useful metabolite like ammonia. The overall reaction,

$$\text{Light} \xrightarrow[\text{photosynthesis}]{H_2O \quad O_2} 2H^+ + e^- \text{ (reduced ferredoxin)} \xrightarrow{\text{hydrogenase}} H_2,$$

describes the production of useful hydrogen gas at the expense of only sunlight and water. Its major drawback is its nonoccurrence in normal functioning plants. Any organism which collects energy (photosynthesis) only to throw it away needlessly (evolved hydrogen gas) clearly has a large evolutionary disadvantage. (Some nitrogen-fixing organisms do however leak large amounts of hydrogen; Evans and Barber 1977, and Andersen and Shanmugam 1977.) Some organisms (see also Part II, Chapters 2 and 3) evolve hydrogen some of the time, but hydrogenase—which occurs quite widely, particularly in algae— predominantly catalyzes the reverse reaction (see Figure II.1.11). Hydrogenase, which merely catalyzes an equilibrium, does not represent the control preventing hydrogen gas evolution. Control is instead exerted via the unavailability of reduced ferredoxin for hydrogenase reduction. As this reverse process is of little interest here, discussion of the enzyme hydrogenase will be saved for a section on synthetic biochemical hydrogen production (see Part II, Chapter 5).

Fermentation and Respiration

While the first three subdivisions of this section described dark processes that under natural or unnatural conditions store solar energy (as sugar, ammonia, or hydrogen), respiration and fermentation are processes that release stored energy for immediate use.

Respiration.   Sugar, the direct energy storage product of photosynthesis is normally totally oxidized via respiration, in Man for instance, or combustion to the low-energy molecules carbon dioxide and water (the starting materials for photosynthesis). The stoichiometry of this oxidation process is (Lehninger 1975),

$$C_6H_{12}O_6 + 6CO_2 \rightarrow 6CO_2 + 6H_2O + 675 \text{ kcal.}$$

Such processes became possible with the "late" (geologically) appearance of atmospheric oxygen and the evolution of organisms tolerating oxygen. Oxidation with oxygen yields a great deal more energy than is possible with organic oxidants, and this greater efficiency gave these organisms a selective advantage.

The chemical energy (ATP) generated from glucose respiration or oxidation ("oxidative phosphorylation") occurs in solid cellular membranes, most commonly in the mitchochondina of eucaryotes. Eucaryotes evolved from the much simpler procaryotes which possessed no inner membrane structure, but can in some cases generate ATP on the cell membrane. In each case a complex set of enzymes is immobilized in the membrane.

The multistaged oxidation process begins with the absorption and conversion of food in the form of fats, amino acids and sugars to a common two-carbon product, acetyl CoA, which is the substrate for the following Krebs or tricarboxylic-acid cycle enzymes (Lehninger 1975). The enzymes of that cycle stepwise remove CO_2 and reducing equivalents from acetyl CoA. The reducing equivalents are in turn used for ATP synthesis (oxidative phosphorylation) in the series of electron-transfer steps, the so-called respiratory chain. Each sugar molecule, when completely oxidized, yields 36 ATP molecules for later use as energy packets and 24 protons which are oxidized by oxygen to water.

Fermentation.   In the absence of oxygen (or before the geological appearance of oxygen) energy is still derived from oxidation-reduction reactions, but an organic molecule is the ultimate electron acceptor, not oxygen. This extraction of energy from glucose (and other molecules) in the absence of oxygen is called *fermentation* and is possible only in primitive microorganisms. Due to the absence of oxygen, the fermentation products have the same oxidation state as the substrate, and consequently little energy can be removed. Microbial fermentations can therefore produce a variety of useful energy-rich products from sugars, products which can later be oxidized completely with oxygen to yield the remaining energy.

Fermentation of glucose can follow the two paths shown in Figure

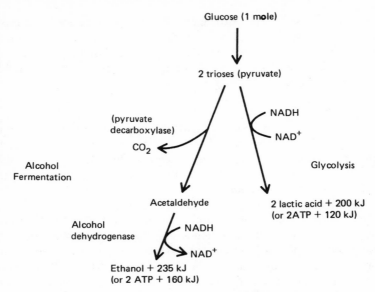

Figure II.1.12. Alternate anaerobic breakdown pathways for glucose.

II.1.12 and thereby produce either ethanol or lactate; the latter is normally called glycolysis (Lehninger 1975). Both lactate ($CH_3CHOCOOH$) and ethanol (CO_3CH_2OH) are fermentation products whose residual energy content cannot be used in the absence of oxygen. The fermentation products are, however, good fuels or respiration substrates containing 93% of the energy originally stored in glucose. As a conversion process, the organisms involved are thus 93% efficient with the remainder released for the organism's survival.

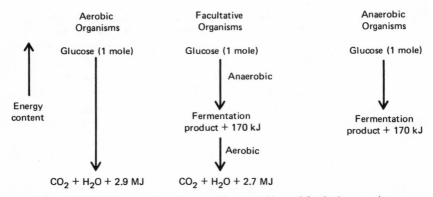

Figure II.1.13. Energy release from aerobic, anaerobic, and facultative organisms.

In some facultative organisms, fermentation occurs under anaerobic conditions, and the fermented products are respired under aerobic conditions by the same organism. This relatively unusual scheme plus the more common metabolisms of glucose are summarized in Figure II.1.13. Many other fermentations exist including those producing higher alcohols and methane, and of course many substrates other than glucose can be used. Some of these, such as hydrogen, will be discussed later (see Part II, Chapter 3).

LIMITS TO PHOTOSYNTHETIC PRODUCTIVITY

Between sun and sugar lies an extraordinary number of energy losses. Many losses are physically unavoidable (e.g., atmospheric absorption and reflection), and others result from thermodynamic limitations, but several result from immediate conversions to low-grade heat due to low-nutrient concentrations. Typically less than one percent of the incident energy results in carbon fixation, i.e., photosynthesis. This high-quality energy storage is eventually re-released via metabolism, combustion, or fermentation and re-emitted from the earth as low-grade heat energy.

Some of the multifarious factors limiting photosynthesis are listed below:

1. Genetic Factors
   • stomatal or CO_2-diffusion resistance
   • differences in CO_2-fixation pathways and resulting photo-respiration
   • photosystem composition, arrangement, concentration, and activity
   • shape, structure and number of chloroplasts
   • leaf anatomy, optical properties and distribution
2. Environmental and Internal Factors
   • radiation intensity, frequency composition and duration
   • CO_2 supply
   • leaf and air temperatures
   • water supply and internal distribution
   • photosynthate concentration and translocation rate
   • mineral supply, uptake and distribution
   • pathological state
   • endogenous diurnal cycle
   • seasonal cycle
   • age of plant and leaf

Of the many factors listed above, only those most significant, most common, or controllable will now be discussed.

Solar Energy Sinks to Ground Level

Beyond our atmosphere, the incident solar radiation is 0.135 W/cm^2, 8.1 J/cm^1-min, or 11.3 KJ/cm^2-day and is of course constant until eclipsed by the earth or moon. The solid angle impinging on the earth is insignificant for all but us on this planet: at a distance of 1.5×10^8 km and a projected area of 1.28×10^8 km^2, the earth intercepts a solid angle of 6×10^{-9} steradian. According to a familiar analogy, the earth seen from the sun is like an apple on a tree seen at a distance of 0.7 km. It can be no surprise that we receive only 2.4×10^{-6}% of the sun's radiation.

The above figures are of interest only to proponents of orbiting solar collectors or for those calculating the atmospheric component in the earth's total energy budget. Of greater interest is the radiation at ground level: of the incident radiation, 35% is reflected and 17.5% is absorbed by the atmosphere and clouds, and only 47.5% remains at the biosphere level (Oort 1970).

Non- and Pre-photosynthetic Solar Energy Sinks

Photosynthesis plays a nearly insignificant role in the earth's overall heat balance by absorbing less than 1% of the available radiation. More exactly, one can sum the gross sea and land biomass production (1.2×10^{11} and 0.7×10^{11} tons biomass, or 2×10^{21} and 1.1×10^{21} J/yr, respectively; Hall 1976) and compare it to the 3×10^{24} J/yr solar energy incident on the earth. Only 0.1% of the incident solar energy is stored photosynthetically.

Of the remaining incident energy 30% is reflected (mostly long wavelength radiation), 49% "lost" as heat of water vaporization to drive the hydrologic cycle, and 21% as low-grade heat which due to differential absorption causes the wind currents and the one-world ecosystem (Gates 1962).

For long-term climatological considerations the 3×10^{24} J/yr received by the earth should be compared with the present world's energy consumption of just under 3×10^{20} J/yr. While world-wide energy consumption is only 0.01% of the solar input, cities such as London can dissipate up to 20% of their solar input and cause local urban-rural temperature gradients of 10°C. Assuming energy sources are available, various projections show world consumption crossing the 1% solar input level by the middle of the next century. A 1% level will result in a 0.7°C average world-wide temperature increase, an increase large enough to cause major climatological changes and possibly the melting of polar icecaps (IFIAS 1975).

The distribution of incident solar energy throughout the world averaged for a 24-hour period is shown in Figure I.9.4. It is this energy which is available for plant growth, but the areal photosynthetic productivity is further deprived of solar energy due to several simple physical factors.

1. Transmission of visible light through the normally translucent leaves can be 10-15% (see Figure II.1.14), but the dense plant canopy of mature plants reduces the net transmission loss to near zero. Canopies are often described in terms of the leaf area index (LAI = the ratio of leaf area to ground area) which is often 4-5 for normal crops but as high as 20 in tropical forests (Golley 1972). Vertical leaf canopies are most effective in reflection and transmission prevention.

2. Reflection from a single leaf is angular-dependent and can be as high as 10%, but as for light transmission the plant canopy absorbs much of the reflected and diffused light and reduces the net reflection (Golley 1972). An algal pond will, however, have a high reflection of 20%, but is even more angular-dependent than a crop canopy.

3. Coverage losses in natural ecosystems are less common than in crop systems where both excessive plant spacing and seasonal canopy variations reduce the effective solar-energy conversion efficiency. The latter factor is the more significant, especially during the time between harvest, replanting and development of an effective plant canopy (this point should be remembered when comparing yields later in this chapter, see "Biomass Productions," page 185) and accounts for most of the annual vs daily growth yields (Schneider 1973). For instance, if at time t the percentage of ground covered by plant canopy is assumed to be $C = C_{\max} (1 - e^{-t/T})$, where

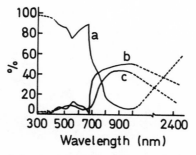

Figure II.1.14. Percent absorption (a), reflection (b), and transmission (c) of light by green leaves (by permission of Porter and Archer 1976).

$C_{max} = 1$ and T is the time required for 63% ground coverage, then the yield for period t_0 to t is the definite integral,

$$Y = \int_{t_0}^{t} N(DI)C dt \ / \int_{t_0}^{t} DI \ dt.$$

For constant daily solar insolation (DI) and conversion efficiency (N), this equation gives,

$$Y = N \left\{ 1 + \frac{T}{t - t_0} (e^{-t/T} - e^{-t_0/T}) \right\},$$

or for $t_0 = O$ (i.e., the total accumulated yield from planting time)

$$Y = N \left\{ 1 + \frac{T}{t} (e^{-t/T} - 1) \right\}.$$

Increasing t to the end of the growth period, e.g., $2T$, gives a seasonal yield of 56% of the daily rate, and an annual yield of 20% the daily rate if the season is a third of the year. Such results are consistent with estimates that the earth's total photosynthetic productivity has, in fact, decreased due to cultivation—including fertilization effects—and indicate the need for time-staggered inter-cropping whenever possible.

4. Absorption by nonphotosynthetically active plant components varies greatly with species, but a useful average is 10%.
5. Photosynthetically active radiation (PAR, i.e., wavelengths 400 − 700 μm) comprises a mere 45% (slightly dependent on atmospheric conditions) of the total incident solar radiation whose power spectrum is shown in Figure II.1.15. PAR is related to the chlorophyll absorption spectra shown earlier, but is more usefully represented as the action spectrum, i.e., the efficiency or "quantum yield" (see

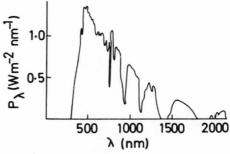

Figure II.1.15. Solar radiance per unit wavelength (by permission of Porter and Archer, 1976).

section under next heading) plotted vs radiation wavelength as in Figure II.1.3. Note that all low-energy radiation, i.e., longer wavelength than 700 nm, is lost as heat due to its inability to create the required potential difference (voltage) in the photosystems (see pages 130).

Photosynthetic Quantum Efficiency

The conversion efficiency of photoabsorbed light quanta to stored energy can be calculated to the point of reduced bound ferredoxin—the end of the light reactions—or to fixed CO_2, the net result which includes metabolic energy consumed in fixation and plant maintenance.

Either calculation begins with the data in Figure II.1.3 showing that no photon whose wavelength is longer than $700\mu m$ can photoinduce the required charge separation. As $\lambda = 700$ μm is the minimum energy required $(E = hc/\lambda)$, shorter wavelength (higher energy) quanta are equally effective, but the additional energy above the minimum is wasted in the collection pigments (see pages 55, 126) and results in lower efficiency. Such inefficiency results from the quantum nature of the photosynthetic process, that each photo-excitation requires a single photon irrespective of energy content beyond a minimum threshold level (700 μm). This is the compromise solution which has evolved to at least make use of most of the quanta above the threshold value.

The discussion of the photoacts at the beginning of this Chapter demonstrated the need for two photo-excitations coupled in series to lift the electron from the low-energy form (water) to the high-energy reduced-ferredoxin form, that is two photons of wavelength less than 700 μm per electron transferred. How many electrons are required can be determined from the overall carbon fixation equation,

$$CO_2 + H_2O + Nh\nu + 2NADPH + 2H^+ + 3ATP \rightarrow$$
$$(CH_2O) + O_2 + 2NADP^+ + 3ADP + 3P_i.$$

As each $NADP^+$ requires two electrons from the photosystems, i.e., from water molecules, a total of four electrons are required which at two photons per electron gives $N = 8$ quanta. This number represents the theoretical minimum, the experimental confirmation of which remains somewhat in dispute. A compromise number $N = 10$ is often used (Lehninger 1975).

Assuming a fixed quantum yield $N = 10$, the energy efficiency calculation next requires an integration over the PAR for the spectral density weighted average wavelength (Kok 1968).

$$\lambda_{av} = \int_{400}^{700\ \mu m} I(\lambda)\ d\lambda = 575\ \mu m.$$

Correspondingly, the average effective (i.e., 700 μm) photon energy is

$$E_{av} = N_0 hc/\lambda_{av}$$
$$= 210\ \text{KJ/Einstein}$$

where,

$$N_0 = 6.02 \times 10^{23}\ \text{mole}^{-1}$$
$$h = \text{Planck's constant}$$
$$= 6.6 \times 10^{-34}\ \text{J}$$
$$c = \text{velocity of light}$$
$$= 3 \times 10^8\ \text{m/sec}$$

Note that E_{av} should be compared with the 172 J/Einstein for the lowest energy light (700 μm) which is still PAR. Therefore 20% of the incident PAR energy is wasted due to the absorbed quanta being more energetic than required, while 55% of the total radiation is wasted as having too low energy (i.e., nonPAR). As ten quanta are required for CO_2 fixation, the input energy (10 hv) corresponds to 2.05 MJ/mole CO_2 fixed. The energy stored in the produce (CH_2O) can in turn be calculated from the carbohydrate chemical energy content (0.47 MJ/mole C) to give an overall PAR efficiency of 23%. As only 45% of the radiation is photosynthetically active, the total solar radiation conversion efficiency is reduced to 10.4%.

Alternatively one can calculate the efficiency of the photosystems themselves, i.e., the efficiency with which water is split and an electron elevated to X or reduced ferredoxin. This excitation represents a potential difference of ca 1.25 V, or for a single electron an energy difference of 1.25 electron volts (eV). As this potential results from the two series coupled photocells, two photons are required per electron, or a solar energy content of 2 $\times$ 1.83 eV (for red light). The conversion efficiency of red light is therefore, $\epsilon = 1.25/3.66 = 34\%$. Some ATP is produced between the two photocells (about one ATP per two quanta, Bassham 1977), hence raising the efficiency to ca 37%. This number must be reduced as before by 20% due to light more energetic than red light and by 55% for nonphotosynthetically active radiation. The resulting overall efficiency is 13.5%.

Rate-Limiting Factors

Light Saturation and Photoinhibition.   Saturation and inhibition of the light absorption/conversion processes are treated together here as they may well be, in principle, synonymous. Photoinhibition and photode-struction of photosystems are also probably synonymous (Lien and San-Pietro 1976).

These problems all result from plants' evolutionary adaptation to variations in daylight intensity and spectral distribution. The absorption surface must be both maximized for area and include pigments of varying colors, i.e., the mixed-pigment antenna system described earlier in this chapter. As daylight is maximal only on the canopy top at noontime on clear days, antenna systems have evolved to accommodate more common, lower sunlight intensities and hence contain ca 200-300 molecules per reaction center. The result is, however, a more than tenfold variation in normal reaction rates demanded of the reaction center. The antenna system size has been optimized for the daily average canopy intensity at the expense of high light-intensity efficiency. Thus, a clear noontime summer light intensity provides photo-oxidized electrons at a rate faster than the reaction centers can process; the same problem refers to those leaves (or algal cells) highest or least obstructed in the plant canopy.

The rate-limiting reaction lies between the two reaction centers and has the result that at maximum light intensity, conversion efficiencies are low and considerable energy is wasted. This again is particularly relevant for those most exposed leaves or the upper algal layer in a pond.

The fate of the excess energy is the heart of the related problems of photoinhibition and destruction; inhibition most likely represents the delay for repair and establishment of new equilibria following photode-struction (Jones and Kok 1966). Moderately excessive light intensity is probably dissipated via certain accessory pigments, but, if too great, photo-oxidative pigment bleaching occurs. In the latter case, oxygen is consumed during light exposure, not released as with the normal function. Both photosystems can suffer from photoinhibition, but with the exception of very intense and prolonged exposure to light, normal repair mechanisms are able to reverse these effects (Lien and San Pietro 1976).

Photosynthetic quantum efficiency is known to decrease, i.e., saturation begins, already at moderate light intensities, well below noontime maximums (predominantly for those leaves receiving direct sunlight). The exact threshold is somewhat variable and has motivated attempts to increase the threshold in "shade" plants to that found in desert plants. Two potential approaches are:

1. decrease the size of the photosynthetic unit (i.e., the number of antenna pigment molecules) to prevent overloading of the rate-limiting reactions during high light intensity;
2. increase the rate of the limiting reaction.

Preliminary indications are that the second alternative is more relevant in Nature (Myers and Graham 1975) and more promising for manipulation. Avenues for manipulation include both cultivation under high light intensity and genetic breeding, but, while the former has been the major experimental approach, practical improvements would require genetic alteration. Results are only preliminary and as summarized in a recent review (Lien and San Pietro 1976), "we currently lack sufficient information to make any statements about the genetic manipulability of the unit size and maximal turnover rate of the (photosynthetic units)."

Carbon Dioxide Uptake. Carbon dioxide is commonly limiting under conditions of intense sunlight for most common plant varieties. Provision of additional CO_2 at peak growth periods can result in a 45% increased growth at midday or daily maximum improvements of 10-20% (Lemon, Stewart, and Showcroft 1971). The enzymatic implications will be discussed later, but there is some speculation that the atmospheric CO_2 concentration, now only 0.03%, was greater during the earlier periods of plant evolution.

The CO_2 requirement gives some indication of the problem's magnitude: if 20 grams/m^2-day dry organic matter are produced, 29.2 grams/m^2-day CO_2 are required. With only 0.6 mg CO_2/liter of air, a 50-meter column of air is required over that square meter ground just to provide CO_2! Peak growth rates for maize can be four times this value or equivalent to a 200-meter column of air! On a still sunny day a sharp CO_2 gradient can result with the plant canopy and further inhibit growth.

The special problem of CO_2 for water plants (algae) will be discussed in Part II, Chapter 2.

Photorespiration: A Comparison of C$_3$ vs C$_4$ CO$_2$ Uptake Pathways

Photorespiration is simply defined as the light-stimulated oxidation of photosynthetic intermediates to CO_2:

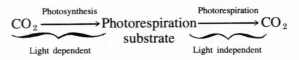

This phenomenological definition and the above figure, however, reveal nothing of the conundrum photorespiration is. Normal dark or mitochondrial respiration is also essentially the reverse of photosynthesis, that is the breakdown of fats and sugars to release the stored energy, but this respiration is small and necessary. Photorespiration is an enormous waste and apparently purposeless.

Controversy abounds on the subject, but agreement extends at least to the following points (see also Bassham 1977, Chollet and Ogren 1975, Gifford 1974, Björkman and Berry 1973, Tolbert 1977):

- Photorespiration wastes up to 50% of the stored energy.
- It primarily occurs in C_3 plants (see below), but is not observed in C_4.
- Glycolic acid is (probably) the substrate (although its synthesis mechanism is unknown).
- Photorespiration is associated with oxygen's inhibitory effect on photosynthesis.
- It is detected as a rapid release of CO_2 immediately following the darkening of a leaf and decays after a minute to the dark (mitochondrial) respiration rate.
- It increases proportionally with photosynthesis and hence light intensity and CO_2 concentration.
- It increases faster with temperature than photosynthesis, possibly due to altered enzyme (RuDP carboxylase) kinetics.
- It is apparently driven by the photorespiratory pool which accumulates during photosynthesis.
- CO_2 concentrations between 0 and 300 parts per million do not alter photorespiration rates, i.e., the difference between true and net photosynthesis.
- The extent of photorespiration can be characterized by the CO_2 compensation point: the CO_2 concentration at which photosynthetic CO_2 uptakes equals photorespiratory CO_2 release. This value depends strongly on temperature and linearly on O_2 concentration up to 100% O_2: CO_2 fixation is 45% greater in 2% O_2 than in 21% O_2 (at 300 parts per million CO_2, 25°C).
- Dark respiration and refixation of respired CO_2 complicate accurate measurements of true photorespiration rates; the rate of mitochondrial or dark respiration in the light is not known.
- Photorespiration at atmospheric conditions (300 parts per million CO_2, 21% O_2 at 25°C) has a rate approximately one-sixth the rate of photosynthesis.

The complexity of the photorespiratory problem, its magnitude, and attempts to avoid it require a discussion of the two alternative CO_2 uptake and initial enzymatic systems.

C_3 Pathway

Most plants take up CO_2 according to the scheme in Figure II.1.16 (Björkman and Berry 1973). CO_2 diffuses in through the variable

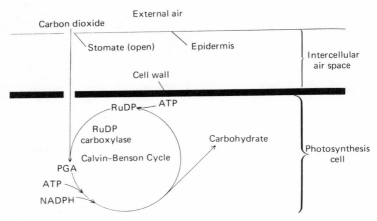

Figure II.1.16. Co$_2$ uptake and utilization in C$_3$ plants (by permission of Björkman and Berry, 1973).

stomatal openings and reacts with a phosphorylated five-carbon sugar (ribulose-1.5 diphosphate, or RuDP, the reaction catalyzed by RuDP carboxylase) giving two three-carbon molecules (phosphoglyceric acid or PGA), hence the classification C$_3$. PGA is reduced to triose phosphate and on to sugars and regenerated RuDP to complete the so-called Calvin-Benson cycle. Somewhere in the cycle the two-carbon acid, glycolate, is involved, but the mechanism is not known. Proposals abound, but acceptance of any mechanism must be consistent with several observations:

- Massive amounts (ca 50%) of the photosynthetically fixed CO$_2$ must flow through glycolate;
- Glycolate synthesis and photorespiration are stimulated by oxygen and repressed by CO$_2$ whereas photosynthesis is stimulated by CO$_2$ and reversibly inhibited by CO$_2$, and
- O$_2$ stimulates (linearly up to 100% O$_2$) biosynthesis of glycolate and its photorespiratory oxidation to CO$_2$ without saturation while dark respiration saturates at 2% O$_2$.

Photorespiration-free Mutants?   These combined CO$_2$ and O$_2$ photorespiration effects represent an enormous potential for increased photosynthetic production. How such a wasteful process evolved, as no purposeful role has yet been found, remains in doubt. Suggestions range from an ancient symbiosis involving glycolate to a protection mechanism against photodestruction during water stress and high irradiance or simply a natural consequence of the RuDP carboxylase reaction. Yet another suggestion is possible photorespirating ATP generation for the reduction of nitrate to ammonia. The latter suggestions allow little hope for photorespiration-free species or mutants.

The search for photorespiration-deficient mutants uses as a diagnostic test a reduction in the CO_2 concentration at which photosynthetic CO_2 uptake equals photorespiratory CO_2 release. A decrease in this so-called compensation point indicates an increased CO_2 affinity and/or a reduced affinity toward O_2. The lack of success in breeding experiments suggests that photorespiration is a property inherent in the system involved: 200,000 mutagen-treated soybean seeds (Menz et al. 1969), 6000 oat cultivars, and numerous wheat cultivars were screened without detection of any with reduced CO_2 compensation concentration (Moss, Krenzer, and Brun 1969). While this finding can hardly be called conclusive evidence, it does reinforce the pessimistic conclusions based on the enzymatic mechanisms (Tolbert 1977).

Photorespiratory Control in Algae. Chlorella, Englena gracilis, and a few other algae lack a light-induced release of CO_2 and possess near zero CO_2 compensation concentrations, although they are classical C_3-fixing species (Chollet and Ogren 1975). They may simply excrete much of the glycolate they are known to synthesize, refix the evolved CO_2, or perhaps possess a restriction in the glycine to serine pathway, the proposed point of CO_2 evolution. Each alternative has its proponents, but the evidence is lacking for any decision.

C_4 *Fixation Pathway.*

In the early 1960s sugar cane photosynthetically fixing CO_2 was shown to incorporate most ^{14}C into the four-carbon malic acid (hence "C_4") instead of phosphoglyceric acid (a three-carbon acid or "C_3") as in most plants (Björkman and Berry 1973). The list of important C_4 plants now includes several other grasses besides sugar cane: corn (maize), sorghum, and noncrop grasses such as crab grass and Bermuda grass. The species number in the hundreds, nearly 100 genera, and 10 plant families, but no algae, bryophytes, lower vascular plants, gymnosperms or primitive angiosperms have the C_4-fixation pathway. As some genera have both C_3 and C_4 fixation, the alternative CO_2-fixation mechanism seems to have evolved separately several times. Although the C_4 species exist over the entire plant kingdom, they share characteristics other than the production of malic acid (Björkman and Berry 1973):

- All possess a wreath-like leaf (''kranz'') structure in which the vascular tissue is surrounded by a concentric layer of bundle-sheath cells containing starch-filled chloroplasts; this layer is in turn surrounded by a ring(s) of chloroplast containing mesophyll cells.
- Large quantities of certain enzymes are found in C_4 plants which are present in only small quantities in C_3 plants.
- The relative amounts of ^{12}C and ^{14}C are different in the two types of plants, e.g., as in sucrose of sugar cane (C_4) vs sugar beet (C_3).
- CO_2 compensation concentration is essentially zero, i.e., an enclosed C_4 plant will deplete the atmosphere of all CO_2, even to one or two parts per million.
- C_4 plants have a higher maximum photosynthetic rate, normally at high light intensity and temperature.
- They have a tendency to inhabit areas of high light intensity, high temperature and limited water supply.
- C_4 plants show no O_2 photosynthetic inhibition.

The above features are explained by the biochemical CO_2-fixation mechanism, the details of which divide the C_4 plants into three types, but

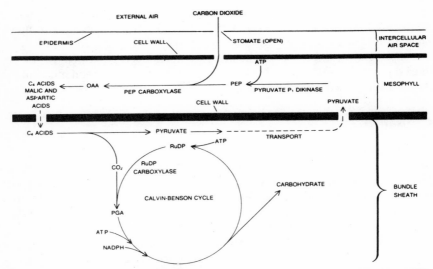

Figure II.1.17. CO_2 uptake and utilization in C_4 plants (by permission of Björkman and Berry 1973).

all basically follow the scheme shown in Figure II.1.17. In each case the CO_2 enters through the stomates as in C_3 plants, but now is fixed by phosphoenolpyruvate (PEP) carboxylase in the mesophyll cytoplasm, the resulting four-carbon oxaloacetic acid (OAA) being reduced to malate (maize, sugar cane, sorghum or crab grass) or aspartate.

The C_4 acids are transported into the bundle sheath chloroplasts where they are oxidatively decarboxylated to pyruvate (it returning to the mesophyll to regenerate PEP) plus CO_2. At this point RuDP carboxylase fixes the locally evolved CO_2 to C_3 compounds. The bundle sheath is apparently impermeable to CO_2, and hence RuDP carboxylase is provided with a CO_2/O_2 concentration ratio higher than in C_3 plants.

A dispute remains whether the lack of O_2 inhibition and lack of photo-respiration results from the PEP carboxylase "pumping" mechanism or the refixation of any potentially escaping CO_2 by the same mechanism. The results of the two mechanisms would be the same.

Before comparing the relative productivities of C_3 and C_4 species, three important parameters other than CO_2-uptake ability must be included. These three parameters—temperature, light intensity, and water-use efficiency—are functionally and morphologically related, but the inter-dependence is complex.

Temperature. C_4 varieties typically, but not exclusively, have a temperature optimum between 10° and 20°C higher than C_3 varieties; the geographical regions of natural occurrence of these two varieties reflect their climatic preferences. Figure II.1.18 shows typical temperature dependencies for C_3 and C_4 grasses; their photosynthetic maxima differ by a factor of two and occur at temperatures differing by 25°C. Such differences were confirmed even for C_3 and C_4 varieties of the same genus, *Atriplex rosea* (C_4) and *Atriplex patula* (C_3) (Björkman and Berry 1973).

The relation between temperature and CO_2-uptake mechanism is not known, but exceptions to the high temperature maximum suggest there may be no causal link. *Spartina townsendii,* a salt marsh grass from the British Coast is a C_4 plant whose temperature optimum is 7–10° lower (30°C) than other C_4 grasses and yet fixes CO_2 at 5°C at rates comparable to other C_3 grasses (Long, Incoll, and Woolhouse 1975).

Light intensity.   As will be discussed in greater detail below, the advantages of C_4 varieties occur with high light intensity and disappear at intensities below 5×10^{-2} J/m²-sec. Such conditions are mostly encountered in the tropics while in most areas the advantage averaged over a

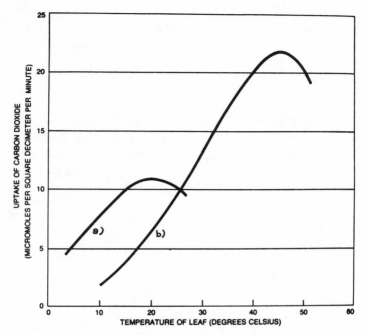

Figure II.1.18. Temperature dependencies of (a) a temperate C_3 grass (*Deschampsia caespitosa*) and (b) a C_4 desert shrub (*Tidestromia oblongifolia*) (by permission of Björkman and Berry 1973).

long period is much smaller. The light dependencies for *A. patula* (C_3) and *A. rosea* (C_4) are shown in Figure II.1.19.

Water-use efficiency. Plants metabolize only a small fraction of the water absorbed by the root, while much more is transpired through the same leaf stomata to dissipate excessive heat. Closed stomata prevent water loss during drought conditions, but the concomitant increase in resistance to CO_2 uptake prevents growth in C_3 plants. This is not the case with C_4 plants: *Tidestromia oblongifolia,* a herbaceous perennial from the US Southwest and Death Valley's floor (annual rainfall 4.2 cm), has its maximum photosynthetic productivity under normal day conditions at over 47°C when the stomata are closed to prevent water loss (Björkman and Berry 1973). The "additional" pumping mechanism in C_4 plants still provides the plant with the necessary CO_2 which as a result has as much as double the water-use efficiency (ratio of CO_2 fixed to water transpired) of C_3 varieties. The same pattern is seen in *A. rosea* (C_4) and *A. Patula* (C_3) for which productivity for a given stomated resistance is always greater for the C_4 variety.

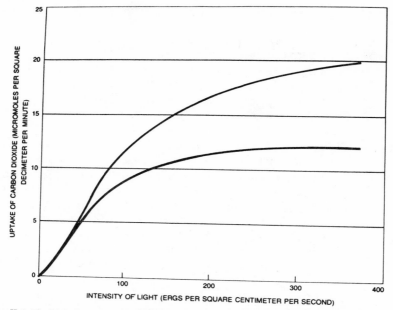

Figure II.1.19. Light-intensity dependencies of *Atriplex patula* (C$_3$, lower curve) and *A. rosea* (C$_4$, upper curve) (by permission of Björkman and Berry 1973).

C$_3$ vs C$_4$ *Productivity*

The relative photosynthetic productivities of plants with C$_3$ and C$_4$ CO$_2$-fixation mechanisms are reviewed here at different levels of measurement to determine if the C$_4$ varieties' advantage at the primary reaction level remains as a final economic yield. The essential questions are how much and under what conditions will photorespiration, which can waste one-half of the fixed CO$_2$, reduce net photosynthetic yield, and how significant are the advantages of the C$_4$ mechanism (Gifford 1974):

1. The primary CO$_2$-uptake (carboxylation) reactions of the C$_3$ and C$_4$-fixation mechanisms involve the enzymes RuDP or PEP carboxylase, respectively. A comparison of the reaction rates for the two enzymes (or K$_m$'s, also described as the substrate (CO$_2$) concentration at which the reaction velocity is one-half the maximum) reveals that for the same reaction velocity, the C$_4$ system has a 60-70-fold advantage over the C$_3$ system. Recent work disputes the "accepted" K$_m$ for RuDP carboxylase suggesting that enzyme modification has resulted before the *in vitro* measurement and that

in fact the C_3 system is only a factor of two worse than the C_4 system.

2. In intact mesophyll cells the relative rates of RuDP and PEP carboxylase can be hidden or perturbed by other factors:

   • Cellular pH may not be optimal for enzyme activity, nor need the cells of C_3 and C_4 plants be the same.
   • *In vitro* reaction rate measurements may differ *in vivo*.
   • Carboxylase concentrations may not be the same.
   • Other enzymes in the cycle can be limiting.
   • Diffusion of other substrates or metabolites may be limiting. Some combination of the above factors (or the disputed K_m for RuDP carboxylase above) results in the 60-fold C_4 advantage over C_3 systems being reduced to a fivefold advantage as measured by the mesophyll resistance. (This term, also called the residual or intracellular resistance, is the difference between the mesophyll CO_2 concentration and the compensation concentration divided by the CO_2 exchange rate per unit leaf area; its inverse is roughly photosynthetic potential.) Unfortunately, these measurements were performed at the same temperatures rather than at corresponding positions with respect to their temperature optima; this measurement may underrate the C_4 advantage.

3. For an intact leaf under full sunlight, atmospheric CO_2 concentration is well below saturation. The CO_2 diffusion-limited concentration gradient across the cell membrane, one limitation of CO_2 fixation, is quantified as the gas-phase resistance, i.e., the sum of boundary layer and stomatal resistance. In C_3 plants larger stomatal openings reduce this resistance to about half that of C_4 plants and reduce the latter's advantage from fivefold to one and a half to threefold. Presumably this is related to water conservation.

4. Instantaneous photosynthetic rates: In a pure crop stand with closed canopy all the factors discussed in earlier sections contribute: leaf area index, spatial radiation and CO_2 gradients, water, and nutrient conditions. Since the C_4 mechanism is most advantageous under intense light conditions, the CO_2 and radiation gradients result in a complex canopy summation of photosynthetic rates. The relative C_3 and C_4 production is then a complex function of conditions, and meaningful comparisons are difficult. Figure II.1.20 is a comparison of wheat (C_3), rye (C_3), and maize (C_4) under identical conditions of growth, leaf area, and age. As expected the advantages of the C_4 mechanism occur only under the most intense light condi-

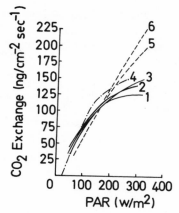

Figure II.1.20. Dependence of net photosynthesis (ng CO_2/cm²-sec) on photosynthetically active radiation (PAR, Watts/m²) for wheat (C_3, *T. aestivum*; curves 1, 2 and 3 for leaf area indices 4.2, 4.6, 4.2, and temperatures 14°, 14°, and 17°C, respectively), rye grass (C_3, *Lolium perenne*; curve 4 with LAI = 4.5), and maize (C_4, *Z. mays*; curves 5 and 6 with T = 27ǫC, and LAI = 4.3 and 4.2, respectively) (by permission of Gifford 1974).

tions (400-450 W/m² is the natural maximum) while the slight disadvantage at low light intensity may indicate a lower energy efficiency in the two-stage C_4, CO_2 fixation mechanism. This would explain the decrease for a canopy average as few leaves experience maximum solar intensity.

5. Short-term growth rates average variations in the daily solar radiation include dark respiration. A comparison of *Atriplex hastata* (C_3) and *A. spongiosa* (C_4) at three weeks' age shows similar growth rates, the former developing a greater leaf area to compensate for a lower photosynthetic potential. A wider summary of C_3 and C_4 plant growth rates under varying conditions and leaf areas shows comparable daily averages (see Table II.1.4); cloudy and night conditions remove any C_4 advantage.

6. Long-term and economic yields depend on nutrient, water and light conditions, the plants' various responses to these factors, and the length of the growing season. The latter factor is dominant in a comparison of annual forage yields for C_3 and C_4 grasses (Table II.1.5). C_4 grasses commonly grow year round where, incidentally, the response to intense light may also be advantageous; C_3 grasses, common to temperate regions, have a shorter season where temperature enhancement of photorespiration is not a problem, and

Table II.1.4   Maximal Values of Short-Term Crop Growth Rates of
C_3 and C_4 Species.
(Values expressed per unit of land area) (Gifford 1974)

Species	PAR (MJ m^{-2} day^{-1})	Crop Growth Rate (g m^{-2} day^{-1})	Comments
C_3 species:			
Helianthus annuus	4.5	76	7-day mean; with roots; LAI = 5.8
	6.4	68	28-day mean; with roots; LAI = 4.3–7.1
Agrostemma githago	7.2	57	6-day mean; with roots; LAI = 8.7–11.0
Phragmites communis	8.6	57	130-day mean; with roots; in hydroponic culture
Oryza sativa	10.62	55	8-day mean
Typha latifolia	10.4	53	24-day mean; shoots only
C_4 species:			
Zea mays	9.4	78	7-day mean; shoots only; LAI = 11.5
	9.4	67	28-day mean; LAI = 8.7–11.5
Pennisetum typhoides	9.6	54	14-day mean; shoots only
Zea mays	13.8	52	12-day mean; with roots; LAI = 15–20
Sorghum vulgare sudanense	13.0	51	35-day mean; shoots only

perhaps they have a slight efficiency advantage over C_4 plants. Thus, C_4 species have a long growing season under conditions most suitable for their growth while photosynthetic potential plays but a small role. For example, Napier grass (*Pennisetum purpureum*) in El Salvador has daily rates ranging from 7.0 grams/m^2-day in the February-May quarter to 36.1 grams/m^2-day in the May-August quarter, the average being 22.5 grams/m^2-day. These are not beyond the C_3 daily potential under *its* optimal conditions, e.g., a hydroponic stand of reed swamp grass (*Phragmites communis*) in Czechoslovakia had a 57 grams/m^2-day average over its 130-day growth season (Table II.1.4).

Table II.1.5   Maximal Annual Forage Yields Recorded for C_3 and C_4 Species (Gifford 1974)

Species	Annual Forage Yield (kg ha^{-1})	Country
C_3 species		
Lolium perenne	29,000	United Kingdom
	26,600	New Zealand
Medicago sativa	32,500	California, USA
C_4 species		
Pennisetum purpureum	85,200	El Salvador
	84,700	Puerto Rico

In summary, one can say that while plant varieties with the two alternative CO_2-uptake mechanisms do demonstrate different functional dependencies on light, temperature and CO_2, the initial 60-fold advantage of C_4 over C_3 systems at the enzymatic level is essentially eliminated at the short-term growth level. Season duration and the suitability of the local climate determines the annual yield, and invariably a C_4 plant finds its optimal climate coincident with a long growth season. Note the C_4 advantage under water-stress conditions is not emphasized here, but can under certain circumstances have major significance.

Incorporating C_4 Features into C_3 Plants.   Hybridization experiments have been attempted with *A. rosea, A. patula,* and three other C_3 *Atriplex* species (Björkman and Berry 1973). First generation hybrids are intermediate in their morphology and variable in PEP carboxylase concentration. A few second generation hybrids resemble the C_4 parent (*A. rosea*) in leaf anatomy and biochemical characteristics, but none had a complete C_4 metabolic pathway. The PEP carboxylase mechanism has evolved many times in Nature, but the number of trials far exceeds the limited attempts (and perhaps possibilities) of laboratory hybridization. Breeding of C_3 plants is not expected to produce C_4 characteristics in the near future, but may select varieties for a relatively lower rate of photorespiration (Wilson 1972).

A more fruitful approach to producing C_4 features may be by breeding C_4 plants into agriculturally useful species, e.g., the C_4 grain plant amaranth (a relative of *tidestromia*) once used by Central American Indians. Alternatively, C_4 plants should be employed with agricultural practices adapted to exploit their potential, e.g., breeding plants for high water-use efficiency, probably *not* a result of current breeding programs.

PLANT NUTRIENTS

Nitrogen

Nitrogen Cycle.   Nitrogen as a plant nutrient is required only to the extent that proteins are synthesized. Protein on the average contains 16% (by weight) nitrogen, almost all of which is complexed as the amino group ($-NH_2$, an ammonia derivative) of amino acids,

$$
\begin{array}{c}
H \\
R-C-COOH \\
| \\
NH_2
\end{array}
$$

"R" in this formula represents the side group which is unique for each of the various 20-odd naturally-occurring amino acids. These side-groups are principally small hydrocarbons, but occasionally contain oxygen, sulfur and sometimes nitrogen. Up to several thousand amino acids can be linked together according to genetic plans into long protein chains,

$$
\begin{array}{ccccccc}
R_1 & O & & R_2 & O & & R_n & O \\
| & \| & & | & \| & & | & \| \\
NH_3^+-C-C & -N-C- & C-N-C-CO^- \\
H & & H\ H & & H\ H
\end{array}
$$

amino acid: 1   2   n   ,

which can serve either catalytic (enzymatic) or structural functions. As a result protein content of plants can vary from negligible quantities as for sugar cane to over 60% for algae. The nitrogen removed with each harvest corresponds to the protein content and can easily be calculated using the approximation mentioned above:

kg nitrogen annually removed per hectare
= 16% of the total protein yield
= (16%) (% protein) (annual total dry plant yield).

The world's three leading grain crops—rice, wheat and maize—contain 8, 12.1, and 10.5% protein, respectively. As the maximum yields for each can reach 13 tons/ha-yr, the above formula gives nitrogen harvests of 167, 250, and 218 kg/ha-yr. Typical yields are about half these values (see page 185).

Although the air we breathe is 79% nitrogen, food production is limited foremost by the inability to replace this lost nitrogen and water. This sad paradox results from plants' (except microorganisms, see below) inability

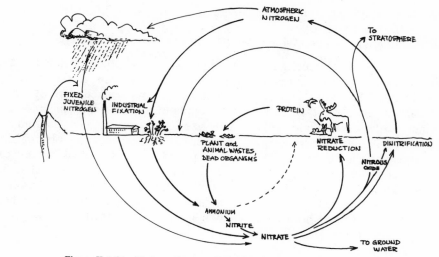

Figure II.1.21. Modern nitrogen cycle (redrawn from Delwiche 1970).

to "fix" nitrogen, i.e., combine it with hydrogen to form ammonia (NH_3). Production of plant protein requires instead that nitrogen have the form of ammonia, nitrate (NO_3^-) or nitrite (NO_2)—only these forms can be absorbed and provide the nitrogen nutrient requirement. Animals including Man have even more stringent nitrogen requirements: protein in Man and other animals can only be synthesized from ingested protein or amino acids, not from any simple fixed nitrogen source as plants do. Thus, all animals must eat plants or other animals, not just for energy but for protein. This nitrogen cycle is summarized in Figure II.1.21.

Nitrogen fixation. From the plant protein production point of view, the conversion or fixation of atmospheric nitrogen into ammonia is the most essential step. This process, chemically summarized as,

$$N_2 + 3 H_2 \rightarrow 2NH_3,$$

can occur in one of four ways, only three of which are natural:

1. industrial fixation from petroleum sources (40 million tons per year);
2. biological fixation at the expense of photosynthetically-stored energy (Eriksson et al. 1976):
   a. terrestrial fixation (140 million ton/yr)
   b. aquatic fixation (90-130 million ton/yr);

3. atmospheric fixation due to ionizing radiation (7.6×10^6 ton/yr; Delwiche 1970);
4. fixation in active volcanoes (0.2 million ton/yr; Delwiche 1970).

Many of the above figures are rough estimates subject to change as until very recently natural biological fixation has been grossly under-estimated. Biological fixation is without exception mediated by microorganisms, most of which are nonphotosynthetic and metabolize energy-rich organic molecules. Such fixation processes have continued since early evolutionary periods and represent the constant supply for all photosynthesis. In the absence of competing processes breaking down fixed nitrogen to nitrogen gas again, fixation would continue unabated and eutrophication of all open waters would be but the first disaster. Processes mediated by a variety of organisms do exist which complete the nitrogen cycle and hold each component in a well-tuned balance (Delwiche 1970, 1977).

Soil nitrification. Soil nitrification is the bacteria-mediated oxidation of the ammonium ion, ultimately to nitrate (NO_3^-). Each of the two stages of oxidation provides sufficient energy for the survival of the oxidizing bacteria:

1. $NH_3 + 3/2 \, O_2 \xrightarrow{\text{Nitrosomonas}} H^+NO_2^- + H_2O + 66 \, \text{kcal}$, and

2. $K^+NO_2^- + 1/2 \, O_2 \xrightarrow{\text{Nitrobacter}} K^+NO_3^- + 17.5 \, \text{kcal}$.

Denitrification. Denitrification is the reaction which produces nitrogen or nitrous oxide gas from nitrates and is mediated (half terrestrially and half marine) by bacteria (e.g., *Pseudomonas denitrificans*) under anaerobic conditions; nitrate or nitrite replaces oxygen as the electron acceptor in these metabolic oxidations, but is not an energy source. Energy is derived from sugars via reactions whose yields are nearly as great when oxidized with nitrate/nitrite as with oxygen. The reactions are:

glucose + nitrate $\rightarrow CO_2 + H_2O + N_2$ or $N_2O + \sim550 \, \text{kcal}$.

This denitrification reaction, together with protein harvesting, is the chief loss of fixed nitrogen for plant use.

Ammonification. Ammonification is the microbial conversion of protein amino acids to ammonia, e.g.,

CH_2NH_2COOH (glycine) $+ 3/2 \, O_2 \rightarrow 2CO_2 + H_2O + NH_3 + 176 \, \text{kcal}$.

The nitrogen cycle is thus a multicomponent cycle with only the first stage, nitrogen fixation, requiring energy. An imbalance anywhere in the cycle immediately causes a nutrient surplus in some other branch of the cycle with resultant prolific growth. The most well-known example of this nutrient surplus is caused by fertilizer runoff or sewage disposal, both of which are rich in ammonia or nitrates. Algae thrive in open waterways due to this enrichment and initiate processes which ultimately can lead to eutrophication or death of the waterway. Such accidents are prevented in Nature by the rapid doubling time for bacteria, in this case denitrifying bacteria which are the prime balance against natural fixation processes. The enormous perturbations due to massive and often geographically concentrated use of chemical fertilizer show signs of straining these natural balances. Denitrifying bacteria can compensate to some degree, but even their normal product, nitrous oxide, has been shown to produce harmful side effects: nitrous oxide can be oxidized to nitric oxide (NO) in the stratosphere where it reacts with ozone (Crutzen and Ehhalt 1977, Knelson and Lee 1977). A few percent increase in atmospheric nitrous oxide has been observed and may result in a corresponding decrease in ozone (McElroy 1976). Ozone absorbs solar ultraviolet radiation and protects biological organisms from this genetically damaging radiation.

The microorganisms responsible for organic decomposition are particularly essential for nutrient cycling, including nitrogen. As long as organic litter is available for this bacterial growth, these microorganisms and hence the ecosystem are quite resilient to small changes in inorganic nutrient levels or even major accidents to bacterial population. Even small losses ($\sim$10%) of organic litter can, however, greatly alter plant growth and decomposer (bacterial) populations. Recovery of the population to such a perturbation can take 20 years while a simple microbial population perturbation recovers after a few months (Dudzik et al. 1975). The difference is, of course, basically due to plant vs bacterial biomass doubling times. If these conclusions from computer modeling are correct, careful use of a large fixed nitrogen pool (via recycling) may be less destabilizing than interrupted decomposition processes.

Runoff or leaching of nitrates and nitrites to ground water is the most harmful, wasteful and destabilizing effect of large-scale nitrogen-fertilizer use. Nitrates are particularly mobile in soil due to their negative charge, but while the positively charged ammonium ion is less mobile, past evidence suggested it was not as easily absorbed by plant roots. While both have been found contaminating drinking water, nitrites are far more toxic than nitrates, but several other nitrogen compounds are also toxic to human and animal life. Some of these are carcinogenic, but the most immediate threat to life is the reduction of hemoglobin to methemoglobin by nitrates or nitrites (methemoglobinemia); death occurs by suffocation due to the decreased oxygen-carrying capacity of the blood (Magee 1977). Children, particularly infants under six months, are most susceptible.

Runoff occurs equally whether manure, compost, or synthetic chemical-type fertilizers are used, in each case the use efficiency is only 50%. Recent work with trickle irrigation, where both water and soluble nutrients are delivered according to a carefully planned schedule to individual plants or rows of plants in perforated hoses, has increased use efficiency to 90-100% (Ingestad 1977). Much of this increased efficiency results from its more local distribution to the plant roots, but at least as important is the nutrient application *when* it is needed. Trickle irrigation allows repeated applications in small amounts programmed to preserve the plants' internal nutrient status constant. Traditional fertilizing techniques result in growth inhibitory overdosage followed by deficit stress which accounts for both waste and suboptimal growth.

Nitrogen fertilizer is undeniably a sound economic investment for both crop yield and soil health, but the demand for nutrients is substantially increased for new plant hybrids which are selected for their response to fertilizer. These hybrids often fail to yield as much as traditional varieties unless fertilizer is supplied in excess of 80 kg/ha per crop, and multiple crops per year require multiples of this 80 kg/ha figure. For comparison, the imbalance in regional fertilizer consumption is shown in Table II.1.6.

Estimates for world nitrogen-fertilizer production for 1977 are 50-60 million tons (124-million tons as ammonia fertilizer), but this figure is based on use, not need (Bolin and Arrhenius 1977). Equal use of 80 kg/ha-yr (US maize uses 130 kg/ha-yr) on the world's 1.5-billion arable hectares increases the requirement to 120-million tons per year while multiple cropping can double or triple this figure.

Even the 80 kg/ha-yr nitrogen assumed above is less than optimal as crop yields continue to increase with nitrogen-fertilizer use, although

Table II.1.6   Nitrogen Fertilizer Use (FAO 1973).

Region	kg/ha-yr Arable Land
Europe	70.6
North and Central America	31.6
Latin America	12.5
Far East	11.1
Near East	10.4
South America	7.4
Africa (developing)	2.0

marginal yields decrease at high application levels: marginal yields in developing countries are almost 20 tons grain per ton fertilizer while at industrialized agriculture rates the marginal yield drops to ten. Figure II.1.22 shows more detailed data for maize yields vs nitrogen fertilizer applied: a maximum in the *marginal* return occurs at 120 kg/ha, but the *total* yield, of course, continues to increase to about 200 kg/ha (Pimentel et al. 1973). As Figure II.1.22 assumes fixed phosphate, the maximum yield may even exceed that shown. Furthermore, recent evidence with carefully programmed "trickle" fertilizing does *not* show diminishing marginal returns until the sharp optimum is reached (Ingestad 1977). This massive amount of fertilizer, well beyond that amount which is normally fixed by soil bacteria, inhibits bacteria from making even their normal contribution.

Industrial Nitrogen Fixation. Increased use of hybrid crops which respond profitably to large amounts of fixed nitrogen explain the doubling in industrial fertilizer demand every five years. This rate of increase has continued unabated in the industrialized agricultural sector since the Second World War. A brief look at the industrial fixation process makes clear the fuel and capital limitations for realizing the fertilizer-based Green Revolution in developing countries.

Virtually all ammonia is synthesized via the Haber Process, a catalytic

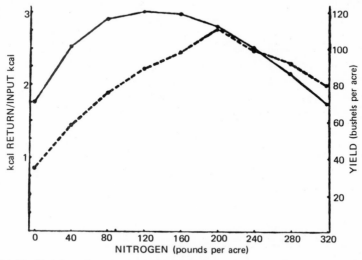

Figure II.1.22. Variation of maize crop yields (dashed curve) and energy return (solid curve) on nitrogen fertilizer (by permission of Pimentel et al. 1973).

synthesis developed 60 years ago. As with biochemical fixation, this process needs to lift a nitrogen molecule from its nearly inert state to its energetic reduced form (ammonia, urea, or nitrates).

Natural gas (methane) provides both the reaction energy and the heat required for the initial endoergic ''reforming'' reaction (Safrany 1974),

$$CH_4 + H_2O \rightarrow CO + 3H_2.$$

Carbon monoxide and hydrogen are then separated in another high temperature ''shifting'' reaction,

$$CO + H_2O \rightarrow CO_2 + H_2.$$

The final step is the actual Haber Process in which molecular hydrogen and nitrogen gases are combined catalytically at high temperature (500°C) and pressure (400 atm) to form ammonia. This final step is exoergic and releases 38 KJ, but such a statement conceals both the high energy content of hydrogen in the first reaction and the high process temperature. The Haber Process is therefore an extremely energy-intensive synthesis. Its high efficiency leaves little room for improvement as most of the high hydrogen-energy content is stored in the ammonia. It is this energy which in fact sustains nitrifying bacteria.

Energy investment in the actual nitrogen-fixation process is sketched in Figure II.1.23 where the ammonia product is referenced to the low energy equilibrium substances CO_2, H_2O, N_2. Such a description emphasizes the fact that ammonia is a fuel form which is produced quite efficiently but at

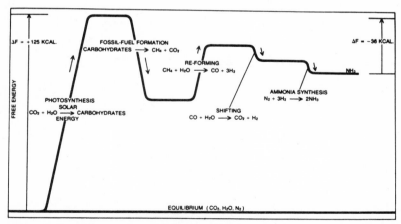

Figure II.1.23. Thermodynamics of the Haber nitrogen fixation process (by permission of Safrany 1974).

high capital and energy costs. As the original energy source is a hydro-carbon (methane), its cost has increased drastically, and the supply is limited.

The concept of ammonia and all materials as fuel or energy equivalents also allows a more complete energy analysis for its production, capital and feedstock costs in energy terms. Capital and fixed economic costs are translated into energy units (see Part I, Chapter 3) and for fertilizer factories has been calculated as follows (Leach 1976):

new buildings	$1134 \ 10^6$ MJ
plant and machinery	1177
replacement parts for machinery	770
all other materials (excluding feedstocks)	4702
fish, mefillers	29
Total capital + fixed costs	$7812 \ 10^6$ MJ

Dividing this total equally among the outputs gives an input for ''capital'' and ''fixed costs'' of some 870 MJ/ton which is about 3 to 4% of the input/ton for most products. To this should be added packaging costs of 410 MJ/ton product and transport costs of 300 MJ/ton. Direct energy inputs are however over 90% of the total energy costs of between 39.7 to 57.6 MJ/kg ammonia. Note that ammonia nitrate is 34.5% nitrogen, ammonium sulfate is 21% nitrogen, urea is 46.6% nitrogen, and liquid ammonia is 82.4% nitrogen. Fertilizer use will be used per kg elemental nitrogen for which the energy cost is taken as 80 MJ/kg.

If one views both fertilizer and maize in terms of their energy equivalents, one can determine an optimum application yield separate from the normal economic evaluation. Reference back to Figure II.1.22 shows such a calculation: energy efficiency has a clear maximum at about 120 kg/ha, but energetically any given farmer could justify increased fertilizer use as long as the output/input ratio is greater than one.

Biological Nitrogen Fixation.   Biological nitrogen fixation (see page 137 for biochemical discussion) is the reduction of atmospheric nitrogen by a photosynthetically-produced reductant whose reduction potential is near that of hydrogen; this process is totally analogous to the industrial Haber catalytic process, but is mediated by an enzyme system (nitrogen-ase) at ambient temperature and atmospheric pressure. The overall reaction, however, is the same,

$$H_2O \xrightarrow[\text{photosynthesis}]{\text{light} \quad O_2} H \xrightarrow[\text{nitrogenase}]{N_2} NH_3.$$

Nitrogen-fixating organisms include all photosynthetic bacteria, some blue-green algae, some aerobic soil bacteria (e.g., *Azotobacter*), some facultative bacteria (e.g., *Klebsiella* and *Achromobacter*) in addition to about 13,000 species of leguminous plants and 250 nonlegumes with symbiotic nitrogen fixation (Stewart 1977). As the latter two groups depend on the symbiotic nitrogen-fixing bacterial genus *Rhizobia*, all biological nitrogen fixation except blue-green algae occurs in bacterial systems; blue-green algae, a procaryote, lies between the bacterial and plant worlds.

Most of the advantages of biological nitrogen fixation are obvious:

- fixation occurs without capital investment;
- solar energy is the driving force;
- fixation occurs in place, precluding the need for transport and spreading (economically and energetically important);
- excessive ammonia production is automatically prevented;
- no fossil fuels are required.

But biological fixation is not without disadvantages:

- fixation does not occur with principal crops (e.g., wheat, maize, and rice);
- fixation control mechanisms generally prevent maximum plant productivity;
- legumes have a special taste, not universally popular;
- legume yields are lower than most cereal crops;
- fixation has a large and often limiting energy requirement.

Biological nitrogen fixation is not a universal phenomenon in Nature due in large part to the high energy requirement and extreme oxygen sensitivity of the enzyme system (see page 139) involved. As oxygen in a natural product of photosynthesis, the incompatibility of the energy source and the energy-consuming process desired (nitrogen fixation) has required a number of novel evolutionary developments. Organisms have evolved into those which maintain their independence and those which live in symbiosis or close association (associative symbioses) with a photosynthetic organism. These three groups together with their representative members are shown in Table II.1.7.

Free-living microbial nitrogen fixation.   Truly free-living nitrogen-fixing organisms are rare, or at least not very productive, due to the dual requirement of an energy-rich carbon source (e.g., sugars) and a low (or zero) oxygen tension to prevent enzyme (nitrogenase) inactivation. Those which fix nitrogen aerobically, e.g., *Azotobacter,* are believed to reduce

Table II.1.7   Present-Day N$_2$-Fixing Plants (Stewart 1977).

1. Free-living organisms	a) Heterotrophic bacteria, e.g., *Azotobacter, Clostridium, Spirillum, Beijerinckia, Klebsiella*
	b) Autotrophic bacteria, e.g., *Rhodopseudomonas, Rhodospirillum, Thiobacillus*
	c) Blue-green algae, e.g., *Anabaena, Calothrix, Nostoc, Plectonema, Mastigocladus, Gloeotrichia*
2. Associative symbioses	examples are *Paspalum notatum-Azobacter paspali, Digitaria decumbens-Spirillum lipoferum*
3. Root nodule-forming symbioses	a) *Rhizobium*-legume associations, e.g., *Glycine max, Phaseolus vulgaris, Vicia faba, Trifolium repens*, etc.
	b) *Rhizobium*-non-leguminous angiosperm association: *Trema cannabina*
	c) *Actinomycete*-non-leguminous angiosperm associations, e.g., *Alnus glutinosa, Myrica gale, Hippophaë rhamnoides, Casuarina equisetifolia*, etc.
	d) Cycad-blue-green algae associations e.g., *Bowenia, Cycas, Encephalartos*, etc.

the oxygen tension locally by having a very high metabolic rate. A further decrease in free-living fixation rates occurs because nitrogen is fixed only when required; free-living bacteria therefore produce less fixed nitrogen than those bacteria who live in symbiosis/association with an ammonia-requiring host. Free-living heterotrophs fix only nominal amounts of nitrogen and at low efficiencies: about 1 kg nitrogen/ha-yr at an efficiency ratio of N$_2$ fixed to carbon compound consumed of 5-20 (Stewart 1977).

Of somewhat greater local importance are the photosynthetic bacteria (nonsulfur bacteria and purple and green-sulfur bacteria). These organisms evolve no oxygen because they photosynthetically reduce various inorganic compounds other than water. They are strict anaerobes and occur in sulfur ponds or in the anaerobic layer below algae, particularly in polluted waters.

Blue-green algae are the most effective free-living nitrogen fixers due to their unique ability to photosynthesize normally with water as the electron donor. The most common form of blue-green algae is the filamentous type that differentiates some normal vegetative cells into nonvegetative and hence nonoxygen-evolving cells which nevertheless possess one active photosystem (Stewart and Rowell 1975). These cells, called heterocysts, are able to fix nitrogen for themselves and exchange ammonia via pores to adjoining vegetative cells for their photosynthetically-produced sugars (see Figure II.1.24). Oxygen is thus excluded from the heterocyst's nitrogenase which is reduced by its one active photosystem; the ultimate electron donor for this photosystem is not known. The fact that up to 50% of the fixed nitrogen leaks out into the medium (Stewart 1977) and the ability to inhibit the normal regulation

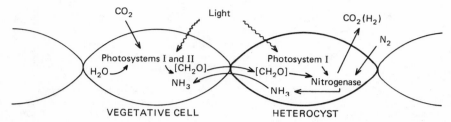

Figure II.1.24. "Symbiosis" in heterocystous blue-green algae (redrawn after Benemann and Weissman 1976).

mechanisms, makes filamentous blue-green algae attractive for "*in situ*" fertilizer production (see Part II, Chapter 2).

Nonheterocystous filamentous algae are of somewhat lesser interest; while they can grow aerobically, they only fix nitrogen in the absence of oxygen as in some salt marshes and swamps. Unicellular blue-green algae (e.g., *Gloeothece* or *Gloeocapsa*) more rarely fix nitrogen with an as yet undetermined oxygen protection mechanism.

Root nodule-forming symbiosis. Obligatory symbiotic relations exist between host plants and several nitrogen-fixing bacteria, algae and *Actinomycetes* (Brill 1977). The group of symbionts producing 40% of the world's naturally-fixed nitrogen and hence of dominant agricultural interest are the legumes, a class of pod-bearing plants which includes soybeans, peanuts, chickpeas, string-type beans, cow peas, and pigeon peas. Areal rates of nitrogen fixation for these organisms also far exceed all others with the possible exception of blue-green algae: typically 100 kg/ha-yr nitrogen is fixed, but as shown in Table II.1.8, nine times this amount is documented. Symbiotic nitrogen fixation in legumes almost invariably involves the bacterial genus *Rhizobium* which infects legume root hairs and there forms anaerobic nodules. The absence of oxygen is essential for the nitrogen-fixing enzyme's (nitrogenase) activity.

The *Rhizobia*-containing root nodules are linked to the plant vascular system from which they receive a constant portion of the host's photosynthesis. Root nodule demands for the host's energy reserves are quite substantial and represent a selective disadvantage when fixed nitrogen is available. Of the 9-10 kg carbohydrates required per kg nitrogen fixed, half are complexed with the exported product (as amino acids), 3.6 kg are respired by the nodule and 0.4 kg is for nodule growth; about 4 kg of the 10 kg consumed by the nodule are required for the actual fixation process. Recent evidence shows that energy-use efficiency in legume root nodules is not constant: 30% or more of the energy used by the nitrogenase

Table II.1.8   Nitrogen Fixation in Various Leguminous Plants (Nutman 1975).

Species	N Fixed (kg ha^{-1} ann^{-1})
Arachis hypogaea	124
Cajanus cajan	224
Calopogonium mucunoides	202
Canavalia ensiformis	49
Centrosema pubescens	259
Cicer arietinum	103
Cyamopsis tetragonolobus	130
Desmodium intortum + D canum	897
Enterolobium saman	150
Glycine max (soya)	103
Lens culinaris	101
Lespedeza sp	95
Leucaena glauca	277
Lotonosis bainesii	62
Lotus corniculatus	116
Lupinus sp	176
Medicago sp	199
Melilotus sp	199
Mikanea cordata	120
Phaseolus atrapurpurea	291
Phaseolus aureus (green grain)	202
Phaseolus aureus (mung)	61
Pisum sativum	65
Pueraria phaseoloides	99
Sesbania cannabina	542
Stylosanthes sp	124
Trigonella foenumgraecum	342
Vicia faba	210
Vigna sinensis	198

enzyme system is wasted as evolved hydrogen gas (Evans and Barber 1977, Andersen and Shanmugam 1977). The same investigations have, however, isolated *Rhizobial* strains which do not leak hydrogen. Cow pea is one such example and apparently has a hydrogenase which recycles any evolved hydrogen gas.

The limitation of *Rhizobial* nitrogen fixation to a certain few plant species has prompted repeated attempts to culture *Rhizobia* for *in vitro* nitrogen fixation. Suggested dependence on a host "diffusable factor" (even nitrogenase was a considered candidate) has recently been eliminated with the discovery that two carbon sources, a pentose and a dicarboxylic acid, are required for *Rhizobial* growth (Pagan et al. 1975). Both of these sugars are normally provided by the host. Host-species selection is thus not related to any nutritional factor, but the factors which do trigger nodule formation remain unknown.

Host-bacteria specificity is known to occur at several levels. Thus, although the topsoil (ca 17 cm) bacterial population can be 500 kg/ha, their colonization on host-root surfaces is not assured. Specificity at this level is rather low, but growth does vary with pH and soil water content. Infection on the other hand is highly specific for host and *Rhizobium* species: the two seem to be linked together via a protein (trifoliin) which connects two antigenically similar binding sites. The protein that binds to *Rhizobium* species infecting soybeans will not bind to species infecting other legumes (Brill 1977). Awareness of even this incomplete recognition model indicates the importance of matching bacterial and host species for effective fixation.

Infection and nodule formation are the two initial hurdles, but by no means assure successful nitrogen fixation. The magnitude of nitrogen fixation which does occur depends upon a wide range of factors (Brill 1977, Hardy and Havelka 1977, and Bergersen 1977):

1. Temperatures outside of a 15°C temperate range (15°-30°C) strongly inhibit fixation rates. The fixation rate drops to half its maximum value at 8°C (Figure II.1.25a).
2. Excessive oxygen inhibits nitrogenase activity and stability while insufficient oxygen reduces the availability of metabolic energy (ATP via oxidation phosphorylation; Figure II.1.25b). The nodule membrane reduces the soil pO_2 by four orders of magnitude to an internal level maintained by the oxygen-binding enzyme

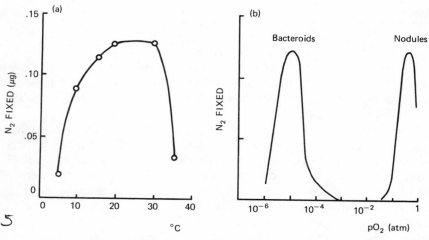

Figure II.1.25. Nitrogen fixation rate dependence of *Rhizobia* nodules in soya on (a) temperature, and (b) oxygen tension (redrawn from Bergersen 1977).

leghaemoglobin, in much the same manner as haemoglobin in blood.

3. The volume of the bacteriod-containing tissue determines in part the number of fixing organisms.

4. The specific bacterial nitrogenase activity determines the rate at which each organism fixes nitrogen.

5. The duration of active fixation varies seasonally, and in the case of soybeans ceases for unknown reasons at senescence when a fixed nitrogen supply is still required.

6. Mild water deficits cause reversible inhibition of nitrogenase while severe stress causes irreversible inhibition. Excessive water inhibits fixation by interfering with the soil-nodule gas exchange.

7. As 32% of plant photosynthates are passed to root nodules where half are used for bacterial nitrogen fixation and nearly half are returned to the host complexed with ammonia, fixation rates vary with photosynthetic rates.

8. Ammonia or nitrates in the soil inhibit fixation and nodule formation except for certain *Rhizobia*. Recently fixed nitrogen is normally quickly converted to glutamine, but if not, accumulated ammonia will inhibit fixation (Burris 1977).

9. CO_2 concentration affects nitrogen-fixation rates almost linearly up to 1500 parts per million (Hardy and Havelka 1975). This is a photorespiration effect. A tripled CO_2 tension under experimental conditions increased fixation fourfold (from 90 kg/ha to 425 kg/ha), decreased soil nitrogen uptake (from 225 kg/ha to 75 kg/ha), and doubled the crop yield. A large part of the increased nitrogen fixed also results from a delay in senescence which may or may not be a photorespiration effect.

10. Stage of pod growth strongly affects fixation rates: From the flowing stage (thirtieth day) fixation rates increase exponentially (8% day) to a maximum when the pod is little more than half-filled (ninetieth day). Senescence is rapid, and little nitrogen is fixed during the final crucial week of development.

Clearly a wide variety of factors affects the rate of nitrogen fixation in legumes. To these should be added (although many are the same) the previously reviewed factors affecting photosynthesis in general. The special role of nitrogen, however, has removed the plant breeders' major trick: a selection of hybrids for their special responsiveness to fertilizers. Legumes, of course, still require soil ammonia or nitrates (under normal fertilized conditions they fix only 35% of their need of 85 kg/ha-yr), but additional fertilizers simply decrease fixation activity; nitrogen fixation is energy-intensive, and plants only fix if necessary. Addition of trace nutrients molybdenum and sulfur (also copper and cobalt) can occasionally have dramatic results, but, in general, legume yields (see Table II.1.9) have not followed cereal crops' responses to hybrid selection and heavy fertilizer use. Breeding has succeeded in extending the northern boundary where legumes grow, and other research possibilities include:

1. 50% increase in yield if photorespiration is eliminated.
2. Elimination of the 50% loss of photosynthetically-produced electrons which occurs via hydrogen production from nitrogenase.
3. A near doubling of nitrogen fixed if nodule senescence (cessation of fixation) is delayed one week.
4. Preliminary encouraging results in fields with *Rhizobial* strains superior in infection, nodulation, or fixation properties, should be continued. An optimal matching of host to existing soil bacterial species is the other alternative.
5. Bacterial strains derepressed in nitrogen-fixation control mechanisms may be more easily inoculated and would continue to fix in the presence of fertilizer. *Rhizobium japonicum,* for example, shows suppression of neither nitrogenase activity nor synthesis by ammonia. An alternative may be regulatory points, but these pose greater dangers.
6. Methods to harvest protein remaining in the fields for fodder use: a third of the fixed nitrogen is never harvested, but remains as roots, stems, and low-lying pods.

Table II.1.9   World and US Grain Legume Acreage, Production and Yield (Evans et al. 1975).

	WORLD			US		
	10^6 ha	kg/ha	10^6 M Tons	10^6 ha	kg/ha	10^6 M Tons
Soybeans	38.5	1378	53.0	18.5	1886	34.9
Dry beans	22.3	489	10.9	0.56	1458	.82
Groundnuts	19.7	859	16.9	0.60	2469	1.49
Chick peas	10.5	637	6.72	—	—	—
Dry peas	9.3	1154	10.2	0.06	2034	.12
Cow peas	5.0	254	1.26	0.033	606	.02
Broad beans	4.7	1137	5.33	—	—	—
Pigeon peas	2.6	665	1.73	—	—	—
Vetches	1.9	1087	2.04	—	—	—
Lentils	1.7	689	1.18	0.027	1292	.035
Lupins	1.0	721	0.75	—	—	—
Other pulses	6.3	564	3.57	—	—	—
Total	123.7	918	113.58	19.78	1890	37.39

7. Optimal plant husbandry for highest efficiency: plant densities have been decreased by a factor of ten from recommended density (300,000 plant/ha to 25,000) and yet produced the same yield. Soya is particularly adaptive: a decrease in pods simply increases number of seeds per pod for the same yield (Bergersen 1977).

8. All methods should be investigated which increase the available photosynthate to the nodule. Leaf canopy, for example, is not limiting: 50% has been removed without decreased yield.

Nonleguminous root-nodulated plants which fix nitrogen are all woody perennials. Several subgroups can be described (Stewart 1977) of which the most important are the more than 160 species of actinomycetes. Many of these organisms are primary colonizers in isolated regions of poor soil quality. A remarkable exception—and one which indicates the perhaps yet unknown/underexploited potential of these organisms—is *Alnus rubra* (red alder), an important hardwood tree which fixes 200-300 kg nitrogen/ha-yr.

Nonroot-nodulated symbioses and associations. These nitrogen-fixing systems have traditionally been underestimated due to technical difficulties in measuring fixation rates. Oxygen inhibition of the nitrogenase system during measurement has become avoidable with the so-called acetylene-reduction technique which, in turn, accounts for the steadily increasing estimate of world biological nitrogen-fixation rates.

One group of bacterial associations of great recent interest are those with grasses and grains. Symbiosis in this case is not significantly different from legumes except that nodules are not formed. The bacteria do, however, infect the root system at specific sites where they achieve access to the host plants' photosynthates in return for the nitrogen they fix. This association is somewhat less stable than legumes, but the nitrogenase activity is several times more important than fixation by free-living soil bacteria. Active fixation depends on several known variables and probably several more which are as yet unknown (Dobereiner, Day, and Bulow 1975, Dobereiner 1977):

1. Only mature, well-established roots can be infected.

2. The bacteria involved (*Spirillum lipoferum*) are microaerophilic with pO_2 optima between 0.01 and 0.015 atm (air has a $pO_2 = 0.21$ atm).

3. Temperature optimum is very high (31°C) for both fixation in isolated *Spirillum* cultures and in association with maize. Little decrease in activity is observed up to 40°C (much higher than for legumes), but falls rapidly below 25°C. A fivefold difference in

fixation rate with plant type (maize vs Digitaria) is observed at 22°C.

4. Ammonia concentration in the soil is severely inhibitory above 200 parts per million.
5. Soil humidity is significant only at the wilting point when fixation ceases.
6. Supply of photosynthate limits at various points of the growth cycle and season.

Of the factors listed above, soil temperature and ammonia concentration are operationally the most significant whereas soil pO_2 is often very near its optimal value. Fertilizer application must be limited for effective nitrogen fixation, but 20 kg nitrogen/ha applied several times is not inhibitory.

Soil temperature may place a geographical limitation on effective bacterial-fixation, but comparisons of fixation rates must also include the high temperature characteristics of the C_4 species studied. The coincidence of associative nitrogen fixation and its occurrence in C_4 plants may also be related to the CO_2-fixation pathway, or both may result independently from their high temperature optima. Differences (factors of five) in fixation rates with host plants and *Spirillum* species greatly encourage breeding experiments: maize—*Spirillum* breeding yielded a variety with a nitrogenase activity 20 times the original. Reasonable fixation rates have even been observed in Northern US.

Fixation rates observed vary widely with host plant species and variables including those listed above. As far as is known, forage grasses and C_4-type grain plants have the most effective associations with nitrogen-fixing bacteria.

Recent studies have shown some associations comparable with good legume fixation rates: one maize strain fixed 2 kg/ha-day while an elephant grass (*Pennisetum purpureum*) association fixed 1 kg/ha-day (Dobereiner et al. 1975, Dobereiner 1977). Maximum fixation rates are not sustained for more than a few days (the silking stage in maize), and except for about a one-month period little nitrogen is fixed. Fixation does, however, occur during the period of maximum growth and as mentioned can be supplemented with synthetic fertilizers up to 20 kg nitrogen/ha. Similar experiments in temperate climates (Florida, USA) demonstrate not only a tolerance but also a need for some "priming" nitrogen fertilizer. As shown in Figure II.1.26, about 20 kg nitrogen/ha were required to achieve equal yields on *Spirillum lipoterum* inoculated fields of pearl millet (*Pennisetum americanum*) as on uninoculated fields (Smith et al. 1976b). At 40 kg nitrogen/ha-yr the yield from inoculated

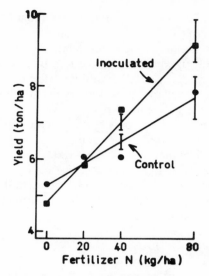

Figure II.1.26. Pearl millet yield with and without inoculation of *S. lipoferum* (by permission of Smith et al. 1976b).

fields gave yields equivalent to those which received 80 kg/ha—i.e., *Spirillum* inoculation saved 40 kg nitrogen/ha. Similar results were obtained for guinea grass (*Panicum maximum*).

A significant contribution toward the fixed-nitrogen need does seem to be possible with bacterial associations in maize, sorghum and millet in addition to forage grasses *Paspalum, Digitaria,* and *Pennisetum purpureum* (elephant grass). The advantages seem to be greatest in tropical climates, but the data is as yet too tentative for definite conclusions. What is clear, however, are some of the more pressing research needs:

1. identify the infecting bacterial species;
2. understand the relevant factors in the infection process;
3. investigate specificity using pure cultures;
4. accurately determine the nitrogen budget in relevant ecosystems;
5. study fertilizer (nutrient) requirements for optimal fixation;
6. determine relation of fixing rates to environmental factors;
7. breed fixing associations for cooler temperatures and maximum fixation rates;
8. determine if field inoculation aids infection rates.

Blue-green algae (usually *Nostoc* or *Anabaena*) can also form symbioses with various eucaryotic plants including fungi, liverworts, and a particularly interesting water fern, *Azolla* (Peters 1977). The latter (nonobligatory) symbiosis has been used for centuries as a source of 100-160 kg nitrogen/ha-yr in the rice paddies of China (Stewart 1977). Some amount of ammonia may leak out while the plant is alive, but most is freed together with fixed carbon as the rice plant canopy develops and the *Azolla* plant dies. As in other symbioses, the constant drain of fixed nitrogen to the host forces the algae (*Anabaena* in this case) to fix more nitrogen than it would if freestanding.

Other unusual symbioses can occur, such as the reported occurrence of nitrogen-fixing *Klebsiella* in Papuans eating a high-starch diet (Bergersen and Hipsley 1970). Whether any nitrogen fixation occurs or if it is of any significance is yet unknown. The same is true for certain ruminants where such a functioning symbiosis would of course be a highly attractive means of producing high-quality meat protein from ruminants eating little else but cellulose.

Phosphorus

Although phosphorus is present in neither protein nor carbohydrates, its storage in Nature is the principal cause of soil infertility. The role for phosphorus lies with the complex supplying energy for all enzymatic reactions: adenosine triphosphate (ATP). Without phosphorus ATP cannot be formed, and many enzymatic processes cease.

In natural standing biomass phosphorus is only a tenth as common as nitrogen, for example (C:N:P = 750:10:1), but normal compound fertilizers often contain equal amounts of the two elements. Most phosphorus is in fact applied as phosphate or superphosphate, the form which occurs in ATP, but as is typical of most fertilizers, only 50% is absorbed.

Current usage in intensive farming, e.g., US maize, is typically 50 kg/ha phosphorus. This application rate is ten times greater than the 1945 maize cultivation and a hundred times greater than that used (when used at all) in developing countries. Unlike nitrogen, however, no natural process replenishes any of the phosphorus exported in crop material; if phosphorus is not locally recycled, chemical fertilizer must be added.

Projected usage of phosphate indicates the need for careful waste recycling for two reasons:

1. Phosphate which is not returned to the fields ultimately ends up in the waterways where it is the principal cause of eutrophication. Increased fertilizer application rates can only aggravate this problem unless the efficiency of reuse increases.
2. Phosphate resources, while large, are limited and nonrenewable.

A closer look at the latter point indicates that projected usage, at a modest application rate of 10 kg/ha-yr will be 4×10^7 tons per yr or five times as much if US rates are assumed (Millington 1976). As recent estimates of known and potential phosphate reserves are ca 20×10^9 tons, projected usage would consume these reserves in 100-500 years. Pollution effects will probably require altered recycling patterns before this.

Trace Minerals

Trace mineral requirements for plant growth is a complex and poorly understood question. The list of elements involved includes potassium, magnesium, sulfur, iron, manganese, copper, boron, zinc and molybdenum.

While none of these are photosynthetic reaction substrates or products, their influence results from:

- being bound in or required for the activity of pigments (e.g., magnesium in chlorophyll) and enzymes (e.g., iron in ferredoxins);
- determining membrane permeability;
- regulating stomates;
- altering size and number of leaves;
- affecting life-times of photosynthetic tissues.

Thus, most of these minerals if suddenly removed are not greatly missed, but are required for the machinery of photosynthetic production. Little can be said about supplementations required without soil analysis, but a striking example occurred in Australia where soybeans had never succeeded. The addition of molybdenum, one of the metals essential for the nitrogen-fixing enzyme, nitrogenase, resulted in a flourishing soybean crop!

BIOMASS PRODUCTIONS

The limits to photosynthetic productivity have been discussed at length on pages 147–154. This section will summarize typical and reasonably expectable biomass yields. Upper limits are useful for reference purposes for which an estimate can be made from the solar radiations and an assumed 3% overall conversion efficiency. Yield estimates for several locations are summarized in Table II.1.10 (Cooper 1975). Note that the geographic variations in annual insolation are about 2.5, that the maximum occurs at 15°N or S, and that the seasonal variations are less than a factor of two in the tropics while in the temperate regions they can differ by tenfold. The latter fact is almost of greater importance than solar energy because it determines growing season length and average canopy cover. For regions not listed, the following calculation can be used in conjunction with the

Table II.1.10   Annual and Seasonal Energy Inputs and Estimated Potential Production in Different Climatic Regions (after Cooper 1975).

	Latitude	Annual Input (KJ cm^{-2} yr^{-1})	TOTAL RADIATION* Seasonal Variation (lowest month) (J cm^{-2} d^{-1})	(highest month)	DRY MATTER FOR 3% CONVERSION EFFICIENCY* Whole Year (total)
Temperate					
Aberystwyth, UK	52°N	350	210	1880	52.5
Berlin, Germany	53°N	400	170	2010	60.
Tokyo, Japan	36°N	420	630	1510	63.
Wellington, New Zealand	41°S	480	500	1970	72.
Madison, Wisconsin, USA	43°N	500	540	2220	75.
Subtropical					
Deniliquin, Australia	36°S	650	880	2930	98.
Davis, California, USA	39°N	670	750	2720	101.
Algiers, Algeria	37°N	690	880	2800	104.
Brisbane, Australia	28°S	710	1250	2720	107.
Imperial Valley, California, USA	33°N	730	1130	2760	110.
Tropical					
Manila, Philippines	15°N	540	1130	1920	81.
Hawaii	21°N	650	1340	2180	98.
Singapore	1°N	650	1670	2010	98.
Puerto Rico	18°N	670	1550	2130	101.
Townsville, Australia	19°N	750	1670	3590	112.

*Total Solar Radiation Conversion Efficiency and Assuming Dry Matter Energy of 20 MJ/kg.

solar energy map (Figure I.9.4) of Part I's Chapter 9 to predict annual yields.

The assumed conversion efficiency can also be broken down to allow for percent of canopy cover (C, which itself varies with point in the growth cycle), light absorptivity (A), photosynthetic efficiency (13%), and physiological losses (L, such as photorespiration). An assumed plant energy content of 17 KJ/gram then gives for daily light radiation I (KJ/m²-day; Schneider 1973),

$$\text{yield (grams/m}^2\text{-day)} = 74 \text{ AC } (1\text{-L}) \text{ I}$$
$$= 37 \text{ I } (\text{KJ/m}^2\text{-day})$$
$$(\text{for A} = 0.84,$$
$$C = 1, \text{ and}$$
$$L = 0.4; 1 \text{ gram/m}^2 = 10 \text{ kg/ha}).$$

As mentioned above, both canopy cover and solar irradiation will vary from day to day, and of course one assumes optimal water and nutrient supply. Maximum growth rates can reach 52 grams/m²-day (maize), but half is more typical for most plants; the former, if sustained, would yield 100 tons/ha-yr, but daily maximums are held at most a few weeks. Yields for various types of agriculture will now be discussed.

Agricultural Crops

Even when restricted to traditional agricultural crops, maximum possible and typical yields constitute two separate discussions. Husbandry, in addition to soil, climate, and crop genotype, plays a role which is often difficult to identify separately. Maximum yields, mostly from research studies, are useful for selecting crop genotypes and as a guide for necessary development or alterations in traditional practices.

Table II.1.11 (from Cooper 1975) summarizes annual productivities for a number of common crops in various climates (temperate, subtropical and tropical) and latitudes. Where comparisons are made, note not only the latitude (i.e., solar radiation) but also the length of growing season. Thus it is not surprising that forage grasses from the tropical or subtropical regions produce the highest yields: Napier grass yields 85 tons/ha-yr while sugar cane yields 67 tons/ha-yr. These figures compare quite well with the theoretical yield of about 200 tons/ha-yr.

The effect of latitude can be traced for other C_4 grasses, such as sorghum, Sudan grass and Bermuda grass, all of which produce well up to a latitude of 40°N or S: their yields drop from 40 tons/ha-yr in tropical regions to 20 tons/ha-yr in the subtropics. Latitude, of course, combines several productivity factors (solar insolation, climate, growing season and canopy coverage), but as shown in Figure II.1.27 it can serve as a practical summary (Cooper 1975). As discussed at length earlier, C_3 and C_4 plants show a wide spread in yields, but mostly reflect a tendency of C_4 varieties to thrive under hot, high solar intensity and perhaps even slightly arid conditions. The difference in maximums shown is by no means a hard rule, but rather stems from the widespread occurrence of C_4 forage grasses. Forage grasses owe their high yields much more to a high annually averaged canopy coverage plus long growing season than they do to

Table II.1.11 Crop Productivity in Different Environments (from Cooper, 1975).

Crop		Location	Latitude	Product	Yield (t ha^{-1})	Growing Period (days)	% CONVERSION OF LIGHT ENERGY	
							Year	Growing Period
I. TEMPERATE								
A. Forage crops								
Perennial ryegrass (Lolium perenne)	C$_3$	UK	52°N	Forage	25	365	3.0	3.0
Perennial ryegrass	C$_3$	Netherlands	52°N	Forage	22	365	2.4	2.4
Kale (Brassica oleracea)	C$_3$	UK	54°N	Forage	21	140	2.4	3.8
Sorghum (Sorghum sp.)	C$_4$	Illinois, USA	40°N	Forage	16	140	1.2	2.2
	C$_4$	Netherlands	52°N	Forage	15	88	—	4.4
	C$_4$	Ottawa, Canada	45°N	Forage	19	—	1.6	—
Maize (Zea mays)	C$_4$	UK	51°N	Forage	17	160	1.9	3.0
	C$_4$	Japan	35°N	Total dm	26	—	2.5	—
	C$_4$	Kentucky, USA	38°N	Total dm	22	129	1.7	3.4
	C$_4$	Iowa, USA	42°N	Total dm	16	141	1.1	2.8
	C$_4$	Italy	45°N	Total dm	40	140	3.2	6.4
B. Tubers and roots								
Potato	C$_3$	UK	52°N	Tubers	11	164	1.1	1.6
					12	122	—	2.4
(Solanum tuberosum)	C$_3$	Netherlands	52°N	Tubers	18	162	1.9	2.8
				Total dm	22	162	2.3	3.4

Crop	C₃/C₄	Location	Latitude	Part				
Sugar beet (*Beta vulgaris*)	C₃	UK	52°N	Sugar	8	217	0.8	1.0
					8	153	—	1.5
				Roots	13	217	1.4	1.7
				Total dm	14	153	—	2.5
	C₃	Washington, USA	46°N	Sugar	23	217	2.5	3.0
				Roots	14	230	1.0	1.2
				Total dm	26	230	2.1	2.4
				Total dm	32	230	2.5	3.0
C. Cereals								
Wheat (spring) (*Triticum vulgare*)	C₃	UK	52°N	Grain	5	160	0.5	0.7
	C₃	Netherlands	52°N	Grain	6	153	—	1.1
	C₃	Washington, USA	47°N	Total dm	12	69	0.9	3.7
Barley (*Hordeum vulgare*)	C₃	UK	52°N	Grain	12	—	2.4	—
				Total dm	30	148	0.7	1.1
Rice (*Oryza sativa*)	C₃	Japan	37°N	Grain	7	61	—	2.0
	C₃	UK	51°N	Grain	6	123	0.7	1.6
	C₃	Japan	35°N	Grain	7	160	0.5	1.1
	C₃	Kentucky, USA	38°N	Grain	5	—	1.3	—
Maize	C₄			Grain	14	127	0.7	1.4
	C₄	Iowa, USA	42°N	Grain	9	141	0.6	1.6
II. SUB-TROPICAL (including Mediterranean)								
A. Forage Crops								
Coastal Bermuda grass (*Cynodon dactylon*)	C₄	Georgia, USA	31°N	Forage	27	365	1.8	1.8
Alfalfa (*Medicago sativa*)	C₃	California, USA	38°N	Forage	33	250	2.1	—
Sudan grass (*Sorghum* sp.)	C₄	California, USA	38°N	Forage	30	160	1.9	3.4
Sorghum	C₄	California, USA	33°N	Forage	47	210	2.6	3.6

continued

Table II.1.11—continued

Crop		Location	Latitude	Product	Yield (t ha⁻¹)	Growing Period (days)	% CONVERSION OF LIGHT ENERGY	
							Year	Growing Period
B. Tubers and Roots								
Potato	C₃	California, USA	38°N	Tubers	20	—	1.2	—
				Total dm	22	—	1.4	—
Sugar beet	C₃	California, USA	36°N	Sugar	19	240	1.1	1.4
				Roots	35	240	2.1	2.9
				Total	42	240	2.6	3.5
C. Cereals								
Wheat	C₃	Syria	33°N	Grain	8	—	0.4	—
	C₃	California	38°N	Grain	7	—	0.4	—
	C₃	Mexico	27°N	Grain	7	—	0.4	—
				Total dm	18	—	1.0	—
	C₃	California, USA	38°N	Grain	11	—	0.7	—
				Total dm	22	—	1.4	—
Rice	C₃	NSW, Australia	32°S	Grain	14	190	0.8	1.1
Maize	C₄	California, USA	38°N	Grain	13	130	0.8	1.5
				Total dm	26	130	1.7	3.3
	C₄	Colorado, USA	39°N	Total dm	27	117	1.7	3.7
	C₄	Egypt	30°N	Grain	12	—	0.5	—
				Total dm	29	—	1.4	—
III. TROPICAL								
A. Forage crops								
Napier grass (*Pennisetum purpureum*)	C₄	El Salvador	14°N	Forage	85	365	5.4	5.4
	C₄	Puerto Rico	18°N	Forage	85	365	4.9	4.9
Bulrush millet (*Pennisetum typhoides*)	C₄	NT, Australia	14°S	Forage	22	112	1.1	3.8

190

Crop	Pathway	Location	Latitude	Product		Days		
B. Cane crops								
Sugar cane (*Saccharinum* sp.)	C$_4$	Hawaii	21°N	Sugar	22	365	1.2	1.2
				Total dm	64	365	4.0	4.0
C. Tree crops								
Oil palm (*Elaeis guineensis*)	C$_3$	Malaysia	3°N	Oil	5	365	0.8	0.8
				Fruit	11	365	1.1	1.1
				Total dm	40	365	3.2	3.2
D. Roots and tubers								
Sugar beet (two crops per year)	C$_3$	Hawaii	21°N	Sugar	14	365	0.8	0.8
				Total dm	31	365	1.9	1.9
	C$_3$	Tanzania	7°S	Tubers	16	330	0.9	1.0
				Total dm	31	330	1.7	2.0
Cassava (*Manihot esculenta*)	C$_3$	Malaysia	3°N	Tubers	22	270	1.4	1.9
				Total dm	38	270	2.5	3.8
E. Cereals								
Wheat	C$_3$	Sudan	17°N	Grain	7	—	0.3	—
	C$_3$	NT, Australia	15°S	Grain	11	125	0.5	1.5
Rice	C$_3$	Philippines (dry season)	15°N	Grain	10	122	0.5	1.8
	C$_3$	Philippines (wet season)	15°N	Grain	7	115	0.4	1.5
	C$_3$	Peru	7°S	Grain	12	205	0.8	1.3
				Total dm	22	205	1.5	2.6
Sorghum	C$_4$	Philippines	15°N	Grain	7	80	0.4	1.8
	C$_4$	Thailand	15°N	Grain	7	103	0.4	1.2
Maize	C$_4$	Peru	12°S	Grain	16	103	1.0	2.7
					10	—	0.7	—
				Total dm	26	—	1.7	—
F. Multiple cropping								
Rice	C$_3$	Philippines	15°N	Grain	5	115	—	—
+ Sorghum	C$_4$	Philippines	15°N	Grain	6	90	—	—
+ Sorghum (ratoon)	C$_4$	Philippines	15°N	Grain	7	80	—	—
+ Sorghum (ratoon)	C$_4$	Philippines	15°N	Grain	5	80	—	—
Total for year	—	Philippines	15°N	Grain	23	365	1.6	1.6

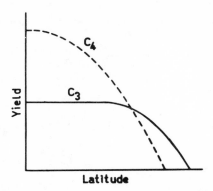

Figure II.1.27. Approximate variation in crop yield with latitude (by permission of Loomis and Gerakis 1975).

their C_4 carbon-dioxide-fixation mechanism. Multiple harvesting with a root system left intact virtually assures a faster canopy closure.

C_4 plants tend to decrease in yield at higher latitudes, but maize is one critical C_4 crop which is uniquely respectable even beyond 40°: it yields 34 tons/ha-yr at 45° N and 17 tons/ha-yr at 52° N. C_4 plants demonstrate much greater dependence on latitude than C_3 varieties.

Among the C_3 plants, Manioc (*Manihot esculenta*) has the highest yield at low latitudes (41 tons/ha-yr at 8.5° S) while sugar beet is superior at intermediate latitudes (34-42 tons/ha-yr at 35° N). Other productive C_3 crops include particularly forages (e.g., rye grass) and grains, while legumes, except for alfalfa (33 tons/ha-yr at 38° N), have typically low yields at all latitudes.

Preliminary estimates for energy farming have considered forage crops, maize, sorghum and elephant grass, and the tuber crop cassava (McCann and Saddler 1976, Kemp and Szego 1975). Very preliminary cost estimates are as low as \$1.10 per 10^9 J (10^6 Btu) for which water availability and slope of terrain are among the main cost determinants along with transport distances.

Forest Productivity

Yields.   Forests as solar-energy converters present a unique set of advantages relative to other photosynthetic systems:

1. They can be harvested according to demand and availability of labor.
2. Weather variations will be averaged over a longer period and except in most extreme cases represent no catastrophic threat.

3. Perennials need not be replanted until after about five harvests.
4. Some trees demonstrate rapid juvenile growth.
5. Marginal soil can be used.

Wood production is not without its unique problems, and among the disadvantages of forest production are:

1. Low energy-conversion efficiency due to small canopy coverage as juveniles.
2. Interference between adjacent plants after maturity limits planting density and harvesting cycle.
3. Conifers grow slowly when young.
4. Maximum solar-energy-conversion efficiency is lower than the best crop efficiencies.
5. Lignin content is high in many species and limits use of biological conversion methods.
6. Slopes of more than 10-12° limit mechanized harvesting.

Forest productivity varies widely according to geography (principally solar insolation) and tree species (principally leaf canopy structure). As expected, tropical rain forests are among the most productive with a global mean of 20 tons/ha-yr and highs of 44 tons/ha-yr (Troughton and Cave 1975). Evergreen broad leaf forests in warm temperate zones have comparable productivities followed by pines and temperate conifers (10-15 tons/ha-yr) and conifers in the boreal zone (10 tons/ha-yr). Deciduous broad-leaf forests in cool temperate zones have lower productivities (5-10 tons/ha-yr) due not only to climatic conditions but also to their short leafy period. In contrast, the advantages of evergreen coniferous and broad leaf forests lie principally in their great, long-lived leaf surface. The distribution of yields is summarized in Figure II.1.28 and Table II.1.12 (from Kira 1975).

Forest productivities quoted above are long-term yields which average out changes in growth rate. The inadequacy of the leaf canopy in juvenile plants and crowding in older forest stands are the limits to a production curve which maximizes when the canopy closes. A closed canopy represents the period of maximum photosynthetic efficiency for a forest stand with a productivity of 25-35 tons/ha-yr. While the growth period required to reach this maximum varies with specie and tree density, it occurs roughly when the standing biomass density is 80-140 tons/ha (Kira 1975). Optimization of harvest for biomass will clearly lag behind the annual yield maximum in order to make up for the initial period of low growth. Note from Table II.1.12 that while the gross production efficiency for closed forests is 2-3.5%, net productivity efficiency is less than half this

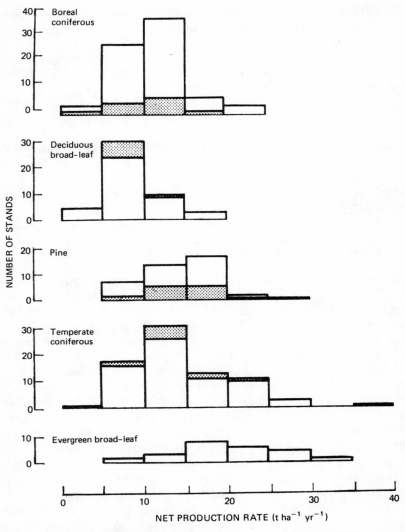

Figure II.1.28. Frequency distribution of annual above-ground net production in 258 forest stands in Japan (by permission of Kira 1975).

amount. The difference is due to biomass (leaves, branches) dropped during the year or metabolic losses from dark respiration.

Energy Return.   As dry wood invariably contains about 20 MJ/kg and ca 45% cellulose (Troughton and Cave 1975), a 20-dry-ton/ha-yr (70-m³/ha-yr) forest yield will have a gross energy yield of 400 GJ/ha-yr (4 ×

TABLE II.1.12 Gross and Net Production Rates in Forest Communities (Kira 1975).

| | | | | RATE (t ha^{-1} yr^{-1}) OF | |
	Localities	Age (yr)	LAI (ha ha^{-1})	Gross Production (ΔP_g)	Net Production (ΔP_n)
Tropical rain forest	S. Thailand	—	11.4	123.2	28.6
Tropical subhumid forest	Cote d'Ivoire	—	3.2	52.5	13.4
Climate warm-temperate ever-green forest dominated by *Distylium racemosum*	S. Kyushu	—	8.8	73.0	20.6
Secondary forest of *Castanopsis cuspidata*	S. Kyushu	11	8.0	45.3	18.7
	S. Kyushu	14	8.9	51.7	22.7
Cryptomeria japonica plantation	S. Kyushu	5	8.6	84.1	29.1
	S. Kyushu	24	7.4	73.3	15.1
	N. Kyushu	28	4.3	54.4	18.8
	W. Kyushu	31	6.8	64.1	16.7
	N. Kyushu	34	—	57.1	16.0
Chamaecyparis obtusa plantation	S. Kyushu	45	5.1	40.9	15.4
Pinus densiflora plantation	C. Honshu	15	—	53.9	15.8
Fagus crenata secondary forest	C. Honshu	30–70	5.7	27.5	15.3
Fagus crenata plantation	C. Honshu	50	7.8	44.1	19.3
Fagus sylvatica plantation	Denmark	8	4.2	13.9	7.5
	Denmark	25		22.3	13.5
	Denmark	46	5.4	23.5	13.5
	Denmark	85		21.4	11.3
Fraxinus excelsior plantation	Denmark	35–45	—	21.5	13.5
Oak-pine secondary forest	New York	—	3.8	26.4	13.5
Picea abies plantation	Denmark	40–50	—	26.5	18.0
Abies sachalinensis forest	Hokkaido	35–40	—	50.2	23.8
Subalpine *Abies* forest	C. Honshu	c.15	—	19.9	7.4
	C. Honshu	—	—	40	11.1
Subalpine *Abies veitchii* forest	C. Honshu	4	5.5	16	8.4
	C. Honshu	25	9.7	45	16.8
	C. Honshu	60	10.6	49	12.8

10^{11} J). Conversion will, of course, be less than 100% efficient and will require in addition some energy investment. These costs are discussed in Part II, Chapter 3.

Energy investments for wood production have been estimated (Cousins 1975) for a forest of 10,000 ha to be about 2% of the assumed 18 tons/ha-yr yield. For the seven stages considered (land preparation, planting, thinning, protection, logging, transport, and shipping), fuel costs constituted about 80% of the energy used, the remainder being attributable to machinery manufacture. Logging consumes 70% of the energy consumed, but if transport distances were increased from the assumed 30 km to 80 km, then transport would consume as much as logging. Logging efficiency and transport distances are the critical energy factors.

Economic Return. Economic product value varies, of course, with end use, but in all cases logging is again the expensive step. Denser hardwood forests are therefore preferred. Traditional forestry in New Zealand has produced radiata pine at $12/m³ for a 23-year rotation, while *Eucalyptus* in a 10-year cycle was expected to involve lower costs due to higher wood density and easier logging (Williams 1975).

An economic estimate has also been made for an energy plantation with a fast growing perennial conifer, the Southern pine (Kemp and Szego 1975). Estimates were made for planting densities of 2000-20,000 plants per ha, harvesting cycles from 3-17 years, and assuming the chipped fuel product is loaded on trucks at the harvesting site. Labor costs (1973 level), fertilizer costs (1974 levels) and 10% capital inventory costs were also assumed. Costs show a distinct minimum for a 12-year harvesting cycle (in contrast to 20-30-year schedules for pulp wood) with a planting density of 13,000 plants. The net cost, ca $3/GJ, results largely from slow growth rates and the standing inventory cost (40%). Inventory costs can be reduced if harvesting is more frequent, such as for hybrid poplars (annual density 8000 plants per ha) harvested every three years. Poplar was also the choice in Sweden (~60° N) for a trial intensive energy plantation to produce 20 dry tons/ha-yr harvesting every one to three years (Sirén 1978).

Grazing Land

Grazing land includes a broad spectrum of climatological and ecological conditions, grazed both by domestic livestock and wildlife. The vegetation is a natural mixture adapted to local conditions and grazing habits but otherwise unaltered by Man. Assuming the animal harvest is not large relative to the natural mineral balance, the annual production will primarily depend upon solar insolation and water conditions. The first factor is directly reflected in a plot of plant yields vs latitude, shown in Figure II.1.29. Several factors vary together with latitude, of course, particularly temperature and season length, but the strong interdependence among these makes latitude the more useful parameter for productivity comparisons.

Annual productivities vary enormously, from 0.5 to 10 tons/ha-yr in arid or semi-arid regions (70% of the world's grazing land) to 6 to 30 tons/ha-yr in more humid regions (Caldwell 1975). Grazing lands have the same potentially high yields as forage crops, but less irrigation and fertilization limits yields. As only 10% of the nutrients ingested by grazing animals is retained and removed from the fields, fertilizer costs would not be excessive if irrigation water is available.

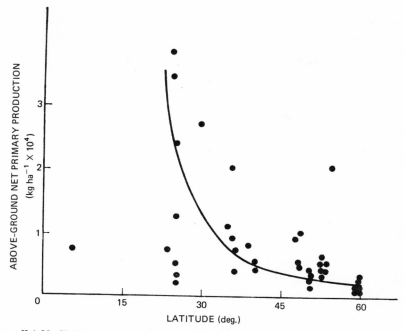

Figure II.1.29. Variation in the above-ground productivity with latitude for grassland and tundra (by permission of Caldwell 1975).

Deserts

Arid regions, those where the potential evapotranspiration exceeds precipitation, constitute a third of the earth's terrestrial surface of about 10-M km². Production in these typically high solar radiation areas is limited not only by the total annual rainfall but by the precipitation schedule and resulting soil conditions.

Vegetation which has evolved to survive arid and semi-arid regions has extreme local importance and little chance for improvement in the absence of water. Artificial provision of water from trickle irrigation in fields of closed greenhouses can, to the limit of water reserves, exploit the abundant solar energy and unused land area in these regions. Saline or covered algal ponds are also appropriate to arid regions.

As up to two meters of water are lost annually in arid regions, water provision or prevention of its loss is the limiting factor in bioproductivity. Using this extraordinary energy and land resource will require very special and capital-intensive techniques. Extensive research is required to determine if biological or physical (or some combination thereof) is most appropriate for solar-energy capture.

Tundra

Tundra, by definition, has permafrost in the soil or an annual subzero average air temperature. Such harshness results not only in low production, but also low biological diversity, low plant density and low system stability. The latter point is intimately related to cyclicity in (small) animal life which, of course, suffers from small changes in the plant system. Isolated meadows in alpine and arctic tundra zones can occasionally produce up to 4 tons/ha-yr and thus have great local significance, but productivities from as low as 30 kg/ha-yr and low solar radiation make most tundra of little practical interest despite the enormous areas.

Wastes

Conversion of organic waste material to fuel or food requires the simultaneous consideration of several factors:

- Economics: Are the collected, transported, separated, dried and treated wastes less expensive than other fuels? Note: A positive credit may result from disposal costs.
- Ecology: Does the waste removal have a positive or negative effect on the ecology? (Decaying waste matter is often crucial for soil nutrition and quality.)
- Energy: Does the conversion of wastes to fuel/food have a net positive or negative energy balance? A negative energy balance is rarely justified, such as when the problem is one of disposal.
- Water Quality: To which conversion processes is the waste suited, from the point of view of both waste content (toxins, separable components of higher value, etc.) and conversion ease?
- Waste Quantity: Geographical distribution and seasonal availability.

The above factors cannot be treated in general as each is highly specific to the country, waste, and time of year. One general conclusion which is possible, however, is that food shortage as a worldwide problem precludes most short-run considerations of energy farms; use of marginal lands, unsuitable for food crop growth, is one obvious justifiable exception. Wastes or by-products from food production, processing, and consumption are more logical starting points whose economic credibility will assuredly increase as oil is depleted.

Wastes are most conveniently divided according to their source which in addition distinguishes them according to their quality, consistency and availability:

Urban Wastes. Urban wastes are in the industrial world an embarassment of wealth and excessive consumption. The USA leads with an annual urban refuse production of a half-ton per capita or nearly 2 kg waste discarded at home per person every day. If all wastes are included, the daily per capita waste production increases to 11 kg (Anderson 1972)!

Waste treatment as an urban problem cannot be treated here except for its two organic components, sewage and waste paper. Inorganic and nondigestible organic wastes (e.g., plastics) are amenable only to industrial treatment, such as pyrolysis, and are insignificant as a fuel/food-producing process.

Waste paper constitutes 40% of (US) urban wastes, and at 75% cellulose represents an easily digested, energy-rich waste. Its value as a collected recyclable waste is worth $8-25 per ton to the paper industry, an ideal use for cellulose refuse already containing potentially harmful inks and dyes. Whatever the end use, economic and low energy-consuming collection systems are required and represent the limiting factor to recycling.

Water-borne sewage, on the other hand, is in modern urban centers collected and transported cheaply and efficiently to central treatment sites. Human excretions of about 1.2 kg (140 grams solids) per day are added to variable amounts of kitchen wastes, a figure which increases with the frequency of garbage-grinding systems. Sewage treatment has traditionally been a problem of purifying water, and even when anaerobic digestion was employed, fuel (methane) production was not sufficiently valuable to warrant optimization. Modern attempts which view sewage wastes as a rich medium for microorganism growth (algae and methanogenic bacteria), discussed in Part II, Chapter 3, combine an efficient collection system (sewage pipes) with energy, water, and nutrient recovery, plus waste disposal credit. Successful realization of this highly appropriate centralized facility rests in the segregation of potentially harmful wastes *before* they enter the sewer system.

Industrial wastes is again a specialized problem which is of limited interest in this discussion except for those wastes originating from food and wood processing. Considerable wastes which are concentrated and of uniform quality are available as waste by-products from meat packers, fish and fruit processers, bakeries, mills, breweries, molasses and bagasse from cane and beet sugar production, plus many others of local importance. These are organic wastes of the highest quality whose food grade level can be maintained via microbiological reprocessing.

Of critical relevance to sewage use in a negative sense are those toxic industrial wastes disposed via the domestic sewage system. Among these wastes are many which make urban sewage totally unacceptable for biological treatment: toxins can kill methanogenic bacteria and/or can become incorporated into algae for re-entry into the food chain. Separate sewage disposal may be required for several industries.

Forest Wastes. Exploitation of forest wastes has potentials and limitations similar to agricultural wastes; in each case a large amount of by-products are left at the point of harvesting and at each processing stage. Discarded wastes account in part for the typically low forest yields: branch and foliage growth is 100-150% bale growth (40-50% of total net production), and yet only bale wood is harvested (Stephens and Heichel 1975). Of the standing crop, branches and foliage represent a harvest waste 12-42% of the total. Nearly all of the foliage and some of the branch growth is lost each year as it falls to the forest floor, but annual harvesting of this diffuse waste is both impractical and detrimental to the forest floor ecology: mineral content, soil quality and water runoff would be impaired.

Forest harvesting wastes not only the 12-42% in foliage and branches, but also discards the upper portion of the bale. Most of these residues (24 million tons in US, 1968) are left in place to rot or are burned. Harvesting of this portion is, of course, possible, but the diffuseness of the waste represents a severe economic and energetic limitation. A costly side-effect of waste harvesting would be the additional fertilizer and soil-conditioner requirements: a disproportionally large fraction of the nutrients (e.g., fixed nitrogen) are found in the rapidly growing and photosynthesizing leaves. Conversion to high-quality protein (see Leaf Protein, pp 314) may improve the economics of forest waste harvest.

An amount roughly equal to logging wastes is discarded in primary and secondary wood-processing plants. These wastes (27 M tons/yr in the US, 1968) are concentrated at manufacturing sites and represent a cellulose potential which is already partly exploited. The major use is for pulp and a lesser amount for fuel. Improvements in forest waste utilization lie principally in increased exploitation of these wastes at the factory particularly by converting the cellulose to the highest possible quality end product: pulp is one important product, but breakdown to sugars for fermentation or microbial growth for food and fodder is even more valuable. Sulfite liquors from the pulping process is a polluting waste which can serve as a growth medium for sulfur bacteria.

Agricultural Wastes.   As with forest yields, agricultural crop yields often seem low due to the neglect of waste production: typically half of the above-ground biomass is waste while the below-ground production can vary from a tenth to double the above-ground production (Cooper 1975). Little can or should be done about the latter portion as it plays an essential role in soil conditioning.

The above-ground wastes, such as those listed in Table II.1.13, increase in availability as total crop production increases and represent an enormous potential for energy production and nutrient recycling. For 10 tons/ha-yr waste production, conversion to methane (assume 50% efficiency) would yield 90 GJ/ha-yr, an energy yield three times the per hectare energy requirement for US maize production.

Cereal straw is by far the world's most abundant crop waste which together with grain husks represents an enormous cellulose resource. Straw has, of course, some competitive uses as a building and bedding material. Sugar cane plantations on the other hand produce prodigious quantities of bagasse (up to 80 tons/ha-yr, of which 50% is cellulose) which has long been burned to fire sugar-mill boilers. It therefore already has a positive value of $6-13/ton, but is considered for microbiological conversion to more valuable end products.

Most agricultural wastes are diffusely distributed and are never harvested. Much is burned to prevent the spread of disease which wastes not only the energy but also nutrient (ammonia) value. Plowing under at least serves as a soil conditioner, promotes activity among soil bacteria and recycles nutrients. An end to this resource waste can be expected with an

TABLE II.1.13   Low Value Agricultural Wastes.

Crop	Waste
Sugarcane	Bagasse
Sugar beet	Pulp
Coffee	Pulp
Carob	Husks
Maize	Cobs and husks
Dates	Pulp
Potato	Peelings
Tomato	Pulp
Grass seed	Straw
Wheat	Straw and husks
Rice	Straw and hulls
Peanut	Hulls
Cotton	Hulls

extensive FAO survey of waste availability (in progress) and the spread of decentralized microbiological techniques for conversion to energy, food, and/or fertilizer. Centralized facilities are precluded in all but special cases by the waste's low energy value.

Animal Wastes. As animals retain only 10% of the food mass they ingest, they excrete large amounts of metabolic wastes and incompletely digested organic matter (10-30% cellulose). This inefficiency is acceptable when low-quality plant products are digested by ruminants, but from the start the dubiousness of animals fed human-quality food must be stated. Grazing animals, however, digest cellulose, and their wastes form an important link in the local, natural recycling process with a relatively low resultant net soil-nutrient loss.

Confined animals, however, produce a prodigious amount of excreted wastes which can alternatively be a severe disposal/pollution problem or a rich source of energy and nutrients. The conversion of these wastes will be discussed in Part II, Chapter 3, but the enormity of this resource should be appreciated: in the US two billion tons of fresh manure (15% solids) or 200-400 million tons dry organic wastes are produced annually, 50-80% from confined animals (Anderson 1972). An increasing fraction of these animals are being confined in mammoth feedlots containing over 30,000 cattle, 1000 dairy cows, up to a million poultry, or several thousand pigs per lot. The often disturbing problem of recycling these wastes even on small farms is exacerbated in such a centralized lot where its 100,000 cattle would annually produce a million tons manure (150,000 tons dry organic waste). Depending upon its location, a feedlot has several alternatives for its manure:

- sell it as fertilizer (each ton of unprocessed cow manure is worth about 2 kg nitrogen in fertilizer) at $1.50 per ton unprocessed or $0.60 per 40-lb. bag ($0.32/kg) of dried manure; reprocess it chemically to a high-protein (35%), fine-textured, sterile fodder for $50 per ton;
- convert it anaerobically to methane (280-560 m^3 per ton dry organic waste) plus sludge containing all nutrients.

All of these alternatives are now being tried, but only the latter two are truly justifiable; conversion to expensive garden or lawn fertilizer is a luxury ill-afforded in a world of food shortages. A $4-million anaerobic digestion plant is being constructed for the 2½ km^2 Monfort, Colorado, feedlot and will process 750 tons manure per day from 100,000 steers. The high water content in manure makes it most suitable to biological processing at the production site, but manure will not be easily digested in its entirety—if it were, the animal would have digested it. Transport quickly becomes energetically and economically inefficient and is most suitable to small-scale use in remote areas. If the above system can return nutrients to neighboring farms at low energy cost, it may prove to be a viable compromise.

2

Algae—A Special Case

WHAT AND WHY?

Algae are ubiquitous in habitat and variable in size and morphology; they have been found in the ice of polar regions and in 90°C hot springs, in both fresh water and brine lakes, and they cover a range from unicellular organisms several microns in size up to kelp 50 m in length. Their cell walls can vary from the rigid, thick silicon impregnated walls of diatoms to fragile walls susceptible to osmotic lysis; their reproduction can be sexual or asexual or a combination of both, and they can be highly mobile or totally passive. While taxonomic classification is not always obvious, algae include all photosynthesizing eucaryotic (cells with nuclear membranes) protists (unicellular or undifferentiated multicellular organisms). To this group need only be added the procaryotic (cells without nuclear membrane) algae, the blue-greens.

The two algal groups of principal interest here are the green (*Chlorophycophyta*) and blue-green (*Cyanophycophyta*) algae. Each type has much of the variety mentioned above, but the former, a eucaryote, stores energy as starch (like higher plants) while the latter, a procaryote, stores energy as glycogen.

Algae have been selected as a special case due to their unique open water habitat and a number of advantages which make them an attractive "crop" in certain areas:

- high yields are typical of the best terrestrial plants such as sugar cane;
- transport distances of nutrients and photosynthates are very small, i.e., cellular dimensions;
- nutrient diffusion from the medium, a surface phenomenon, is facilitated by algae's orders of magnitude more surface area per biomass than terrestrial plants;

- nutrients and water are held in a confined space and transpiration is eliminated;
- protein content is very high (40-60% of dry weight);
- nitrogen fixation occurs in some varieties;
- low-quality land can be used;
- low-quality water can be used and thereby improved;
- low-quality waste heat can be used to mutual advantage of e.g., power plant or diesel motors and increased photosynthesis;
- energy costs are low for biomass, protein and caloric yield;
- doubling times are low, on the order of hours;
- a wide variety of species exist which are suitable to vastly differing conditions of salinity, temperature and nutrients;
- certain species, especially blue-greens have a high temperature optimum (about 32°C). This can also be disadvantageous;
- climatic condition variations are of less importance than for terrestrial plants.

In summary, algae (grown on wastes), when compared with normal terrestrial plants, will require 1/50 the land area, 1/10 of the water, 2/3 the energy, 1/5 the capital and 1/50 the human resources for an equal amount of useful organic matter (Oswald and Golueke 1968).

Algae are, however, not preferable to terrestrial plants in all respects, and growth conditions may easily exacerbate the severity of several disadvantages:

- reflection from the water surface is about 30% of the incident light;
- evaporation of up to 1-2 m water per year can occur under arid conditions;
- low cell densities of 0.2 to 8 grams dry weight per liter makes cell concentration and drying both difficult and expensive;
- similarity of nutrient requirements and growth conditions for most algae results in great difficulty controlling the desired species, except under ideal, sterile growth conditions;
- high temperature optima (greater than 30°C) for some species, especially blue-greens;
- toxins are generated by some (especially blue-greens) algae which can be fatal for fish (see page 228);
- parasites, especially rotifers, can easily attack an algal culture;
- untreated algae are poorly digestible and require post-harvesting treatment;
- CO_2 is required in such vast amounts that nonwaste CO_2 is economically nonviable.

MASS CULTURING—GENERAL TECHNIQUES AND GROWTH LIMITATIONS

Technical details of mass algal culturing must, of course, be suited to the end use (food, fodder, fuel, fertilizer) for the algae, which also determines to a large degree the growth medium used (synthetic vs natural, fresh water or saline). Practical systems, when capital costs and technology must be minimized, invariably employ earthen ponds and differ only in the nutrient source and post-harvest processing.

Large-scale open ponds are generally earthen, dug to a depth of 30-100 cm (effective light intensity is near zero beyond 20 cm), and where the earth is porous the bottom is lined with a double polyethylene sheet. If cost allows, concrete can be used instead as plastic sheet has only a one-year lifetime. Vast improvements are needed for an inexpensive, preferably locally produced durable pond bottom. Pond size is freely variable with neither advantage nor disadvantage, and volumes in excess of 10^9 liters already exist. Common variations to the simple open pond include terraced ponds or ponds with narrow ($\sim$2m) raceways, each complication being designed to aid stirring and aeration.

Figure II.2.1 summarizes many of the components and alternatives for mass algal cultures where each variation is an attempt to meet the five basic requirements for good algal growth:

- illumination of sufficient intensity;
- CO_2 supply;
- temperature optimum;
- mineral supply;

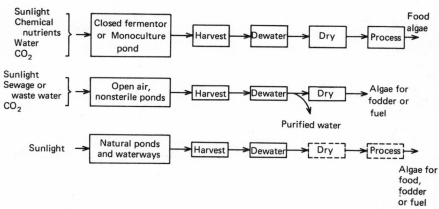

Figure II.2.1. Alternative methods of algae production.

• agitation to prevent sedimentation and promote nutrient and light accessibility.

Light

Light is the limiting factor below the first few centimeters yet is in excess in the first centimeter. This sadly paradoxical situation results from the observation discussed earlier (see page 152) of saturation of photosystem I even at intensities 5-10% of noontime sunlight; furthermore the rate-limiting step has a time constant of 10 msec which is too fast to alleviate by stirring (Lien and San Pietro 1976). That this enzymatic rate constant may well be temperature dependent might find confirmation in the observation that at 40°C the blue-green algae *Anacystis nidulans* does not saturate at high light intensity (Goedheer and Kleinen Hammans 1975). The improvement in algal yield if photosaturation were eliminated could be considerable as the top saturated layer is not only an inefficient solar converter, but filters light from lower cells. The resulting distribution of growth rate vs depth is shown in Figure II.2.2 to be about 40 cm in a natural lake (Talling 1975). For dense algal cultures, little, if any, light reaches such depths, and the maximum is shifted to less than 10 cm.

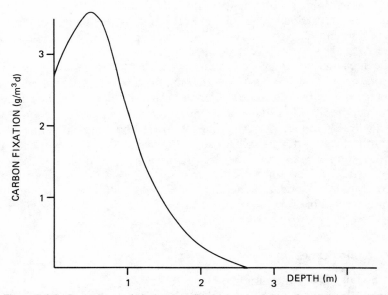

Figure II.2.2. Dependence of algal carbon fixation on depth (data from Mathiesen 1970).

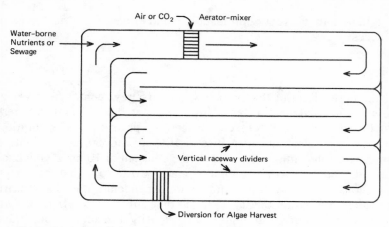

Figure II.2.3. Meandering raceway algae pond (redrawn from Shelef et al. 1976).

The uppermost scum layer, in addition to being photosaturated, tends to dry and requires agitation which also mixes nutrients and prevents sedimentation. Terracing with a slow flow of the culture down a slope is also effective, but complicates pond construction and will add greatly to energy costs if the water must be lifted again to the starting level. With a retention time of a few days and flow velocity of 30 cm/sec one can easily show that pumping requirements will far exceed—by orders of magnitude in some configurations—the solar energy stored in the algae. No terracing system known today is so advantageous as to pay for the added pumping obligations.

The meandering raceway formation shown in Figure II.2.3 provides two alternatives for mixing and aeration: a mixing and aeration pump can emit a jet of compressed air, or alternatively a paddle wheel placed in one corner can be used to provide the required 5-30 cm/sec circulation velocity (Shelef, Moraine, and Meydan 1976). Each system is amenable to windmill power with an auxiliary source for windless days. Circulation in a large, simple open pond can be provided by two pumps in opposite corners (2 × 1/3 hp for 120 liters/min for a 140,000-liter pond; Ryther 1975), but in general the methods of choice will vary with location, materials available and the type of algae (e.g., unicellular or filamentous).

Carbon Dioxide (CO₂)

Although light is the ultimate factor limiting areal yield, CO_2 concentration is the most severe, controllable limiting element. CO_2 is particularly limiting at midday under bright conditions when the CO_2 fixation rate far exceeds the supply rate unless special measures are taken. While algae are not strictly C_4 plants, they do not demonstrate photorespiration; the observation of photo-oxidative death under low CO_2 may, however, account for death of algae during the summer (Abeliovich and Azov 1976). CO_2 concentrations for algae are hence mostly determined by the need to supply CO_2 at a rate matching its consumption. Gas consumption can easily be calculated from the rate of cell mass production—per year, day, or hour—using the conversion factor 1.8 grams CO_2 consumed per gram dry algae grown. A reasonable maximum hourly rate of algal CO_2 fixation is ca 20 grams/meter²/hr or 40 mg/liter/hr (assuming 0.5 meter depth). This amount of CO_2 is contained in air (0.03% CO_2) flowing at the tremendous rate of 70 liters of air per liter of culture per hour; as not all CO_2 in air will be removed, the actual rate of flow would have to be considerably larger. These flow rates are impossible for several reasons and simply point out the need for CO_2-enriched air.

CO_2-enriched air, say 5% CO_2 instead of the normal 0.03% CO_2, decreases the required gas-flow rate both because the gas carries more CO_2 per liter and because the soluble CO_2 equilibrium value increases. The first factor decreases the rate by more than two orders of magnitude ($\times 170$), while the second factor increases both the fraction of CO_2 removed from the gas and the medium's CO_2 storage capacity. Equilibrium with air is only 0.5 mg/liter (pure water at 25°C) while with 5% CO_2 enriched air it is about 70 mg CO_2/liter water, or about two hours' CO_2 requirement.

CO_2 solubility is far more complex than can be adequately dealt with here, but a few important points should be noted. First absorbed CO_2 is found in equilibrium with several other forms:

CO_2 (gas) $\rightleftarrows$ CO_2 (dissolved)
CO_2 (dissolved) + H_2O $\rightleftarrows$ H_2CO_3 $\rightleftarrows$ H^+ + HCO_3^- $\rightleftarrows$ H^+ + CO_3^{2-}

Equilibrium values are determined by gas CO_2 content, water medium temperature, pH, and salt content (Meynell and Meynell 1965). The first two factors for CO_2 solubility in water (mg/liter at 760 nm pressure) are summarized in Table II.2.1.

CO_2 solubility is somewhat dependent on pH and more so on salt concentration, but the amount potentially convertible to nonvolatile and highly soluble bicarbonate (HCO_3^-) is most strongly dependent on pH. Table II.2.1 and Figure II.2.4 demonstrate these equilibrium changes and emphasize the advantages of a high pH growth medium, as for *Spirulina*. Even within the more normal physiological range the variability of available CO_2 in solution (as CO_2 or bicarbonate) is large: at pH 6 (37°C) and

Table II.2.1   CO$_2$ Solubility in Water (mg/liter at 760 nm pressure; Meynell and Meynell 1965).

T (°C)$_2$	100% CO$_2$	5% CO$_2$	0.033% CO$_2$ (air)
0°C	3346 mg/liter	167 mg/liter	1.100 mg/liter
10	2318	116	.765
20	1688	84.5	.557
30	1257	63	.415
40	973	48.6	.32
50	761	38.0	.25

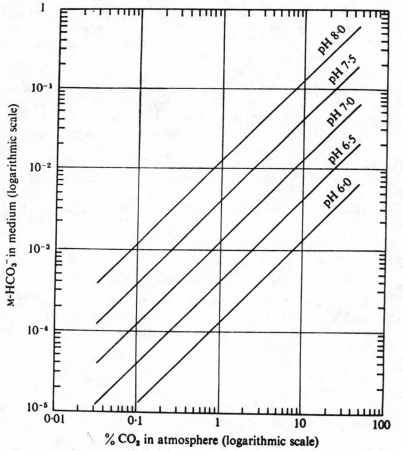

Figure II.2.4. Dependence of bicarbonate on pH and atmospheric CO$_2$ concentration (by permission of Meynell and Meynell 1965).

5% CO_2 gas, 90 mg CO_2/liter are potentially available (72.5% CO_2 and 27.5% as HCO_3^-), while at pH 8 nearly 3.4 grams CO_2/liter are available (97% as HCO_3^-). Any decrease in pH, often accompanying photosynthetic CO_2 absorption, can cause large amounts of CO_2 to be volatilized and points out the need to hold the CO_2—HCI_3^-—CO_3^{2-} buffering system poised at the optimum pH.

As bottled CO_2 is excessively expensive, various other schemes have been attempted:

1. In Israel (Richmond 1975) calcium carbonate is readily available from which CO_2 and hydrated lime (from calcium oxide) are the combustion products:

   $$90\% \text{ calcium carbonate } + 10\% \text{ oil} \xrightarrow{\text{combustion}} CO_2 + \text{calcium oxide.}$$

   Mixing with air for 1-2% CO_2 concentration and bubbling the gas in the medium gives near optimum conditions for *Spirulina* growth.

2. Carbon dioxide is the principal gaseous product of any combustion process and generally—perhaps after a bit of particle removal—is treated as an easily disposed of waste product. Combustion or exhaust gases are, however, ideal from another point of view: they are heated, pressurized and contain predominantly CO_2. A well integrated system for algal growth can exploit these factors to provide the required CO_2 for algal growth and extend the growth season in cool climates. On a larger scale, conventional power plant flue gases can be combined with large open algal ponds (Benemann et al. 1976), while for smaller scale ponds CO_2 from the confined combustion of biogas can be used. Cleaning of the gases may be required for food algal growth, but this has been shown feasible for diesel exhaust gas using first a (sea) water trap for water soluble gases (SO_2 and NO_2) followed by an activated charcoal filter for nonsoluble organic gases (Eisa, Zeggio, and Jensen 1971).

3. The easiest and most widely practiced large-scale production of CO_2 for algal growth is from the aerobic bacteria digesting organic wastes in waste treatment oxidation ponds. Restated from the sewage treatment point of view, the algae are used to provide O_2 for the bacteria in the oxidation pond. The happy symbiosis will be discussed in detail in the section entitled GROWTH MEDIUM on page 216 of this Chapter.

Temperature

The third growth factor, temperature, is rarely controlled, but determines in part the species which naturally bloom and where algal ponds are practicable. Oswald (1973) states that waste treatment algal ponds are practicable wherever ponds don't freeze for more than one or two months. Uses discussed here mostly have more stringent requirements such as for the interesting blue-green algae whose temperature optima lie over 30°C.

As more than one meter water can in certain areas evaporate per year and cause evaporative cooling, a transparent plastic cover may in some cases aid in maintaining optimum temperature. If one assumes, for example, an annual water loss of one meter (10^7 liters/ha-yr), then covering the pond would save not only water but ca 23×10^6 MJ/ha in low-grade heat energy, or 35 times the energy in the algae (assuming a dry algal yield of 50 tons/ha-yr). Low-grade heat energy can be used only with low efficiency, but this waste heat (ca 95% of the incident light energy) is readily available if the temperature needs to be raised in the pond or for example in associated microbiological digesters. Clearly, conductive losses to the earth and air will become limiting in a covered pond and will increase with increased pond temperature, but depending on the application, some insulation techniques may be justified (Dickinson, Clark, and Wonters 1976). Cooling at night may not be disadvantageous as dark respiration increases with temperature.

GROWTH MEDIUM

The fourth growth factor, nutrient requirements and their method of provision, are treated separately here because they divide algal culturing into four major divisions to be discussed in detail here: "clean" algae grown on synthetic medium (for food), algae grown on wastes in conjunction with oxidizing bacteria (for food, fodder, fuel or fertilizer), algae harvested from natural or semi-natural blooms (for food or fodder), and nitrogen fixing algae grown alone or in association with other plants (as nitrogenous fertilizer). The first three of these alternatives were summarized in Figure II.2.1.

Synthetic Medium for "Clean" Algae

The cultivation of clean algae for human consumption differs from the other systems to be discussed here only in the source of CO_2 and minerals; all concepts of pond construction, temperature and harvesting can be identical with simpler systems. Economically, the justifications for these extra costs lie in possibly obtaining greater yields due to medium optimization, the greater possibility to control the algal specie, and a higher retail price (assuming algae from wastes is only suitable as fodder or would require expensive sterilization before human consumption). Presumably, the higher product value justifies a greater investment in ponds and harvesting procedures, but evidence to date is not economically encouraging. The German effort (Soeder 1976) can produce food quality protein from *Scenedesmus* projects in India, Thailand, Peru and Israel (the latter project uses sewage effluent instead of a synthetic medium, see below). This algal protein grown on synthetic media has quality comparable with soya protein, and if the cost of the latter rises a bit, algae may become economically competitive.

Among the major algal species contending for commercial exploitation are *Chlorella vulgaris, Scenedesmus acutus, Coelastrum proboscideum* and *Spirulina maxima,* the most active work being found in Japan (*Chlorella* and *Spirulina*), France (*Spirulina*), Czechoslovakia, Bulgaria and Germany; the latter three all use *Scenedesmus.*

Scenedesmus has the advantage of being prolific, but its small size has continually plagued the economics of harvesting (see page 224). A unique alternative, one appropriate to equally unique conditions is *Spirulina,* a helicoidal blue-green (*Cyanophyte*) algae which is advantageous for several reasons (Pirie 1975a):

- its size (0.2 mm long) and helical shape causes the formation of entangled clumps which are easily removed by filtration;
- its alkaline (pH 10) growth optimum allows little competition from other microorganisms, assures abundant CO_2, and allows use of alkaline lakes common to arid regions whose waters are useful for little else:
- its protein content (64-70%) is greater than any other natural product and is of higher quality;
- it has high digestibility and mild taste;
- it has been used for centuries as a food;
- it has an annual dry weight yield of 50 tons/ha-yr under optimum conditions (35°C, pH 9, 23 grams/liter salinity) to 10 tons/ha-yr under semi-natural conditions.

One such semi-natural medium is found in Mexico, where Sosa Texcoco, S.A. (a Mexican company extracting sodium alkalies) and the French Institute of Petroleum operate a one-ton/day pilot plant using local lake water containing 30 grams solids per liter (Santillán 1974, or Pirie 1975a). The yield from their semi-natural solar evaporator is 10 dry tons/ha-yr (pH 9.8). *Spirulina* is harvested from this cultivation lagoon (normally $0.1 - 0.3$ gram cells/liter) by first concentrating the algae on inclined filters to 5-10 grams/liter and then to 15-25 grams/liter using rotary filters; both stages are designed for high filtration efficiency and low energy consumption. Vacuum filtration and washing reduces the inorganic matter and provides a paste of 15-20% solids. Handling (viscosity) and digestibility of this cell paste is improved by mechanical cell rupture (proprietory type) prior to spray or drum drying at 130°C for six seconds. If flour rather than flakes are desired, grinding can follow the drying.

Natural "Media"—Sea Farming

Natural algae and water plant densities are normally too dilute in natural waterways to justify harvesting except in some polluted waters. Cultivation of natural waterways, or sea farming, is however highly attractive in many respects:

- oceans cover 71% of the earth's surface;
- half of the ocean area lies within 30° of the equator;
- little competition exists for sea use;
- ocean plants are efficient solar-energy converters like most aquatic plants;
- ocean farming is not subject to normal climatic variations;
- terrestrial farming already employs the best arable land and fresh water.

Limitations have, however, heretofore prevented exploitation of ocean farming:

- the ocean bottom lies at depths where light intensity is nil and too deep for the growth of attached seaweeds, and
- well-illuminated surface waters have essentially no nutrients (Figure II.2.5).

This unfortunate situation has potential if a raft mesh is held 15-25 meters below the surface by cables to the sea bottom. Seaweeds are attached to this mesh, according to this proposal at three-meter intervals to provide a density of about 1000 plants per hectare (Wilcox 1976). To overcome the inadequacy of dissolved nutrients shown in Figure II.2.5, the cool sea-bottom waters, rich in phosphorus, potassium and fixed nitrogen, would be upswelled with ca 11 hp/ha pumping power derived from wind, wave or ocean thermal gradients; wave power, believed to be economically and energetically viable for electrical generation, is particularly promising. Additional nutrients can be provided from processing wastes on the shore or ship facilities.

The plant tentatively selected for ecological and efficiency reasons is the giant (up to 100 meters) brown or California kelp, *Macrocystis pyrifera* from the coasts of California, Mexico and New Zealand. *M. pyrifera*, one of the fastest growing plants (15 cm/day or up to 60 cm/day for young

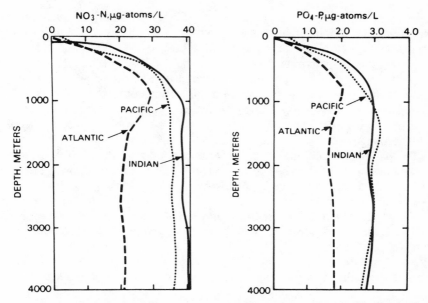

Figure II.2.5. Nutrient gradients in open seas (Sverdrup, Johnson, and Fleming, The Oceans: Their Physics, Chemistry, and General Biology, copyright 1942, renewed 1970, pp. 241 and 242. Reprinted by permission of Prentice-Hall, Inc., Englewood Cliffs, New Jersey).

fronds) has multibladed fronds lying along the sea surface and lives six months before being replaced by new growth from below. As the plant itself does not age, its doubling time of six months allows continual harvesting every three months, replacement being required only by storm, disease or fish damage.

Kelp products have long been used as thickening agents and colloid stabilizers in the food, textile, cosmetic, pharmaceutical, paper and welding industries. Years of experience have shown that kelp can be harvested and partly processed with special ships to aid in locally recycling nutrients and minimizing transport costs. Treatment of harvested kelp begins with removal of the surface coat of fucoidan by washing with warm water and then following chopping, water and salts are removed by cell rupture in weak acid. Separation technique and end-product choice is widely variable from this point.

The most ambitious kelp project, estimated to cost two billion dollars by the year 2000, is a highly integrated scheme designed to combine many product options (Wilcox 1976). While methane via anaerobic digestion is to be the main product, high-quality food in addition to fodder, fertilizers, sugar syrups and several organic chemicals will be by-products significantly reducing the energy costs. Combined marine aquaculture with confined grazing fish is also being considered in what is ultimately hoped to be a carefully integrated, wave-powered 45,000-ha complex.

Ecological dangers must be studied as the harvested area increases, but they are predominantly avoided by the artificial feeding required for survival: the kelp require the cool nutrient-rich water provided by upwelling and will not survive nor spread if this supply is turned off. This property could allow growth in tropical and semi-tropical climates without the danger of infection in unwanted areas.

The ICES (International Council for Exploitation of the Sea), however, is not convinced of the ecological safety and refused to support a French project planned for the coast of Brittany. The viable temperatures of 2-20°C are available in large areas of Western Europe from North Africa to Norway, and ICES felt a real danger existed for navigation, salmon and lobster fishing and defence. Such problems are not expected to be relevant in California.

Water hyacinth, which though not an algae, is a prolific freshwater plant under consideration for exploitation. It commonly blocks polluted streams and waterways and produces up to 60 tons/ha-yr. Cultivation is tentatively being considered, but no projects have reached any serious dimension. An alternative viewpoint is exploiting water hyacinth and other water plants for *in situ* sewage treatment, i.e., the removal of nutrients from polluted waterways where algae size is of critical importance for harvesting (Yount and Crossman Jr. 1970, Wolverton and McDonald 1976).

ALGAE CULTURED ON WASTE

Mass algal culture on waste effluent has traditionally been viewed from the water purification perspective from which the algae's principal function is the production of oxygen for the bacterial oxidation of organic wastes. In addition, the algae functioned as a pollutant sponge for the released nutrients (P, K, and NH_3). The energy for bacterial growth is derived from the oxidation of organic wastes, whereas the algae use solar energy to reassimilate the nutrients freed by the bacteria into a more valuable high-energy form. Such treatment prevents the pollution problem of raw sewage in, for instance, rivers where the bacteria oxidizing the organic wastes deplete the oxygen and release CO_2 and NH_3 to the air and nutrients to the water. Enormous algal blooms thrive on the nutrient-rich water downstream, and the oxygen produced by the algae is in turn wasted. The confined algal oxidation pond combines algal oxygen production and bacterial CO_2 production symbiotically for maximum break-

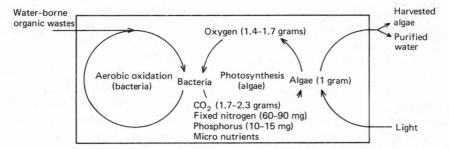

Figure II.2.6.   Confined oxidation pond (redrawn from Oswald 1973).

down of organic wastes and reorganizes the nutrients into algae, leaving relatively clear water (Benemann et al. 1977). These complementary processes are best understood from Figure II.2.6.

As soon as the lack of one nutrient begins to limit algal growth, the waste water can be returned to the natural waterway while the algae can be discarded as a used up sponge or processed further. No nutrient is ever reduced to zero concentration, but Table II.2.2 shows that the limiting nutrient, phosphorus, is reduced by 90% (Shelef et al. 1976). These results are typical except that water scarcity in Israel causes nutrient concentrations to be greater than for example in the US.

A wider and more recent view of this accelerated waste water treatment system is to see algae as a valuable product, a source of fuel, food, fodder and possibly fertilizer. In arid zones, particularly, this process is a step in

Table II.2.2   Removal of Various Substances in the Technion, Haifa, Photosynthetic Pond under Favorable Operational Conditions (Shelef et al. 1976).

Substance	Raw Sewage mg/liter	Pond Effluent mg/liter	Flotator Effluent (alum dossages 80–90 mg/liter) mg/liter
Suspended matter	240	268	15
BOD_5: total	330	106	10
dissolved	—	12	5
COD: total	750	670	148
dissolved	—	64	46
Nitrogen: total	86	71	20
dissolved	—	18	12
Phosphorus: total	16	10	1.4
Coliforms/100 ml	6×10^7	3.5×10^5	8×10^3

water recycling, not just waste treatment. The technical problems for these systems always remain the same as the above more limited view, but the economics are changed dramatically: cost credits for food and fuel produced and the lowered waste disposal costs (or from the other point of view, disposal credits) reduce the net protein costs. The cost advantage over synthetic media is considerable, but in this case optimization must also include maximum algal yield and recovery, not just "pollutant" (nutrient) removal. Pressing shortages of protein and all nutrients supporting food production must make obsolete the concept of waste disposal. Which function is maximized, however, will depend on local needs, but as none is in itself sufficient for economic success, algal oxidation ponds must be viewed as a multipurpose component in a well-integrated system.

Kinetics of Algal Production

Aside from some extreme cases, the rate of algal production is limited by incident radiation rather than temperature or nutrient concentrations; in part this statement results from the ability to vary the detention time such that nutrients do not limit.

Nutrients become available for algal uptake at the rate of bacterial organic waste decomposition. As this decomposition process is an oxidation, this bacterial reaction can be measured by the rate of oxygen consumption, or what is commonly called the biological oxygen demand (BOD). Several nutrients are released simultaneously in this process but carbon dioxide is by far the most important to consume at the rate it is produced—others will accumulate, but carbon dioxide will be lost to the atmosphere due to its limited solubility. Algal growth rate can be determined from the rate of photosynthetic oxygen production, nutrient incorporation (i.e., 1.6 grams O_2 and 65.9 mg N per gram algae), or by the rate of algal harvesting. Whatever the growth rate determinant used, the rate of oxygen produced by the algae (1.7 kg O_2/kg algae) must equal the bacterial biochemical oxygen demand (BOD) to prevent O_2 or CO_2 loss to the atmosphere; in this symbiotic system, oxygen loss is the more serious as it is about fifty times less soluble than CO_2. To cite a practical example, human wastes have a BOD of ca 75 grams O_2 per day, and therefore 40 grams algae per person per day will be produced. Alternatively, the wastes of 1500-4500 persons can be treated per hectare per day (Oswald 1973).

Several parameters including pond depth, light intensity, the nature and loading rate of the organic wastes, and temperature must be considered to achieve a "well-tuned" system. Pond depth should increase with increased sunlight, and hence both seasonal and geographical variations

Table II.2.3 Probable Values of Visible Solar Energy as a Function of Latitude and Month (from Oswald 1973).

Latitude		Jan.	Feb.	Mar.	Apr.	May	June	July	Aug.	Sept.	Oct.	Nov.	Dec.
							MONTH						
0	max	255+	266	271	266	249	236	238	252	269	265	256	253
	min	210	219	206	188	182	103	137	167	207	203	202	195
10	max	223	244	264	271	270	262	265	266	266	248	228	225
	min	179	184	193	183	192	129	158	176	196	181	176	162
20	max	183	213	246	271	284	284	282	272	252	224	190	182
	min	134	140	168	170	194	148	172	177	176	150	138	120
30	max	136	176	218	261	290	296	289	271	231	192	148	126
	min	76	96	134	151	184	163	178	166	147	113	90	70
40	max	80	130	181	181	286	298	288	258	203	152	95	66
	min	30	53	95	125	162	173	172	147	112	72	42	24
50	max	28	70	141	210	271	297	280	236	166	100	40	26
	min	10	19	58	97	144	176	155	125	73	40	15	7
60	max	7	32	107	176	249	294	268	205	126	43	10	5
	min	2	4	33	79	132	174	144	100	38	26	3	1

will occur, but depths of 30-40 cm are typically optimal for regions whose climates permit year-round operation. Table II.2.3 lists the annual sunlight variations at different latitudes and indicates that areas within 35° of the equator are ideal for algal growth, but even growth in remote regions (e.g., 60°) is possible. Light utilization efficiency is shown in Figure II.2.7 to decrease rapidly with increasing intensity above about 40 J/cm²-day) which coincides with the maximum for the product of efficiency and productivity shown in Figure II.2.8.

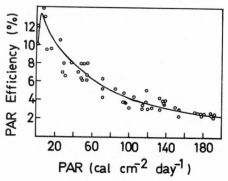

Figure II.2.7. Dependence of micro-algae's light conversion efficiency on photosynthetically active radiation (PAR) intensity (by permission of Oswald 1973).

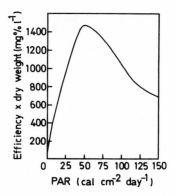

Figure II.2.8. Variation of the product of efficiency and dry weight production with PAR (by permission of Oswald 1973).

Daily oxygen and algal production yields can be calculated using the following simplified relations and the data in Table II.2.3 and Figure II.2.7 (Oswald 1973):

$$\text{Cell yield (kg/ha-day)} \quad = 0.17 \, FS$$
$$\text{Oxygen yield (kg/ha-day)} = 0.28 \, FS$$

where,

$$F = \text{the conversion efficiency (Figure II.2.7)}$$
$$S = \text{the incident solar energy (Table II.2.3)}$$

Figure II.2.8 shows that 4000 cal/liter-day (16.7 kJ/liter-day) gives the optimal algal productivity. This optimum can be attained by adjusting the pond depth to compensate for the light intensity from Table II.2.3; this depth is typically less than 0.5 meters. The optimal productivities for various latitudes and assuming different solar-conversion efficiencies were calculated for Table I.6.1. An annual efficiency of 3% is the typical present maximum.

The third controllable factor, the dilution rate (or its inverse, the retention time), also determines the concentration of algal cells. Figure II.2.9 shows that a clear optimum exists for maximum production efficiency: the decrease at low retention times, less than a day, is due to culture washout while at too long retention times (greater than 4 days) higher cell density prevents adequate light penetration. Optimum detention period varies with light intensity and depth, but three to four days is a typical average (Shelef et al. 1973).

The fourth component determining the rate of algae production is of course the sewage or waste which provides the nutrients for algal growth.

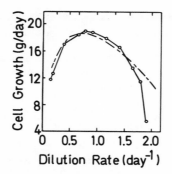

Figure II.2.9. Variation of *Chlorella pyrenoidosa*'s net production rate with dilution rate (detention period^{-1}; solid curve—experimental data, broken curve—theoretical calculation) (redrawn from Shelef, Schwartz, and Schechter 1973).

As each gram of algae requires 0.5 gram carbon, 0.1 gram nitrogen and 0.01 gram phosphorus, the relative nutrient balance in the sewage is of prime importance. If the system is optimized for algal *production* rather than sewage treatment, supplements of the limiting nutrient may be justified for total nutrient usage, particularly ammonia. Ammonia is the nutrient required for protein synthesis and is the most costly energetically, but in excess of 2mM or at pH values over 8.0 it can severely inhibit algal growth (Abeliovich and Azov 1976). Potential nutrient losses include phosphate precipitation under some conditions (flocculation) and ammonia evolution due to the high surface pH on sunny days.

The rather well-balance nutrient characteristics of domestic (US) sewage are shown in Table II.2.4. Table II.2.5 summarizes the additives required for maximal use of other wastes. Vegetable wastes, for example, are typically limiting in nitrogen and phosphorus, while animal wastes are limiting in carbon (carbon dioxide); if CO_2 is provided, 5-10 kg algae can be grown from the nutrients in each kilogram of human or other animal wastes.

Depending on the end use of the effluent water, a marine system may be of interest. At Woods Hole Oceanographic Institution in the US (Ryther 1975), secondary treated urban effluent was mixed with equal amounts of sea water to produce a near monoculture of a marine diatom with yields comparable to fresh water systems (see page 218). If the effluent is to be disposed of, this scheme is highly advantageous for water conservation. More commonly the treated water should be kept salt-free for reuse.

An alternative "waste nutrient" source, the sludge from an anaerobic digester, is neither as energy-rich nor does it have as high a BOD as raw

Table II.2.4 Domestic Sewage Characteristics (Oswald 1973).

pH	9.4
Total solids	553 mg/liter
Volatile solids	212 mg/liter
Fixed solids	342 mg/liter
OH^-	1.6
CO_3^{2-}	48
HCO_3^-	141
Total N	44.5
NH_3	25.5
NO_3	0.1
Organic N	18.9
Total C	80.7
PO_4^{3-}	38.8
H	27.2
MG	11.6
Ca	7.4
K	25.5
S	7.1
Fe	0.5
C-BOD (20°C 5 day)	140
Ultimate BOD	218

Table II.2.5 Potential Nutrient Sources for Mass Algal Culture (Oswald 1976a).

Sources of Water and Nutrients	Required Supplementary Nutrients	Potential Use (subject to toxicological and epidemiological evaluation)
Man-made algal nutrient solutions	None	Human food
Potato-processing wastes	None	Human food
Sugar, tomato, winery wastes	N and P	Human food after heat drying
Natural bodies of water, brackish or fresh	N and P	Human food after pasteurization
Spring and wells, including geothermal waters	N, P, C	Human and animal food
Domestic sewage	None	Animal feed after pasteurization
Animal manure	None	Animal feed after pasteurization
Meat-processing wastes	None	Animal feed after pasteurization
Reduction-plant wastes	C	Animal feed after pasteurization
Petroleum-refinery wastes	P	Animal feed
Irrigation and land drainage	N, P, Fe	Animal feed after pasteurization
Storm waters	C, N, P, K	Animal feed

sewage and hence will not support as much bacterial growth. All the nutrients are present in a simpler, soluble form, but the anaerobic bacteria have transformed most of the energy from fats, protein and carbohydrates to methane (CH_4). Some CO_2 will evolve, however, from aerobic bacteria in an algal pond fed anaerobic-digester sludge as the former break down the more inert organic particles, but optimal growth required additional CO_2; some low-energy CO_2 sources were discussed earlier in this chapter of which CO_2 from the digester gas (ca 40% of the gas) is the most readily available.

SPECIES CONTROL

All open-pond algal growth systems with the exception of the proposed kelp plantations accept whatever species spontaneously dominate with little (successful) effort to alter Nature's choice. Two motivations exist for desiring a given specie: first, if the algae is intended for human food, toxological tests for each specie are required. *Spirulina* is favorable in this case as its optimum medium is suitable for little else. A second case involves selecting filamentous algae because of their ease of harvesting and the resulting tenfold reduction in cost (see following section on "Harvesting").

Filamentous algae rarely dominate except in instances where nitrogen is limiting, and then naturally nitrogen-fixing blue-green species have a selective advantage; this is, of course, one method of assuring their dominance and exploiting their ammonia-producing capacity, but for the purpose of waste-water treatment the advantages of simple harvesting make filamentous algae—in the presence of ammonia—attractive.

Pilot experiments have been conducted using a rotary microstrainer (discussed in the following section on "Harvesting") to harvest filamentous algae grown in large open ponds (Benemann et al. 1976). A fraction of the harvested algae are used to reinoculate the pond to displace the specie(s) which normally would dominate. Recycling favors the selected specie by effectively lengthening its residence time in relation to the faster-growing species.

Such techniques have been applied to encourage the growth of algae on waste water. *Oscillatoria*, a filamentous blue-green algae, was inoculated in an open pond, but the filtering and recycling scheme also favored a colonial green algae (*Microactinium*) which later dominated even without continued recycling. Although the inoculated specie was not established, the goal of obtaining a filterable algal specie was attained. If this technique proves successful, the economically and energetically costly harvesting methods for unicellular algae can be avoided. Similar techniques are to be tried to sustain nitrogen-fixing blue-green algae, the energy requirements and growth rates of which rarely allow their natural dominance except when nitrogen is limiting (Benemann et al. 1976).

HARVESTING

For microalgae, harvesting is the major economic constraint and prevents the use of naturally occurring algal blooms whose densities rarely exceed 3 mg/liter. The minimum density which can be economically harvested is approximately 250 mg/liter, and artificial algal ponds can create densities up to 1 gram/liter. Several stages are generally required for harvesting:

1. The initial concentration step has in principle several alternatives, but in practice none is yet entirely satisfactory:
   a. Centrifugation is the technically ideal method, removing 84% of the algae at 1500 liters/min. Power costs are, however, prohibitive: 2.7×10^3 KWh/ton or $60 per ton algae (at 300 mg/liter; Oswald and Golueke 1968). Capital costs are even more prohibitive: $50,000 (1969 prices!) for a 1600 liter/min capacity unit. Centrifugation is a practical laboratory concept, but holds no applicability on mass scale.
   b. Auto-flocculation is a natural process which occurs on sunny afternoons when the increase in pond temperature, depleted CO_2, and increased pH (up to pH 9.8) causes algae to clump together and sink to the bottom. To exploit this method would require a separate flocculation pond of 8-15 cm depth where the supernatant could be decanted and returned to the growth pond. Economically this is an excellent method, but its dependability remains to be established.
   c. Chemical flocculation or coagulation of microalgae can be induced with the addition of aluminum sulfate (or alum, $Al_2(SO_4)_3$), lime ($Ca(OH)_2$), ferric chloride ($FeCl_3$) or a few other chemicals. The addition of, for example, 70-120 mg/liter alum at pH 5.5-6.0 is followed first by a sequence of rapid and gentle stirring and then by air or electroflotation (Shelef et al. 1976). As shown in Figure II.2.10, the flocculated algal froth (ca 5-10% solids) is skimmed off after a few minutes in the flotator, and after the addition of acid the now soluble alum is separated in a low-power centrifuge from the algae. At least 50% of the alum and acid can be removed for recycling, while the 15-25% solid algal paste is dried or processed further. While flocculation is the harvesting method most widely practiced today, one re-

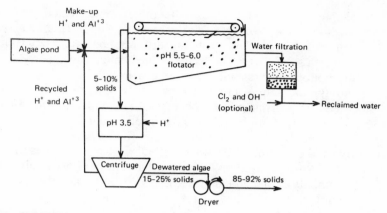

Figure II.2.10. Harvesting by recycled alum flocculation (redrawn from Shelef et al. 1976).

maining problem is the presence of about 4% alum in the algae. This alum greatly limits the algae's end uses, but is not a problem when used as fish fodder due to the near neutral pH of the fish digestive system; a 30% algal diet has been shown to replace 85% of the fish meal supplement (15% of the fish diet; Shelef et al. 1976).

d. Magnetic harvesting of algae which itself is nonmagnetic is possible via the chemical attachment of algal cells to waste magnetite from mining operations (Kolm, Oberteuffer, and Kelland 1975). The complex is then passed through a magnetic pole gap filled with a three-dimensional stainless steel grid and there trapped. The algae can then be removed from the magnet-bound magnetite by caustic soda (or sonication) which hydrolyzes the cells and allows recovery of the protein. Nearly 99% of the algae is removed from the medium of which 84% is hydrolyzed and recovered. Magnetite can also be recovered, but at $15/ton is not yet economically justified. The overall cost for a magnetic separator is approximately $50,000 giving an optimistic separation cost estimate of about $0.004/kg protein, including chemical additions (Mitchell 1977). Smaller-scale units employing permanent magnets are expected to reduce costs substantially, but this will remain a high-capital system.

e. Harvesting by raking partially submerged or floating algae is the least expensive method of harvesting, but unfortunately this is applicable only to filamentous algae. Filamentous algae dominate an algal population only under special conditions or when special efforts are made for its dominance (see page 223 of this chapter). In the latter case the more sophisticated "raking" method or rotary microstrainer shown in Figure II.2.11 is used. The microstrainer is a rotating drum supporting a 25-100 micromesh screen which collects the filamentous algae introduced at the bottom and releases them into a trough after 180° rotation through the help of pressurized water. Costs are estimated to be a tenth that of chemical flocculation (Benemann et al. 1976).

f. Harvesting of algae or aquatic weeds with herbivorous fish, especially bivalves with their built-in filter system, is both effective and economically attractive, but of course assumes that fish is the desired product. This important and traditional algal use, commonly called aquaculture, is discussed separately in the following section "Uses for Harvested Algae."

2. Dewatering is the second harvesting step, and while a centrifuge can be employed, it is economically uninteresting. Filters for the larger microalgae tend to clog easily, but are highly effective and inexpensive for filamentous algae. Microalgae are best dewatered and dried in one step.

3. Drying to less than 12% moisture content is required if long-term storage is desired. Several methods are available (Oswald and Golueke 1968):

a. Mechanical methods are best, but are energy- and capital-intensive. Fast drum drying (5-7 sec at 135°C) in very thin layers costs about $20/ton dry algae, while spray-drying produces a better-quality product due to shorter drying time and lower temperature.

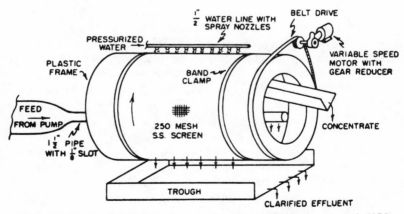

Figure II.2.11. Continuous rotary separator (by permission of Benemann et al. 1976).

b. Sun-drying on a lightly oiled surface is effective beginning with a 10% solids slurry, ca 0.25 cm thick. After a day the moisture content is below 15%, and the flaked solids are easily collected; an area equal to about 5% of the growth area is required.
c. Sand-bed drying can combine dewatering and drying, beginning with 8-12 cm thick slurry. The water quickly drains away while full drying takes three to five days, a rather long period leading to the loss of some vitamins. As in the sun-drying method, the algae curls and flakes and can be separated from most of the sand, but the residual sand content remains a problem for this economically attractive method.

None of these methods, however, simultaneously meets the criteria of high effectiveness and both low capital and energy costs. The awaited development is a simple and effective solar dryer or oven whose effectiveness as far as loss and contamination are concerned, could be greatly improved at little cost. Many schemes at varying levels of complexity are conceivable, and adaptations could possibly permit application to other crop drying problems.

PROCESSING

Processing can range from simple mixing of dried algae in normal food or fodder, grinding to flour, to a complete solvent extraction of the protein. The minimum processing required is the breaking of cell walls, not only to increase digestibility but also to release some of the 10-20% nitrogen content found in cell walls. Cell-wall thickness and hence digestibility varies from *Dunaliella,* which has no actual cell wall and is thus very osmotically fragile, to *Chlorella,* which has a very thick, difficult to break cell wall. Nonruminants cannot even digest the relatively fragile *Scenedesmus* unless the cellulose-like cell wall is ruptured mechanically (homogenization or sonification), chemically (urea soaking or solvent extraction), or by heating (dry steam or cooking). Chemical extractions have the important advantage of removing most pigments and, more importantly, nucleic acids.

USES FOR HARVESTED ALGAE

Algae for Food

The use of algae as food, with protein content as high as 70% (e.g., *Spirulina*) is extremely attractive, but certain risks and feed trial results must first be discussed. The most obvious risk involves direct contamination in the culture medium, a microbiological contaminant, or the normal production by certain algae of exo- or endo-toxins (Schwimmer and Schwimmer 1968). The latter are particularly a problem with blue-green algae: massive mammal poisoning due to *Microcystis toxica* has been reported in addition possibly to *M. aeruginosa, M. flos-aquae, Anabaena flos-aquae, A. lemmermannii* and *Aphanizomenon flos-aquae*. Other possible intoxicating culprits include *Nodularia spumigena, Coelosphaerium kutzingianum* and *Gloeotrichia echinulata*. A minimum lethal dose of an extract from *A. flos-aquae* killed mice in one to two minutes while a small (2600 molecular weight) cyclic oligopeptide from cells of a contaminated *M. aeruginosa* culture killed in 30-60 minutes. Toxins from nonaxenic *Aphanizomenon flos-aquae* have been shown to be structurally similar to a shellfish toxin from *Gonyaulax calenella,* and a dermatitic factor has been isolated from an impure *Lyngbya majuscula* culture. Whether these toxins are avoidable in pure or axenic algal cultures, or if such cultures are possible, remains unknown.

The above toxic possibilities are supplemented by a suspicion of the algal cell-wall material, particularly the common presence of odd-numbered fatty acids; none are found in *Spirulina*.

For human consumption of algae, there exists no doubt of a dietary limitation due to high nucleic acid content in all single-cell proteins: as high as 25 grams nucleic acids per 100 grams protein is known to occur (Scrimshaw 1975). Nucleic-acid content is roughly proportional to the rate at which the organism's cells multiply; this so-called doubling rate tends to decrease with organism size, and hence nucleic-acid levels are much lower in algae than bacteria but still serious. *Spirulina* has, for example, 7 grams nucleic acids per 100 grams protein vs 4 grams per 100 in liver, the latter being unusually high in nucleic acids for animal tissue.

The difficulty with dietary nucleic acids are their depolymerization by nucleases in pancreatic juice and the following conversion to nucleosides by intestinal enzymes prior to absorption. The metabolism of the nucleotides guanine and adenine to uric acid is a problem only for humans who have suffered an evolutionary loss of the enzyme urease, whose function is to oxidize uric acid to a soluble and excretable form, allantoin. Low uric-acid solubility may result in its accumulation as uriate in tissues and joints with gout-type results. Methods for nucleic-acid removal from single-cell protein do exist, but in the absence of such procedures, human nucleic-acid intake should be limited to about 2 grams per day or about 50 grams *Spirulina* (i.e., 30 grams protein or about 40% of the daily requirement). This is not a severe constraint as rarely would consumption of single-cell protein constitute more than half of the daily protein diet either on the basis of need or taste.

The few limited algal-dietary trials indicate ready acceptance of algae as a food additive but not as a staple food (Scrimshaw 1975, Oswald 1976a). *Spirulina* has been eaten for centuries by Africans on Lake Chad, and athletes in Mexico took 20-40 grams *Spirulina* daily for 30-45 days with good results, whereas two men in the USSR who ate 150 grams *Spirulina* daily developed edema of the face and hands, petechial hemorrhages, cyanosis of the nail beds and peeling of the fingers; little problem was however encountered after alcohol extraction of the color. Long-term (500 days) feeding (25% of the diet was *Spirulina*) revealed no toxicity in (100) rats, however, and in limited amounts (5-15%) provides a good food additive to pigs, poultry and fish; even as a limited additive, algae can replace up to 40% of the soybean and fishmeal proteins for chickens and fish (Shelef et al. 1976).

On the basis of absorbed nitrogen, *Spirulina* is surpassed only by cow and human milk. In general, the food quality of algae varies greatly with growth medium and age of culture. Extremes of protein content are 8-75% of dry weight, lipids are from less than 1% to 86%, carbohydrates from 4-40% and ash from 4-45%. (Algae grown on glucose medium can have high (30-47%) intracellular carbohydrate while old cultures or algae grown on low oxygen or fixed-nitrogen contents can store lipid (28-86%). These extremes are rarely relevant, and more typical compositions are shown in Table II.2.6 together with cow's milk and hen's eggs for comparison (Oswald 1976a).

Vitamin and mineral contents for algae, eggs and cow's milk are also shown in Table II.2.6. Soluble vitamin content in algae is high while vitamins A, D, E and K are negligible or nonexistent. Mineral contents, like vitamins, are highly variable with growth media and conditions, but normally compare favorably with both eggs and milk.

High protein content, 50-75%, is not the only criteria for protein's nutritional value. Of the twenty-odd amino acids naturally occurring in proteins, half must be supplied in the correct chemical form while the other half can be synthesized from these so-called "essential" amino acids. Algal protein, while abundant, is not as good as animal protein, e.g., eggs or milk, due to an unbalanced amino-acid content; low contents of sulfur-containing amino acids (methionine and cystine) are the most significant deficit.

Even with all reservations stated, algae has an extraordinary potential nutritional role, most likely as an additive. As a 10% by weight additive, *Spirulina* can still provide half a person's protein requirement. In times of

Table II.2.6   Summary of Values for the Proximate Composition, Vitamin and Mineral Contents of Milk, Eggs and Microalgae (from Oswald 1976a).

	Cow's Milk	Hen's Eggs	Algae
Proximate Composition			
Protein, gram	28	49	51
Carbohydrate, gram	39	3	27
Fat, gram	28	44	7
Fiber, gram	0	0	6
Ash, gram	6	4	9
Kcal/gram	5.2	6.2	3.6
Vitamins			
Thiamine, mg	.24	.42	1.2
Riboflavin, mg	1.25	1.14	3.0
Niacin, mg	.8	.4	10.0
Pyridoxine, mg	.4	.4	0.2
Folacin, mg	4.7	16.3	3.4
Vitamin B-12, mg	3.2	7.6	25
Ascorbic acid, mg	8	0	40
Pantothenic acid, mg	2.4	.4	0.7
Biotin, mg			26
Carotenoids, IU	1110	4500	52
Vitamin E, mg			2.6
Minerals			
Calcium, gram	.93	.21	0.2
Phosphorus, gram	.74	.78	1.8
Calcium/Phosphorus	1.27	.37	0.2
Magnesium, gram	.10	.42	0.6
Sodium, gram	.40	.46	0.1
Potassium, gram	1.14	.49	0.8
Sodium, mg			187
Iron, mg	trace	9	31

acute food shortages, areal productivity will become one of many crucial criteria; one hundred times more algal protein than wheat protein can annually be produced per unit area, while for animal proteins that figure can rise to 500 or 1000. However used, algae represent an enormous potential.

Aquaculture

As was mentioned earlier (see page 224) harvesting is at present the cost-limiting stage in algae production. Aquaculture, the cultivation of fish to harvest algae and water plants, was there listed as an alternative, special-purpose harvesting method.

Aquaculture, like agriculture, implies an active role by Man in not just harvesting but in cultivating the growth of food. Whereas food gathering developed into agriculture due to the press of increasing population densities, open sea or lake fishing has not evolved into aquaculture on the same scale. The chief reason is not difficult to guess: lack of water. For this reason aquaculture has often been linked to areas with well-developed irrigation, as in Hawaii where aquaculture dates back to the 14th century (Kikuchi 1976). Irrigated fields in some of the Hawaiian Islands were often managed in a manner intermediate between aqua- and agri-culture.

In Asia aquaculture has had an impact and history comparable to no other region (Bardach, Ryther, and McLarney 1972). Fish ponds have formed the major link for recycling plant and animal wastes in these regions—human wastes were cycled by placing the latrines directly over the ponds. Cultivation of algae and higher aquatic plants, plus the stocking of primarily herbivorous fish, thus provided not only waste treatment but also high-quality fish protein. Estimated yields vary widely with the amount of cultivation and feeding involved, but harvests range from ca 350 kg/ha-yr in Hawaii to highs of 18 tons/ha-yr in China. Great care is required in estimating yields as fish, like all animals, are rather inefficient converters of protein. High fish yields attained by the supply of high-quality protein (e.g., soya) is just as extravagant as beef feedlots where only a tenth of the feed is recovered as meat. Similar feeding techniques can yield phenomenal quantities of fish; for example, 29 kg/meter^2-yr (290 tons/ha-yr) harvests were obtained in Japan. Intensive fish farming may, however, prove preferable to the energy intensity of open-sea fishing. Energy ratios (energy in the fish caught divided by the energy input for the catch) and the energy input per weight protein harvested are listed in Table II.2.7 to summarize the problem (Leach 1976):

Table II.2.7   Energy Efficiency of Fish Production (Leach 1976).

	Energy Ratio	Energy Input
Wheat (UK)	3.35	42 MJ/kg protein
Average for UK fisheries	0.05	489
Shrimps—Australia	0.058	366
Gulf of Mexico	0.0061	3450

Ten to one-hundredth times as much energy is required for (caught) fish protein as grain protein! Aquaculture will be attractive only if its energy investments can be reduced signficantly below these levels.

The goal of aquaculture is to lower the energy and feed supplements substantially and yet provide high-quality fish protein. Criteria for success include, but are not restricted to, the yield per hectare. Of at least equal importance is the ability to produce fish protein without competing for resources required for normal agriculture:

- *Land:* Aquaculture is possible on any type of land holding or capable of holding water. This includes sea-water ponds on coastal sand dunes, cleared swamps, bogs, sewage purification ponds, old salt pans, and virtually every river, lake or pond.
- *Feed:* Any type of organic matter, human, animal, or plant waste can be used as fish feedstock. Low-quality agricultural wastes such as rice polishings, grain-mill sweepings, raw-coffee wastes or oilseed residues, are suitable directly, or together with animal wastes as an algal-growth medium. Alternatively all these wastes can be digested anaerobically first (producing methane) and the sludge alone used as a nutrient-rich medium for algal growth. In this latter sense, herbivorous fish, especially shell fish, can be viewed as a natural harvesting method for unicellular green algae. Harvesting and drying are (see page 224 of this Chapter) the energy- and capital-intensive stages of any system which exploits the highly efficient green algae.
- *Water:* Water in rather generous supply, but of only reasonable quality, is required for aquaculture. Water is, of course, not consumed but rather is lost only via evaporation: typically a meter of water (10^7 liters) is lost per hectare per year from standing water and a half to a third that for terrestrial crops. Relative yields are then the deciding factor, and, on a per-protein-yield basis, aquaculture far exceeds traditional agriculture-based animal production. Algae consumes 33-70% as much water per dry weight product as agriculture (500 liters/kg vs 700-1500 liters/kg) but only 5-60% as much produced protein (700-100 liters/kg vs 1600-15,000 liters/kg protein). This advantage is retained in fish production, of course, only if animals are fed grains and fish fed algae. In most cases multipurpose water can be used for aquaculture, such as by growing fish confined in cages in irrigation ditches. Such integration will reduce both attributable water and land costs.

Whatever aquaculture system is used, the specie of fish must be carefully chosen. For the most part, fish common to Western diets are flesh eaters, they eat primarily smaller fish, worms and insects, and therefore are not suitable to aquaculture. Two herbivorous species are the most common choices for both modern and historic fishponds: carp (grass carp—*Ctenopharyngodon idellus,* bighead—*Aristichtys nobilis,* silver carp—*Hypophthalmichtys molitrix,* mud carp—*Cirrhinus molitoretta,* common carp—*Cyprinus carpio* and black carp—*Mylopharyngodon piceus*) and *Tilapia* or St. Peter's fish (Bardach, Ryther, and McLarney 1972 and USNAS 1976). The latter variety, while tropical in origin and hence more sensitive to temperature variations than carp, is advantageous for its superior taste. Both are effective herbivores and hence allow harvesting of the primary product in the food chain, algae.

Alternatively, bivalve mollusks, such as oysters, mussels, octopus, or clams, are uniquely well suited to harvesting unicellular algae. Their filter-feeding mechanism is far more efficient for the small algal cells than even herbivorous fish, and they can double their weight in one week. They should normally not be cultivated in the algal pond directly as the algal cell concentration in the fish pond should be lower (about 10^5 cells/ml), and both the flow and aeration of the water should be greater (Ryther 1975). Accumulated metabolites can also easily reach toxic levels.

While not normally considered as aquaculture, fowl and certain other herbivorous animals are potential algae and water-weed harvesters (NAS 1976b). Ducks, geese and swans commonly forage on water weeds, but if extensive cultivation is considered some grain supplement is normally required (NAS 1976b). The manatee or sea cow (ICMR 1974), water buffalo, pigs and certain species of rodents (capybara and nutria; NAS 1976b) also forage on water weeds and can be of extreme local importance for weed control. None of these animals, however, are of such value nor display such voracious herbivorous appetites that water plants are cultivated for their feed.

Typical Systems. Aquaculture as a system is sketched in the same input-output manner as normal agriculture in Figure II.2.12. Note that the waste water from the fish pond will contain algae and excrement from the fish and would suggest that the fish and algae should be grown in the same pond. Such a system is possible, but it will be optimal for neither and will more resemble the low productivities of a normal healthy lake. Care must be taken not to try for too high productivity by heavy fertilization as eutrophication can occur.

A more logical solution is to still keep the system closed but to return the output "waste" water to the algal pond. Depending on the type of fish or bivalve used some dilution of the algal output will be required, but if

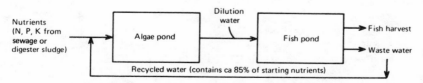

Figure II.2.12. Two-stage aquaculture system.

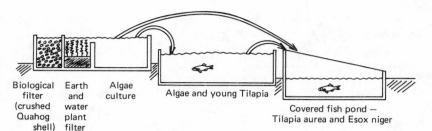

Biological   Earth   Algae
filter   and   culture
(crushed   water   Algae and young Tilapia
Quahog   plant
shell)   filter   Covered fish pond —
                                 Tilapia aurea and Esox niger

Figure II.2.13. Three-stage aquaculture system (redrawn from McLarney and Todd, 1974).

the water added is greater than that needed to compensate for evaporation, it can be used elsewhere in a complete integrated system.

A variation on this theme is shown in Figure II.2.13 (McLarney and Todd 1974) where a series of three-terraced ponds are used. The first is basically an aerobic biological waste-treatment plant (and could be eliminated profitably if an anaerobic digester existed anyway) while the second pond mainly produces the algae. In addition the second pond may produce some small algae-eating animals, such as the crustacea, daphnia, and higher water plants (e.g., water sprite) which will be eaten in this case by tilapia. The final pond, covered to raise its temperature in temperate climates for the tropical tilapia, is for cultivating the fish whose wastes will be recycled back to the first pond.

Yet another variant is to employ aquaculture as a multicomponent water treatment system. Figure II.2.14 shows one variant which employs sea water. This system is designed to produce economically valuable seafoods and water which is clean enough for irrigation or for returning to open waterways. As developed by the Woods Hole Oceanographic Institution in the US (Ryther 1975) the system consists of:

- Waste water received after secondary sewage treatment diluted with sea water (e.g., 1:3 for sewage effluent of 20-25 mg/liter nitrogen and 10-15 mg/liter phosphorus). Urban waste water typically is imbalanced with excess phosphate (N:P=5−7:1) due to detergents, whereas the algae would equally remove nutrients if the N:P ratio

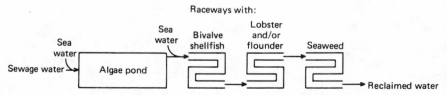

Figure II.2.14. Salt-water aquaculture system for waste-water treatment.

were 15:1. In the former case all phosphate will not be removed, and nitrogen will limit growth.

- Open algal pond (1 meter deep × 16 meters × 16 meters) for mass culturing of phytoplankton, unicellular marine algae; the diatom *Skeletonema costatum* (winter) and *Phaeodactylum tricornutum* were dominant species. Both circulation and aeration are provided by two one-third HP (40 gal/min) pumps in opposite corners whose return jets are just above the surface to aid aeration. Maximum algal production was 12 grams dry weight/meter^2-day while 9 grams was typical for summer and 6 grams for spring and fall periods.
- Bivalve mollusks confined to vertically stacked trays which are fed algae diluted with seawater (1:1-5 for cell concentration of 10^5/ml) flowing rapidly to improve feeding, aeration and removal of wastes.
- Lobster (*Homarus americanus*) and flounder (*Pseudopleuronectes americus*) could also be grown in trays in the raceways (the same or in one following the bivalves even though they are carnivores/omnivores. Feed is provided by small invertebrate detritovores (amphipods, polychaetres, bryozoans, tunicates and mussels) which in turn feed on the shellfishes' solid wastes.
- Seaweed (red algae, *Gracilaria* and *Agardhiella*) confined in raceways (12 meters × 1.3 meters × 1.6 meters deep) served as a final polishing step to remove any remaining nutrients, especially those returned to soluble form by the fish cultures. Seaweed is harvested once a week with nets, dried and potentially salable for commercial extraction of agar or carragenan. Red-algal yields surpass the initial open green-algal yields: maximum dry weight yields were 16 grams/meter^2-day with 13 grams typical for summer and 5 grams for spring and fall. Nitrogen removal is so effective that the final portions of the raceway contain red algae which are pale yellow, due to nitrogen deficiency.
- Water leaving the final red-algal raceway contains little nitrogen due to the 90% removal efficiency, including nitrogen added to the system by the sea water. The flow of nitrogen at the various stages is shown in Table II.2.8. Note that the phytoplankton removed 98% of the sewage effluent plus seawater nitrogen, but the shellfish convert a third of the algal nitrogen to inorganic nitrogen as waste material. Two-thirds of this nitrogen is removed by the red algae, but more complete removal could be achieved by a one-third increase in raceway length.

Fish Yields. Yields of edible fish from the above systems can vary by orders of magnitude, but the desire for increased yields must be bal-

Table II.2.8   Mass Flow of Inorganic Nitrogen (Ammonia, Nitrite and Nitrate) through the Phytoplankton-Oyster-Seaweed System (Ryther 1975).

		Grams of Nitrogen per Day
Phytoplankton pond input		85
Sewage effluent	84	
Seawater	1	
Phytoplankton pond output		1.5
Shellfish raceway input		4.5
Phytoplankton pond harvest	1.5	
Seawater	3.0	
Shellfish raceway output		27
(= seaweed raceway input)		
Seaweed raceway output		9.4
(Final effluent from system)		
Total N removal efficiency (including seawater)		89.3%
Effluent N removal efficiency		93.6%

anced against the capital and energy costs incurred. Feed is a major cost determinant, but, with care conversion efficiencies (net weight fish/net weight food) of 10-20% are attainable. Protein conversion efficiencies are somewhat better, 18-25%, and are comparable to milk (21%) and egg (25%) production efficiencies; beef has a predictably low protein conversion efficiency (12%; Windsor and Cooper 1977). As presently practiced in the West, fish farms produced food-quality fish at a cost over three times that of milk (per unit protein), twice that of eggs and broiler chickens, and about comparable to beef.

Feed partly determines yields, but energy intensity also plays a role. Aeration of experimental high intensity ponds in Israel, for example, increased yields by more than threefold (6.6-9.5 tons/ha-yr to 20-25 tons/ha-yr). A more typical yield for relatively inefficient grass carp and fresh water shrimp fed on the green macroscopic algae, *Chara sp.*, was about 4 tons/ha-yr, with a conversion efficiency of only 2.6%.

The system at the New Alchemy Institute (Woods Hole, Mass., USA) consistently strives for lower limits on the energy and capital consumption spectrum. By employing windmill-driven water pumps and the scheme sketched in Figure II.2.13, they have attained highly respectable yields with minimal energy intensity and feed supplements. The level of energy and capital intensity cannot be decided in general, but the overall feasibility increases dramatically if waste treatment, water recycling and algal and fish production are combined.

Potential Research Developments

Algal Photoproduction of Hydrogen.   Many of the same arguments used elsewhere (see Part I, Chapter 3, page 272) discussing bacterial photoproduction of hydrogen apply here. Of principal importance is the realization that the release of energy and reducing capacity is normally self-defeating especially for a photosynthetic organism. Hydrogen transfer and activation systems ($H_2 \rightarrow 2H^+$ via hydrogenase) in algae have undoubtedly evolved from the absorption and metabolism of hydrogen gas an an energy and electron source. No well-accepted theories as yet exist for normal hydrogen evolution nor does any known ecological niche exist for such organisms, but possible functions include (Kok 1973):

- Under severe anaerobiosis all components of the photosynthetic electron transport chain become reduced.
- Bleeding off H_2—an insoluble gas—converts primary acceptor X to the oxidized state. This allows operation of photosystem I (see Part II, Chapter 1).
- P_{700}^+—a strong oxidant—oxidizes additional intermediates and as a result turns on Photosystem II and O_2 evolution.
- Algae frequently encounter anaerobiosis, especially in mixotrophic growth.
- Leaves never do, need no "priming," and contain no hydrogenase.

Persistent interest in the improbable photoproduction of hydrogen by algae is principally motivated by its attractiveness. If a photosynthetic organism is convinced to evolve hydrogen instead of growing at normal rates, the following process (biophotolysis) would result:

$$H_2O \xrightarrow[\text{algae}]{\text{light}} H_2 + 1/2O_2$$

Storage of solar energy as hydrogen gas at the expense of only water would be a nearly ideal process, the reverse of which releases 285 kJ (68 kcal) per mole hydrogen via combustion to heat or via a fuel cell to electricity. Biophotolysis would ideally employ the two photosystems (see Part II, Chapter 1) of the algal chloroplast to split water (with the evolution of oxygen) and lift the resulting electron about one volt for the immediate reduction of protons to hydrogen gas; some minimal amount of energy would be needed for plant survival, but overall efficiencies of a few percent could be hoped for.

Photoproduction of hydrogen by algae does occur naturally and was first observed 35 years ago (Bishop, Frick, and Jones 1977). Laboratory conditions for sustained algal hydrogen evolution are not exactly natural, however: carbon dioxide and oxygen must first be excluded during a 10-20-hour dark-incubation period. Following this adaptation period, the reactions listed in Table II.2.9 can be catalyzed by hydrogenase containing green algae (e.g., *Scenedesmus*). Note in the reactions in Table II.2.9 that sugars, i.e., cell material, can be synthesized via normal photosynthesis reaction (1), or by metabolizing hydrogen gas with (2), or without (7) the help of light. Sugar synthesis, in the light or dark, requires energy (ATP) processed in a reaction (phosphorylation) which, when chemically inhibited, allows the photoproduction of hydrogen (3).

Green algae are the principal group of organisms known to be capable of hydrogen evolution, but blue-green algae and a few fresh and marine red algae are also among the hydrogen evolvers; blue-green algae are a special case which will be discussed later. The green algae *Scenedesmus* was the first observed hydrogen evolver, but others include *Chlorella (vulgaris, fusca, pyrenoidosa,* and *autotrophica), Chlamydomonas (dysosmos, moewusii,* and *reinhardii), Ulva, Enteromorpha, Codiotum, Acetabularia* and *Dunaliella.* Typical rates of hydrogen production are, however, extremely low: the peak rate is 1% of the theoretical maximum and normally 100 times more oxygen is evolved than hydrogen (Lien and San Pietro 1976). More recently, dark-adapted anaerobically (He) grown *Chlamydomonas reinhardii* produced 1.9 times as much hydrogen as oxygen, very near the stoichiometrically expected ratio of two (Greenbaum 1977). These experiments, which employed single light flashes intense enough to excite all pigments but of short enough duration to

Table II.2.9   Reactions Catalyzed by H_2-adapted Algae (from Stuart and Gaffron (1972).

Light-dependent reactions:

1. Photosynthesis: $CO_2 + 2H_2O + \text{light} \longrightarrow (CH_2O) + O_2 + H_2O$

2. Photoreduction: $CO_2 + 2H_2 + \text{light} \xrightarrow{H_2ase} (CH_2O) + H_2O$

3. H_2 photoproduction: $RH_2 + \text{light} \xrightarrow{H_2ase} R + H_2$

Dark reactions:

4. Respiration: $1/2O_2 + RH_2 \xrightarrow{H_2ase} H_2O + R + \text{energy}$

5. Dark H_2 production: $RH_2 \xrightarrow{H_2ase} R + H_2$

6. H_2 absorption: $H_2 + R \xrightarrow{H_2ase} RH_2$

7. Oxy-hydrogen reaction: $O_2 + 2H_2 \xrightarrow{H_2ase} 2H_2O + \text{energy}$

$CO_2 + 2H_2 + \text{energy} \xrightarrow{H_2ase} (CH_2O) + H_2O$

allow only a single photoact, showed that hydrogen evolution has neither the delays nor intermediate formation characteristic of oxygen evolution (see Part II, Chapter 1).

Although the biochemical mechanisms of algal hydrogen production are not known, comparisons with the nitrogenase mediated hydrogen evolution from photosynthetic bacteria show that an ATP independent hydrogenase is involved (Lien and San Pietro 1976). The absence of an energy requirement for the actual proton reduction promises a potentially higher efficiency for the algal system. Studies with mutants deficient in one photosystem (II) showed decreased hydrogen production in the dark while stimulation of hydrogen evolution by glucose indicates that the photosystems are *not* necessarily directly involved. Photosynthetically produced sugars may be metabolized in both the light and dark, but some argue that the photosystems reduce hydrogenase directly by demonstrating that no phosphorylation (ATP synthesis) is involved, i.e., no generation of reducing power via sugar metabolism is required (Lien and San Pietro 1976).

Much of the uncertainty, low average hydrogen production rates, and the rapid decrease in the initial rate can however be due to oxygen inhibition of the hydrogenase system (see Part II, Chapter 1, page 144). The anaerobically dark-adapted algae produce oxygen immediately when the light-driven photosynthesis begins. Hydrogen is also evolved immediately, but when the oxygen concentration exceeds 0.2% it inhibits the hydrogenase and decreases the rate of production (Stuart and Gaffron 1972). The only potential solutions to the dilemma of hydrogenase inhibition by photosynthetically evolved oxygen are:

- select for oxygen insensitive hydrogenases;
- provide algae with organic substrates;
- separate oxygen and hydrogen evolution, physically or temporally.

Breeding experiments have been started by looking for hydrogen evolving cells at osygen tension slightly higher than normally permissible (2%). *Chlamydomonas reinhardii* was chosen for study due to its constitutively synthesized hydrogenase, but one inspection of 43 algal strains yielded no differences in oxygen sensitivity (Bishop 1975).

The second alternative above, metabolizing sugars to produce hydrogen, is simply not practical; if organic materials (wastes) are to be used, photosynthetic bacteria provide a much simpler conversion system (see Part II, Chapter 3, page 272).

The third alternative above, i.e., the temporal or physical separation of oxygen (photosynthesis) and hydrogen (light or dark reaction) production is exemplified naturally in certain blue-green algae. Physical separation of the two functions occurs in some filamentous blue-green algal species (e.g., *Anabaena cylindrica* and *Nostoc muscorum,* see also Part II, Chapter 1) when some vegetative cells eliminate the oxygen evolving system (Mn^{+2} for biophotolysis is missing), develop thick walls to exclude extracellular oxygen and hence become nonvegetative "heterocysts." Elimination of both extra- and intra-cellular oxygen in these heterocyst cells allows the synthesis and functioning of two oxygen-sensitive enzyme systems: hydrogenase and nitrogenase.

Heterocysts in blue-green filamentous algae normally function by receiving photosynthates (a sugar, maltose) from the vegetative cells which presumably provide all the fixed carbon and reducing power required. CO_2 fixation does not occur in heterocysts due to a missing enzyme system (ribulose diphosphate carboxylase). Only photosystem I is believed to be present, and its function is mostly for photophosphorylation (ATP production), but in principle could also directly reduce ferredoxin which in turn reduces nitrogenase/hydrogenase. Heterocysts will normally fix nitrogen and exchange the sugar from vegetative cells for glycine, the fixed-nitrogen form produced.

Under unnatural conditions, this "symbiotic" cell specialization in blue-green algae can use photosynthetic energy and reducing power to evolve oxygen and hydrogen simultaneously:

$$H_2O + light \xrightarrow[\text{algae}]{\text{blue-green}} H_2 + 1/2O_2$$

Both oxygen and hydrogen can be produced simultaneously because:

- the enzyme system involved (hydrogenase or nitrogenase) is protected from extracellular oxygen;
- no oxygen is evolved in the heterocyst;
- nitrogen, the normal nitrogenase substrate, is eliminated by a purge gas, e.g., argon.

This final point, of course, requires that the cells are first grown normally on nitrogen (fixed or gaseous) for protein synthesis, and then the system is closed and nitrogen removed. Oxygen at normal tensions (18%) is only mildly inhibitory (ca 25%), while nitrogen gas nearly eliminates all hydrogen evolution (Benemann and Weare 1974). Carbon monoxide selectively inhibits the nitrogen-fixing function of nitrogenase but *not* hydrogen evolution, not even in the presence of nitrogen. In principle, 2% carbon monoxide is as effective for hydrogen production as growth under argon.

Most laboratory attempts to produce hydrogen begin with a normal growth period (with 0.3% CO_2) and then replace all nitrogen with argon (plus 3% CO_2). Hydrogen/oxygen production from a two-liter culture has continued for several weeks with conversion efficiencies of ca 0.4% (Benemann and Weissman 1976). The decrease in hydrogen production shown in Figure II.2.15 is believed to be due to filament fragmentation. While a pure photolytic process would produce a hydrogen-oxygen ratio of 2:1, observed ratios vary from 1.7-4.0:1. Such variations are not at all surprising as the link between the two processes is long and complex: hydrogen is evolved only after the CO_2 fixed as sugars in the vegetative cells is transported to heterocysts where the sugars are metabolized and oxygen is released. The rate of reductant supply from vegetative cells limits heterocyst hydrogen production, but efficiencies of a few percent should be possible.

The enzymatic mechanism of hydrogen production in blue-green algae is not known, but both hydrogenase and nitrogenase are known to be present. Either enzyme can evolve hydrogen, but nitrogenase requires ATP while hydrogenase does not. This energy requirement for hydrogen production from nitrogenase is a key factor in the low energy-conversion efficiency and is thermodynamically a totally unnecessary waste. One recent experiment with purified heterocyst cells from *Nostoc muscorum* showed that hydrogen could be evolved in the light (with dithiothreitol) or in the dark (with dithionite), but in neither case was ATP added (Tel-Or

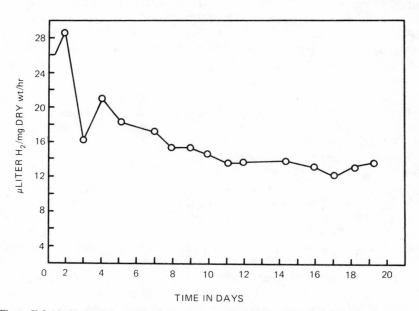

Figure II.2.15. Hydrogen evolution from a two-liter culture of *Anabaena cylindrica* under low light (by permission of Weissman and Benemann, 1977).

and Packer). If substantiated, the ATP-requiring nitrogenase cannot be involved. Further evidence against nitrogenase having a role in hydrogen evolution is that nitrogen was slightly stimulatory, while carbon monoxide was inhibitory. Hydrogenase *is* known to have a role for hydrogen uptake, a competitive reaction which can become serious if the gas is not continually removed.

A second remaining question is the role of the remaining photosystem I in the heterocyst. In theory it could directly reduce ferredoxin which in turn can reduce nitrogenase or hydrogenase, but an electron donor at about zero potential must be added. Alternatively, the photosystem can be used entirely for photophosphorylation with the ferredoxin reduced by sugars metabolized via pyruvate to NADP. The latter reaction certainly occurs in the dark.

A practical system clearly needs to be completely enclosed with constant removal of the photoproduced hydrogen to minimize the competitive uptake reaction, oxygen to prevent inhibition, and nitrogen to prevent nitrogen fixation. In blue-green algae carbon monoxide can profitably be added to inhibit nitrogen fixation selectively. Care should be taken in choosing the gases used to assure their easy removal or suitability for end use. For fuel-cell use the purity requirements for the gas are much more stringent, but for combustion carbon dioxide added in large quantities for photosynthetic carbon fixation (sugar production) will only lower the heating value of the produced gas. Carbon dioxide can of course be removed, and any inert gases used must be removed in any case for reuse. Carbon monoxide added to inhibit nitrogen fixation need not be removed for combustion, as it is also a good fuel.

Ammonia Production by Blue-Green Algae.   As discussed earlier, heterocystous blue-green algae can fix nitrogen for direct fertilizing into rice paddies, production of a dried organic fertilizer, or fed to an anaerobic digester to increase the total amount of fixed nitrogen in a completely recycled system.

Research directed toward increased ammonia yield from blue-green algae takes one of two alternative directions:

1. Increase the relative number of heterocysts. One method, successful on a laboratory scale, uses the natural response to persistently low ammonia levels which causes a fivefold increase in heterocysts if fixed nitrogen is occasionally provided in small amounts for required protein synthesis (Benemann and Weissman 1976). A conceivable alternative is genetic selection of mutants which overproduce heterocysts.

2. Impair self-regulation. This more promising approach deals directly with the limits to nitrogen-fixation rates irrespective of heterocyst frequency. These regulatory mechanisms can either halt nitrogenase synthesis or perhaps cause the reversion of heterocysts to vegetative cells. Temporary regulation, that involving ADP inhibition of nitrogenase activity is probably not alterable. Inhibition due to excess nitrogen is believed to be effected by the amino acid glutamine, synthesized from ammonia via glutamine synthetase, and not due to ammonia directly. Two approaches for deactivating this control are conceivable: ideally a mutant could be found which is not responsive to ammonia concentration, i.e., one which most probably lacked glutamine synthetase. No efforts have yet been successfully directed toward this goal, nor is it known which of the poorly understood regulatory steps is most susceptible to uncoupling. The alternative of chemically poisoning the regulatory system has successfully caused massive excretion of ammonia into the medium (Stewart and Rowell 1975). Apparently, ammonia conversion to glutamine is blocked, and fixation is stimulated by apparent nitrogen starvation. Heterocyst formation is likewise freed from regulation. Practical applications of these recent developments have not yet been considered, but would require removal of the ammonia analogue to prevent ammonia-uptake problems in the crop to be fertilized.

Potential fixation rates can only be estimated by maximum natural rates and a theoretical maximum. The theoretical maximum can only be a crude guess, as the energetics of the fixation process are poorly understood. If one assumes three electrons and 15 ATP molecules per fixed ammonia molecule, then a 10% conversion efficiency produces several tons of ammonia per hectare per year. This figure, taken as a very crude theoretical estimate, can be compared with the maximum observed fixation rates discussed in Part II, Chapter 1, for natural systems. Estimates there extended to 500 kg/ha-yr, but more reasonable extreme maxima are 300 kg. *Azolla* in symbiosis with the blue-green algae *Anabaena* gives a particularly useful figure in the present context of ca 160 kg/ha-yr (Stewart 1977). Of course a great deal of the solar energy is used for plant growth.

A free-living *Anabaena* can produce a biomass of 50 tons/ha-yr which could contain (assuming 40% protein and 16% nitrogen content in protein) 3.8 tons fixed nitrogen, but most of this nitrogen is taken from the medium and not fixed by the organism. How much nitrogen would be fixed by an ammonia-free medium rich in all other nutrients remains to be seen. A natural organism has of course the regulatory mechanisms discussed above intact and does not represent the full potential; *Anabaena* fixation in symbiosis with *Azolla* is only mildly inhibited by fixed nitrogen (Talley, Talley, and Rains 1977). Ideally the biomass would be semi-permanent and tricked by some of the means discussed above into devoting nearly all its efforts toward fixing nitrogen. Potential ammonia yields should exceed a ton, but a more exact estimate is premature.

Algal Production of Alcohol. Nature's adaptability is exemplified in an unusual unicellular algae: *Dunaliella parva* which thrives in high-salinity (from 1.4% salt to saturated solutions) waters typical of arid regions. Osmotic pressure, resulting from the enormous and variable salt gradient across the cell membrane, is balanced by rapid synthesis of glycerol. The glycerol content doubles within one hour after the salt concentration doubles.

Glycerol, an energy-rich derivative of propyl alcohol, has an energy content of (4.3 kcal/gram) and constitutes up to 85% of the cell's dry weight. Glycerol is also an important industrial chemical for lubrication, pharmaceuticals, cosmetics, soaps, and foods.

The *Dunaliella* cell itself lacks the characteristic indigestible cell wall and hence is easily breakable. It can be grown on simple salts at temperatures up to 40°C (its optimum is 33°C) with a density of ca 5×10^{12} cells/meter^3, a doubling time of 30 hours and a production rate of about 50 grams/meter^2-day (Avron 1976). Assuming a daytime insolation of 500 cal/cm^2-day, the efficiency can be calculated to be an optimistic 4%.

Dunaliella can be cultivated in semi-arid regions in natural or artificial ponds using brackish water which supports few other organisms. Whereas harvesting of unicellular algae is normally difficult, *Dunaliella* can be separated using its property of settling in the cold, or alternatively it can be centrifuged at 100 grams for ten minutes. A sudden decrease in salinity of the harvested cells causes the thin membrane to rupture and allows the glycerol to be separated. The cell membranes, containing 40% protein plus some glycerol, make an easily digested energy- and protein-rich fodder without further treatment.

ECONOMIC AND ENERGETIC COSTS FOR ALGAL PRODUCTION

At the present stage of development, analysis of both the economic and energetic costs of mass algal production point to the major burden: harvesting of unicellular algae. Drying and CO_2 costs, if the latter is not provided as a waste by-product, are the next greatest burdens.

Two points must be remembered when discussing algal system efficiency, whether economic or energetic. Firstly, mass algal culture in its modern perspective is in its infancy, nearly all efforts in the past, present and near future are rather scientific and technological in nature, with little effort or motivation to reach any economic or energetic optimum. An exception in this context is the Sosa Texcoco *Spirulina* project in Mexico which is a commercial effort, although one of a rather special character. No major attempts at low or intermediate technology (capital) mass algal culture have yet been made. Secondly, the type of algal culture on which the analysis is based must be clearly stated. Reasonable economic estimates have been made for only three systems:

1. *Spirulina* grown by Sosa Texcoco in brine lakes;
2. *Scenedesmus* grown on artificial media by the various German-(Dortmund)-inspired projects plus one proposed by the Proteus Corporation of Concord, California;
3. Mixed unicellular green algae grown on waste water, predominantly in California, USA.

Spirulina from Sosa Texcoco presently costs about $2.50/kg (in one-ton lots), but while its being of food quality is a distinct advantage, this price still cannot compete with conventional protein sources at one-tenth the cost. Present uses are confined to health foods or as an industrial chemical source.

Figure II.2.16 summarizes calculated costs for large-scale algal growth on inorganic media (for human consumption) and on waste waters (fodder or industrial uses). At 10 tons per day, food-quality algae is estimated to cost about $0.33/kg, while fodder would cost about $0.20/kg (Oswald 1976a). The former begins to compete quite well with competitive conventional foods especially in light of impending food shortages.

Waste-grown algae is the most advanced technology on a large scale and provides an interesting comparison with alternative foods. Soybean-oil meal (40% protein) sells for $200 per metric ton compared with an equivalent cost for algae. As algal protein content is somewhat higher, it may command a higher selling price and thereby become economically viable. On the other hand, algae presently suffers from being an unconventional food, but in the long term its competitive position should improve. Prospective decreases in algae costs are exemplified by a crude breakdown for large-scale waste grown algae in Table II.2.10 (from Oswald 1976a, and Lewis 1976).

Separation and drying are the greatest operational costs and are economically the costs most sensitive to energy costs and shortages. Here lies the greatest hope and need for improvement and re-emphasizes the importance of recent successes with species control (see page 223 of this chapter). Growth of filamentous algae permits an estimated tenfold decrease in the separation costs (Benemann et al. 1976). Drying costs are expected to

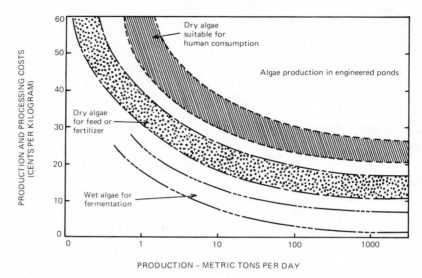

Figure II.2.16. Estimated costs of producing and processing algae of various grades (37° North Latitude; from Oswald 1976a).

remain unchanged, but these are not relevant for algae used in methane production.

Simplified separation techniques with filamentous algae has the additional advantage of not only decreasing the capital costs, but also makes possible practical operation on much smaller scales. Little advanced work has been done on low- or intermediate-scale algal growth, and hence estimates for such installations have been neglected here. This latter area is one deserving highest priority.

Table II.2.10   Estimated Economic and Energy Costs for Algal Production (Oswald 1976a; Lewis 1976).

	Economic Cost (5000 tons/year)	Energy Cost (500 tons/year)
Amortized capital	$110/ton	2.8 GJ/ton algae
Labor	18	—
Mixing costs	7.30	2.8
Separation	80	42
Drying		14
Misc. (taxes, insurance, etc.)	20	—
Sewage medium (credit)	—	−4.3
	$235/ton	57.6 GJ/ton

3

Conversion Processes for the Production of Fuels from Organic Matter

Organic materials, i.e., plant products, consist principally of carbon, hydrogen and oxygen. The same is true for virtually all fuels for which the energy content is inversely related to oxygen content. Oxidation is, of course, the energy releasing reaction. This close chemical relation among all fuels, plant material included, allows in principle the synthesizing of "anything organic from anything organic." Energy may, of course, be added or removed during the transition.

Which conversion process is appropriate is determined by matching the qualities of the starting material with the qualities of the fuel desired. Some of the myriad of energy-conversion methods and products will be discussed further in this chapter and in the next. The conversion methods can be conveniently subdivided into biological, chemical, thermal and physical methods; the latter in the present context are generally only pretreatments, as for shredding, separations or pelletizing, but in general include all physical solar devices, e.g., photoelectric cells. Biological conversion methods are here further divided into those employing intact microorganisms and those employing extracted enzymes. Fuel characteristics will be discussed in Part II, Chapter 4.

The importance of the starting material properties for the various processes discussed here is summarized in Table II.3.1, together with the type energy form(s) produced. High water content in biomass is particularly relevant here and strongly encourages use of biological conversion methods, whereas low degradability and poison content are uniquely inhibiting of these processes. Each of these factors will be discussed where relevant.

Table II.3.1 Characteristics of Various Conversion Methods.

	EFFECT OF FEED CHARACTERISTICS					ENERGY PRODUCTS				
	Water	Heating Value	Noncombustibles	Biodegradability	Poisons	Heat	Gas	Liquid	Char	Other
Fermentations										
to Methane	+	+	0	+	−		×			×
to Ethanol	+	+	0	+	−			×		
Enzymatic hydrolysis	+	+	0	+	0					×
Acid hydrolysis	−	+	0	0	−					×
Fuel cells	−	+	−	0	0			×		×
Conversion to methanol	−	+	−	0						
Pyrolysis	−	+	−	0	0		×	×	×	×
Hydrogenation	−	+	−	0	0		×	×	×	×
Combustion	−			0	0	×				

248

FERMENTATIVE CONVERSIONS

Fermentation to Ethyl Alcohol

Ethyl alcohol or ethanol is an excellent multipurpose liquid fuel and chemical base which can be synthesized by a variety of synthetic and fermentative processes from chemical and biological feedstocks, respectively. As a natural fermentation process, its spontaneous occurrence was observed and then duplicated by numerous ancient civilizations, primarily for beverage use. Other traditional uses are for the making of perfumes, cosmetics and medicines. Additional modern uses are for solvents, beverages, food and feed (via single-cell protein, SCP), petrochemical synthesis (via ethylene), gasoline dilutant and biological energy (ATP).

The fermentation process (also discussed in Part I, Chapter 1) consists of first breaking starch or cellulose chains into individual sugar molecules ($C_6H_{12}OH$) and then fermenting the sugar into ethanol ($C_2H_{12}OH$) and carbon dioxide (Calvin 1974):

$$[C_6H_{10}O_5]_n + nH_2O \xrightarrow{\text{hydrolysis}} nC_6H_{12}O_6 \text{ (glucose)}$$

$$C_6H_{12}O_6 \xrightarrow{\text{fermentation}} 2C_2H_5OH + 2CO_2$$

Energetically, the reaction can be written,

$$\text{glucose} \longrightarrow 2 \text{ ethanol} + 75 \text{ KJ}$$
$$\text{(180 grams/mole, 2.82 MJ/mole)} \quad \text{(46 grams/mole, 1.37 MJ/mole)}$$

The above equation shows that fermentation of glucose to ethanol yields very little energy to the fermenting organism (yeast) and hence makes the process highly efficient from the energy-conversion point of view. The theoretical conversion efficiency from sugar to ethanol is 93% with the ethanol weight slightly more than half that of sugar; the energy density for alcohol as a fuel is consequently nearly twice that of glucose.

Fermentation Substrates.   Four sugars (hexoses) can be fermented by yeasts: glucose, fructose, mannose and galactose (with difficulty). Alternatively any material with these sugars accessible (e.g., starch or cellulose) is also a suitable fermentation substrate. Cereal grains (maize, wheat, barley, sorghum or rye) are most common for beverage alcohol, but other sources include sugar beets, sugar cane, fruit-product wastes, molasses, starch crops (potatoes, rice, cassava), sulfite liquors (paper-pulping wastes), wood, crop residues, and fiber crops. The high starch contents of typical cereal grains (67-70%, Miller 1975) and digestibility makes them equally attractive as food and as a fermentation feedstock. Cellulose, on the other hand, is *not* a suitable substrate for yeast (nor humans) as the enzymes for its degradation are absent.

Process.   The fermentation process is a spontaneous reaction occurring in an anaerobic mixture of sugar and *Saccharomyces cerevisiae* or other yeast varieties. The former, while of unknown origin, is used industrially for both brewing and baking. Reaction times will vary with fermentor conditions and substrate concentrations, but at room temperature a few weeks is not uncommon for complete fermentation.

In the batch mode of operation the yeasting process will continue (observed by CO_2 evolved) until the sugar supply ends or the alcohol content increases above the yeast's tolerance, 11-18%. For maximum alcohol yield, sugar must be kept limiting via the use of dilute solutions. In certain cases yeast mutants tolerating higher alcohol content can be used, but in practice a 15% ethanol solution (by weight) is rarely attained as it would require a thick initial slurry with 30% sugar content. Sugar is therefore normally limiting.

When fermentation has ceased (i.e., sugar has been consumed) the mother liquor must be distilled. For fuel use it is necessary to remove water, but for beverage use the distillation and purification requirements are more stringent; water is of little consequence, but it is particularly important to eliminate higher alcohols.

In principle, distillation is a separation procedure which takes advantage of the difference in boiling points of ethanol (78.5°C) and water (100°C). If the mixture is heated to 78-80°C in even the simplest distillation apparatus, the ethanol will boil, vaporize, rise up in the distiller's vertical column, condense in an unheated side-arm and run off into a separate flask. Some water will evaporize even at 80° and results in an impure distillate. Considerable improvement is possible by using a more complex distillation apparatus consisting of a series of distillers so that at each of several stages, redistillation occurs and a new equilibrium is established. A pure ethanol distillate can never be achieved, however, as an ethanol-water mixture (95.6%:4.4%) is an azeotrope (a mixture with a constant boiling temperature), i.e., the mixture's boiling point is 0.1°C lower than pure ethanol and hence cannot be further concentrated by distillation procedures. The theoretical maximum alcohol content for separation at normal atmospheric pressure is 95.6% ethanol (4.4% water), but alternatives do exist to achieve higher alcohol contents. For instance, the common distillation apparatus similar to those discussed can be used but under reduced pressure. That is, a vacuum pump (e.g., that achievable by simple water flow by the Venturi principle) can be attached anywhere in the distiller. At, for instance, 95 mm pressure (one atmosphere is 786 mm mercury), the boiling points are lowered to 33.5 and 51°C for ethanol and water, respectively. Under such conditions distillation yields 99.5% pure alcohol. Alternatively, absolute alcohol can, as in the past, be made industrially from a 95% mixture of alcohol and benzene. Distillation of this mixture gives first a benzene-alcohol-water fraction (boiling point 68.2°C), then an alcohol-benzene fraction (68.2°C), and finally a fraction with absolute alcohol.

Little advantage accrues with the additional purity attained by these latter two distillation procedures; unless the alcohol product is carefully sealed, water will be absorbed until the one atmosphere azeotropic proportion of 95.6% ethanol and 4.4% water is attained. The additional costs involved are only justified for special purposes within the chemical industry and certainly not for fuel use. While the water content does partially decrease the energy density (ca 5%), the higher alcohols which often contaminate ethanol will increase its energy content.

Chemical synthesis. Ethanol since 1950 (when cheap oil became available) has been mostly synthesized chemically from petroleum and natural gas. The reaction producing ethanol from ethylene, for example, is a catalytic (phosphoric acid) hydration of ethylene with water:

$$\text{ethylene} + \text{water} \xrightarrow{\text{catalyst}} \text{ethyl alcohol}$$

$$CH_2 = CH_2 + H_2O \rightarrow C_2H_5OH$$

For a reasonable reaction time, ethylene must be compressed to 68 atm, mixed with a water stream and reacted in the vapor phase at 300°C. While only 4.5% of the ethylene is converted on each pass, yields up to 97% have been observed via repeated recycling.

While the synthetic process is not relevant for fuel production, it is mentioned here for comparison with the fermentation process that it has increasingly replaced in recent times. The capital and energy intensity of the synthetic process is obvious.

Economic Consideration

Chemical synthesis. Ethylene from petroleum has been in the past an extremely cheap substrate for ethanol production. A stable price of 6.5-8.0¢/kg was, however, doubled in 1972 and has now risen to more than 20¢/kg. Ethanol prices have likewise undergone a sharp cost increase from 11¢/liter in 1971 to 19¢/liter in 1974, and to 30¢/liter in

1976. Process costs alone are 7.1¢/liter for 95% alcohol, but ethylene—now at 30¢/kg—adds an additional 200% (Miller 1976). Ethylene can only increase its present 70% share of alcohol costs as petroleum prices rise. While their percentage role is small, capital costs are considerable: economics of scale optimize at an annual capacity of 200 million liters at a plant cost of $32 million; an optimum of such enormity is not unusual for processes which are extremely capital-intensive. Catalytic syntheses requiring high temperature and pressure conditions are invariably capital- and, of course, energy-intensive.

Fermentation. Industrial fermentation plants of optimum scale produce 70-million liters of 95% alcohol annually and require $15.4-million investment (Miller 1975 and 1976). Processing costs (12¢/liter) are more than double those of chemical synthesis, but a by-product ("distiller's grains") credit for animal feed reduces the net conversion costs to 20% less than the industrial process. The deciding economic factor in each case is the feedstock which has in the past definitely favored chemical synthesis from ethylene. Conversion costs and the effects of maize-feedstock prices are shown in Tables II.3.2 and II.3.3, respectively (from Miller 1976). Note that maize costs are 75-90% of the total but only because base conversion costs are paid for by the "by-product feed." Additional credits in a well-integrated system are possible for CO_2 recovered from the fermentation process, but were not included here.

As wood has a lower fermentable sugar content than maize, its use as a fermentation substrate requires a higher capital cost per yield (ca 10%). More important, however, the cost credit for distiller's grain is lost which defrayed half of the alcohol production costs. Lignin, a by-product from

Table II.3.2 Fermentative Conversion Cost of 190° and 200° Proof Ethyl Alcohol from Maize (exclusive of cost of maize; Miller 1976).

Alcohol	Cents/Liter
190° proof (2.82 gallons/bushel):	
Base conversion cost	11.7
($1.95-million/year, 10 years, 67-million liters)	2.9
By-product feed credit	14.6
(11.7 kg/liter alcohol at $100/ton)	9.0
Net	5.6
200° proof (2.7 gallons/bushel):	
Alcohol (1.048 liters at 5.6 cents/liter)	5.9
Cost of dehydration	0.8
Total cost (exclusive of maize, profit, packaging and sales expenses)	6.7

Table II.3.3   Effect of Maize Cost on Ethyl Alcohol Cost (Basis: 2.7 gallons 200° proof alcohol/bushel; Miller 1976).

Maize Price/Bushel (dollars)	ALCOHOL COST/LITER (cents)		
	Maize	Conversion[2]	Total Base Cost[1]
1.50	14.6	6.7	21.3
1.75	17.8	6.7	23.7
2.00	19.5	6.7	25.2
2.25	22.3	6.7	28.7
2.50	24.4	6.7	31.1
3.00	29.3	6.7	36.4
3.50	34.5	6.7	41.1
4.00	39.4	6.7	45.6

[1]These costs do not include profits, packaging, and sales expenses.
[2]By-product grains credited at $100/ton in conversion cost.

wood processing, is 30% of the wood weight and thus accumulates at a rate of 1.3 kg per liter alcohol. Lignin can potentially be used as a fuel and a chemical base, but it has no value at present due to far cheaper alternatives. Sugar cane is, of course, an ideal feedstock in some areas with annual yields of 9 tons sugar per hectare. This sugar could in turn yield 4.4 tons ethanol, 2.5 tons ethylene and 9 tons bagasse. Cane sugar at a present European price of $0.063/kg is too expensive a substrate for alcohol fermentations, and while raw sugar production increases by 3% per year, supply has yet to meet demand. Only Central and South America, South Africa, and Australia are expected to have significant surpluses in the foreseeable future (Hepner 1976).

After maize, both wheat and cassava are economically equally competitive, but the former will remain in demand for food. Sugar economics for industrial fermentation (e.g., for ethanol) can potentially be transformed by developments in waste utilization. As has been discussed in the relative cellulose-to-glucose-conversion sections, the prices of the resulting sugar is as yet uncertain. Assuming one estimate (Spano 1976) of 25¢/kg sugar produced from wastes, the ethanol would cost 50¢/liter compared with today's (1976) 30¢/liter. Credits for CO_2 produced (3¢/liter), combustible residue (3.7¢/liter), recyclable materials from urban trash (4.2¢/liter), and a disposal credit (7¢/liter) would reduce costs to 36¢/liter ethanol (Spano 1976). Similar estimates have also been made recently for not only maize but cane juice and molasses (Lipinsky 1977). Clearly, exact predictions are difficult, but the price trend of petroleum-based ethanol can only continue upwards, while well-integrated microbiological systems, upgrading wastes to fuel ethanol, can hope to become increasingly attractive.

A village or family-level fermentor/distiller for small-scale fuel ethanol production will, of course, require greater labor to save capital. Very simple systems have been built by thirsty cultures through the years and are a credit to the cultures' technical abilities. Motivation can overcome most barriers. Such simple schemes are similar in principle to those shown earlier, but recall that even simpler versions are possible for fuel-grade ethanol which can contain the higher alcohols removed from ingested ethanol. Ingestion of low-grade alcohol is a severe medical and social problem which must be anticipated if such systems are popularized for fuel purposes.

Energy Considerations.   Fermentation proceeds at ambient temperatures, but is accelerated at slightly elevated temperatures. Industrial fermentors are normally held at 30°C. Fermentation is, however, normally sufficiently exothermic to require little if any heating if adequately insulated. A more serious cost is the distillation process. An 8% ethanol solution will require for distillation about 2.5 kg steam per liter of ethanol produced or about 4 MJ per liter (Finn 1975). As the energy content for ethanol is 22MJ/liter, the distillation energy costs are over 20% of the final energy content in ethanol. Further additions would result from a total energy analysis, but simple solar collectors could ease the high-quality fuel costs.

An important step toward integration of low and high grade fuel processes is a system described by Wilke et al. (1976) to convert cellulose enzymatically to glucose and glucose to ethanol and yeast. Nondigested solids (67% moisture content) are combusted to generate electrical power and process steam (56% of the total steam). Electricity is generated for $0.01/kWh at an efficiency of 25% and constitutes 7.6% of the fuel stock's combustion energy; a third of this electricity is surplus. Ethanol contains 12.5% of the feedstock energy which, when summed with energy of the single-cell protein and electricity, gives an overall effiency of nearly 20%.

The quality of energy required, as well as the quantity, is important. The energy for distillation can be low quality heat from simple solar collectors, while the alcohol is a fine, high quality transportable energy source suitable for combustion, fuel cell, and several other uses. Savings are not to be expected from increased conversion efficiency as the process is already 95% efficient, but some realistic improvements include:

- increasing the sugar (and hence alcohol) content as high as possible, e.g., 14-16% instead of the usual 10%;

- increasing yeast concentrations to shorten the fermentation time (at 30°C) from a typical batch time of 40 hours to two and a half to three hours;
- removing the alcohol continuously to eliminate retardation effects. This has been done by fermenting under vacuum (50 mm Hg), which also serves to combine fermentation and distillation. Fermentor productivity was increased as much as twelvefold (Cysewski and Wilke, 1977). The CO_2 recovered is a valuable side product, and if a water vacuum trap is used, might be combined with an algal pond.

Other Fermentations.   In addition to the fermentations discussed, five others of the hundred most common organic compounds can also be produced microbiologically (Pape 1976). These compounds are acetic acid, isopropanol, acetone, *n*-butanol, and glycerol, but virtually all are synthesized industrially instead. Economics, often petroleum-based, has favored industrial synthesis, and as none of these are of particular value as an energy or food source, they will not be considered further here.

Methanogenesis

Organic waste materials in Nature are most commonly decomposed by one of two different types of microbiological processes: one type is performed by aerobic organisms which thrive whenever plant and animal wastes are exposed to air. By oxidizing wastes to CO_2 they provide the metabolic energy required for microbial growth. Such processes occur in compost piles from which the degradation product, humus, is an excellent soil conditioner and fertilizer. However, a good deal of the volatile ammonia is lost.

The second type of decomposition microorganisms, anaerobic microorganisms, grows only in the absence of air, such as under water or in some animal (ruminant) guts. This type yields peat or manure plus a gas phase of varying composition. A certain group of anaerobic bacteria produces a mixture of methane and carbon dioxide, alternately called biogas, gobar gas (India), marsh, or swamp gas. For these organisms, methane is the final fermentation product, a product which still has a high energy content due to the absence of oxygen as an oxidizing agent. In contrast to the aerobic process, anaerobic digestion releases little heat, and the energy is instead stored in methane. Only when oxidized by oxygen, after the fermentation process, does methane release its full stored energy and produce the low energy compounds—carbon dioxide and water. The various stages of biomass conversion to methane are sketched in Figure II.3.1, but emphasis here will be placed on the digestion process itself.

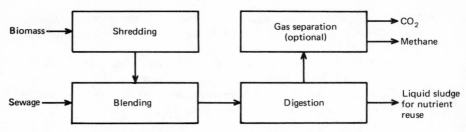

Figure II.3.1. Anaerobic digestion process for methane production.

Process. The process by which organic wastes are converted into methane, carbon dioxide, and a nutrient-rich sludge is a series of reactions involving several organisms. While certain organisms are properly called methanogenic, not all of these organisms are literally methanogenic.

Methanogenic bacteria, available in pure culture only since 1962, grow over a wide range of temperatures, but require near neutral pH, simple (and few) substrates, and the strict absence of oxygen. These bacteria form only the final stage of a methane production sequence which begins with the formation of proper methanogenic bacteria substrates from organic wastes by several other organisms. As the methanogenic bacteria are the most sensitive to temperature and particularly pH changes, an understanding of them is perhaps most crucial. Unfortunately all of the reactions involved in the complex process are not yet completely understood.

Recent experiments with the methanogenic bacteria *Methanosarcina barkeri* and *Methanobacterium thermoautotrophicum* show that acetate in the presence of hydrogen is the substrate most commonly reduced directly to methane (Pine 1971). Electrons from the oxidation of hydrogen can reduce both the carboxy and methyl positions of acetate, reactions which are greatly stimulated by CO_2. Acetate is not the energy source, however, and can be replaced by CO_2/HCO. These two reactions, both mediated by methanogenic bacterial reactions, are

$$4H_2 + HCO_3^- \rightarrow CH_4 + 2H_2O + OH^- + 32.3 \text{ kcal/mole (at pH7)}$$

or for acetate,

$$CH_3COO^- + 4H_2 + H^+ \rightarrow 2CH_4 + 2H_2O + 39 \text{ kcal/mole (at pH7)}.$$

These methanogenic reactions yield sufficient energy to synthesize ATP, but no ATP-yielding reaction has yet been documented for methanogenic bacteria (McBride and Wolfe 1971). On the contrary, some evidence exists which suggests that methane evolution consumes ATP in equal molar amounts. As CO_2 is of course reduced stepwise to CH_4, it is conceivable that certain step(s) require ATP, while the overall process of methane production is energy-yielding.

As hydrogen is the major energy source for methane production, its minimal presence in digester gas indicates a careful balance among the different bacterial species. This balance also preserves a near neutral pH suitable for the sensitive methane bacteria.

Methanogenic bacteria are sensitive not only to pH, but also are competitively inhibited by chlorinated hydrocarbons, such as chloroform, carbon tetrachloride and methylene chloride (Kugelman and Chin 1971). Mercury is also inhibitory to methane formation, but under reducing conditions the more important problem is mercury's methylation to form toxic methylmercury compounds.

Only the final step in methane production, i.e., the step mediated by methanogenic bacteria, has been discussed. These bacteria, however, require highly specific but simple substrates described above, such as acetate, formate, hydrogen, carbon dioxide and ammonia. As photosynthetic plant products cannot be digested directly by methanogenic bacteria, preparation of these simple substrates from organic wastes require the cooperative action of other organisms.

Most of the plant mass has a structural function and has evolved characteristics which maximize its resistance to attack and decomposition. This degradation resistance, discussed more thoroughly in the following section, is particularly true of trees in which lignin protects the more vulnerable cellulose by encapsulation, but is also true of highly ordered or crystalline cellulose. This initial degradation stage is very similar to enzymatic hydrolysis and aerobic decomposition, and as in those cases, degradation is greatly aided by mechanical shredding to decrease order and increase the access to water-borne organisms and enzymes. As expected, the decomposition of organic matter into its soluble organic compounds is the rate-limiting step in methane production.

Organic compounds solubilized by the initial degradation are readily converted by a wide range of bacteria to acetate, various other acids, hydrogen gas and carbon dioxide. While these steps are relatively rapid, they can decrease the pH if the product acids are not further reduced to methane. The pH dependence of these acid-producing bacteria is not particularly severe: their activity is a maximum at pH 6 and slowly decreases to a low value at pH 8. The methanogenic bacteria, on the other hand, have an extremely sharp pH dependence which peaks at about pH 6.8 and is essentially zero below pH 6.2 or above pH 7.6. As methanogenesis does proceed sequentially from insoluble to soluble organic compounds, to acids other than acetate and then on to acetate for the

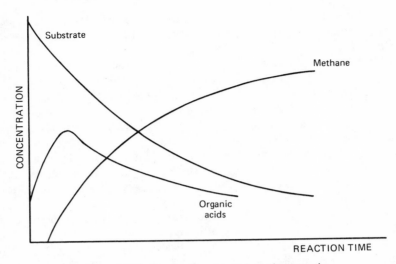

Figure II.3.2. Reaction time course in methanogenesis.

final conversion to methane, any break in this time sequence (Figure II.3.2) can easily lead to digester pH disturbances.

Substrates.   Virtually any type of organic material is a potential substrate for anaerobic digestion. As in all conversion processes, of course, there is a problem of resistance to attack, accessibility of chemical bonds to the bacterial enzymes involved in degradation and presence of inorganic or nondigestible compounds. For these reasons, mixed wastes must be classified in terms of parameters meaningful for digestion yield, not simply the total weight:

- *total solids* (TS)—substrate weight after drying at 100°C (sundried material often contains 30% moisture);
- *volatile solids* (VS)—weight which is combustible; essentially all VS are converted to methane in an anaerobic digester;
- *fixed solids* (FS)—biologically inert material.

Most wastes are quoted in terms of a total weight which includes not only variable amounts of water but also fixed solids. Water content is the largest variable as the difference between total and volatile solids is rarely more than 20%; feces and urine, for example, have 20% and 6% total solids, respectively, and 16% and 4% volatile solids (New Alchemy 1973).

The rate at which a given waste can be digested also varies markedly: variables for animal wastes include age and type of animal, its feed and

Table II.3.4   Daily Animal Waste Yields. (Total Solids—TS—Unless Otherwise Indicated; New Alchemy 1973 and Ensign 1976)

Animal Type	Dry Dung	Urine	Nitrogen (grams)
Cows	3.2–4 kg (2 kg India)	8 liters	52.70 + 60
Humans	50 grams		3–8 grams
Chickens	25–30 grams		1.7
Pigs	0.6 kg (VS)	0.1 (VS)	24 + 4

living conditions, while for plants the degree of cellulose crystallinity or the presence of binding materials such as lignin are most critical.

A second factor affecting gas yields is the relative amounts of substrate carbon and nitrogen in its fixed form (e.g., ammonia). As mentioned above, methanogenic bacteria require ammonia and rely on other bacteria of the mixed anaerobic digester population to metabolize proteins and free the ammonia bound as protein amino acids for their use. Nitrogen content of animal wastes will therefore vary widely particularly with protein content of the feedstock and even with season, time of harvest, etc. Barley, for instance, has 39% protein after 21 days' growth, 12% after 49 days, and 4% after 86 days. (Nitrogen content can be calculated by assuming it is 16% of the protein by weight.) When comparing animal wastes in Table II.3.4 one must remember that ruminant feces, for example, have particularly little nitrogen due to the nutrient requirements of their rumen bacteria. Urine contains most of the animal's excreted nitrogen with the exception of poultry whose urine and feces are combined. Whenever possible, urine should be recycled as quickly as possible via an anaerobic digester although collection is of course impractical for grazing animals. The energy content in urine is minimal, but the ammonia present is required for bacterial growth as a nutrient, not as an energy source.

Most carbon is assumed potentially digestible, but lignin is an important exception. As discussed earlier, lignin poses a compound problem of not only being nondigestible itself, but also inhibits access to digestible cellulose. Lignin degradation is one of the most difficult and important areas for study in anaerobic digestion.

Carbon and nitrogen contents have been discussed here because an optimal substrate for anaerobic digestion will have a C/N ratio of approximately 30 (New Alchemy 1973). A digester with an unbalanced carbon/nitrogen ratio will produce little methane, but a great deal of carbon dioxide plus hydrogen or nitrogen. The effects of these imbalances are summarized in Table II.3.5. Garbage and manure have a near optimal balance, while blood and urine have too much nitrogen, and wood, starch

Table II.3.5 Effects of Carbon/Nitrogen Imbalance (New Alchemy 1973).

		Methane	CO_2	Hydrogen	Nitrogen
C/N Low (high nitrogen)	blood, urine	little	much	little	much
C/N High (low nitrogen)	sawdust, straw sugar and starches such as potatoes, corn, sugar beet wastes	little	much	much	little
C/N Balanced (C/N = near 30)	manures, garbage	much	some	little	little

and sugar wastes have too little. Algae is a useful example to show how a C/N ratio is easily calculated (Golueke and Oswald 1973):

protein content = 60% of the dry weight
nitrogen content = 16% of the protein content
                 = 10% of the total dry weight
carbon content = 40% of the cellulose weight
                 = 85% of the fat weight
                 = 40% of the protein weight

As nitrogen constitutes 10% of the algal dry weight and carbon 40%, the ratio C/N = 3, and hence algae alone is *not* an ideal substrate for anaerobic digestion (but see "Gas Yields" later in this chapter). Algae can be combined with less protein (ammonia)-rich plant wastes (where essentially all nitrogen is bound in protein), to reach the optimum ratio. Paper, for instance, has only trace amounts of nitrogen and must be combined with nitrogen-rich substances for adequate digestion. The ratio C/N = 30 is only an approximate guideline and in practice poses no severe problem.

The problem of poisons was mentioned earlier and is particularly significant where industrial wastes are used. Mercury for instance cannot only be toxic to the methanogenic bacteria, but can be complexed as methylmercury compounds toxic to Man. These compounds will appear in the sludge used later as fertilizer. All heavy metals except iron are toxic in even low concentration (<1mg/liter; Kugelman and Chin 1971).

While many toxic substances are trace nutrients and become toxic only in large concentrations, factors other than simple concentration must be considered. Many factors are not yet well characterized, but are known to play significant roles:

- a toxin can have reduced effect due to *antagonism* by another;
- toxic effects can increase due to a second compound, a *synergistic* effect;
- a toxin can be complexed (e.g., methylated mercury), and thereby its metabolism is prevented *or* promoted;
- toxic effects can be nonfatal through *acclimation* to slowly increased concentrations.

Toxic effects do not end with heavy metals, but rather include light metal cations (calcium, magnesium, sodium and potassium), the free ammonium ion in high concentrations (over 150 mg/liter), and even long chain fatty acids (palmitic, stearic and oleic acids; Kugelman and Chin 1971). Modern soaps are particularly problematic for anaerobic digesters, both those with disinfectant properties and those containing phosphates and EDTA. Little is yet known about toxicity principally because of the complexity added by the antagonistic and synergistic effects of toxins and nutrients in low concentrations.

Digester Design. Among the major non-nutritional requirements of anaerobic digestion are pH and temperature control, and media stirring. Temperature is particularly crucial for digestion and has an optimum typical of all biological processes with one additional complication—there are two types of methanogenic bacteria. One type has a temperature optimum of about 34°C (mesophilic), while the other's optimum occurs at about 60°C (thermophilic).

For mesophilic bacteria both the rate and final extent (at infinite retention time) of digestion increases with temperature. A frequent generalization states that gas production doubles for every 5°C increase in temperature (from 10° to 30°C; New Alchemy, 1973), but inspection of Figure II.3.3 shows this only to be true for short digestion times, i.e., when material remains in the digester for less than 15 days (Loll 1976). If digestion is allowed to proceed to completion, the total gas evolved increases by nearly a factor of two between 10° and 30°C. The digestion time required to reach completion decreases with temperature from about 100 days at 10°C to 25 days at 30°C. Loading rates of 4-5 kg/meter³-day are possible (Loll 1976) under optimal conditions, but these figures can also vary widely with substrate; digestion below 15°C is normally too sluggish to be useful.

Increased temperature may cost more in heating energy than in additional methane yielded, but a second factor must also be considered. Decreased retention times require a proportional decrease in digester volume and hence capital investment. Since temperature control also re-

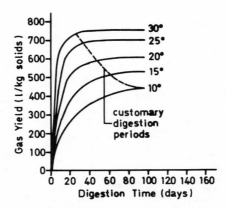

Figure II.3.3. Specific gas yield dependence on digestion time (by permission of Loll 1976).

quires greater capital investment, the optimum must be determined for each installation. Heating from methane or other high-quality fuels is rarely justified, whereas waste, low-grade heat, is abundant if its use is carefully planned.

Thermophilic digestion has several advantages over the traditional mesophilic systems which make it highly attractive (Cooney and Ackerman 1975, Loll 1976):

- the rate of methane production increases 50-75% over mesophilic digestion;
- the extent of waste conversion is more complete, resulting in decreased sludge formation and a lower sludge viscosity;
- pathogenic bacteria are reduced (eight orders of magnitude at 55°C for one hour), and viruses are deactivated (deactivation at 55°C occurs at ca 0.1 min^{-1} or more than two orders of magnitude reduction in an hour);
- loading rates up to 15 kg organic matter/meter^3-day may be possible.

Methane production rates increase for thermophilic digestion partly because digestion rates increase but also because of the increase in the bacterial cell metabolism rate with temperature. The increase in energy for cell maintenance means that more of the consumed substrate is metabolized (converted to methane) rather than converted to bacterial cell biomass. Increased digestion rates allow a decreased retention time and a corresponding decrease in digester volume. Digester volume is, of course, the principal determinant of capital cost.

The extent of conversion, as well as the rate of conversion, increases for thermophilic systems. This increase in extent of conversion corresponds directly to an increase in gas production per unit volatile solid substrate, a decrease in the amount of sludge remaining, and its energy content. The amount of the volatile solids which are converted depends on the digestion rate, retention time, and extent of digestibility—both the first and last of these factors varies with substrate and temperature. In those applications where human wastes are treated, the sterilization at thermophilic temperature is of crucial importance for hygiene and the provision of clean water.

The choice of digester temperature must ultimately be made in terms of all factors: hygiene, economics, energy, or materials and capital required. Rarely can all factors be optimized. Heating and temperature control, for example, require far greater cost, energy, and material inputs in addition to demanding greater skill from the operator. On the other hand, decreased digester volume results in enormous savings and decreased heat losses if the reduced surface area is well insulated. Practical systems should employ only low-grade heat as temperatures of ca 30-60°C are easily achieved with solar heaters or waste heat from other processes. Most commonly, part of the methane produced is used, but use of this high quality fuel is neither energetically nor economically sound and should not be wasted for low temperature heating. Fermentation by methanogenic bacteria is too efficient to depend on the heat of reaction for heating.

Stirring, a second mechanical consideration, is required to promote mixing of reactants, equalization of temperature, degasification, and prevention of both sedimentation and floating scum formation. Too rapid mixing causes a yield reduction presumably due to inhibition by the oxygen entering with each load and the prevention of bacterial cluster formation. In practice some diffusion and convectional stirring does occur, but all but the most primitive digesters should have some form of mechanical stirring even if it is only a hand paddle which is turned a few times per day. More efficient advanced systems require controlled stirring from wind or motor power.

Of the circulation systems shown in Figure II.3.4, reinjection of digester gas has proved to be one of the most effective (Loll 1976). Injected gas bubbles quickly to the surface, causing agitation and the release of small methane bubbles formed on the flocs. In the absence of such turbulence small bubbles adhere to the flocs and are not released until they finally coalesce with larger bubbles. This accumulation prevents adequate metabolite transport and hence digestion rate. Except in the most highly tuned and therefore unstable systems, pH is normally not directly controlled. Acid production (discussed above) is the first step in digestion and forms the substrate for methanogenic bacteria. Any changes in tem-

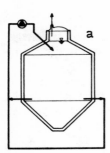 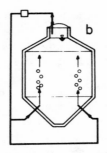 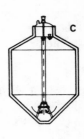

Figure II.3.4. Circulation systems for biodigesters: (a) circulation pump, (b) digester gage injection, and (c) screw-impelled pump (by permission of Loll 1976).

perature, pH, or substrate tend to disturb the methanogenic bacteria more than others, and hence their substrates, the acids, will collect. The resulting increased acidity will further destroy the methanogenic bacteria until all methane evolution stops. Such positive feedback in which a small disturbance can be amplified to cause complete digester failure poses one of the fundamental digester problems. Changes in feed are the most common disturbance, and difficulties rarely occur with a constant feed, for instance with only manure. Alternative and more general solutions include multistage digesters which prevent destruction of all the methanogenic bacteria. Undisturbed stages can be used to reinoculate those which are disturbed. Whether or not a multistage digester is used, early detection of decreasing pH can be counteracted by a cessation in feeding or, if necessary, addition of ammonia. If the digester becomes too alkaline, CO_2 production will increase and automatically reduce the pH. A high technology digester can contain a pH monitor with automatic addition of base to correct any accidents, but such complications are rarely justified.

Anaerobic digesters can be constructed at almost any level of technology. The simplest and most readily available digester is the cow of Figure I.7.1. The cow's rumen automatically heats, stirs and buffers the pH for the anaerobic bacteria, which in turn predigest ingested cellulose. All that is lacking are proper collection techniques to exploit the 0.28 meter^3 (10 ft^3) methane produced each day per cow. A simplified model cow, shown in Figure II.3.5, can be constructed from a rubber balloon with rocks on it for pressurization. Such a design has been constructed from 0.55 mm Hypalon laminated with Neoprone and reinforced with nylon sheet. Sizes range from 6-150 meters^3 and employ batch, continuous, or semicontinuous feeds and cost half the normal designs. If the same design were used with a larger horizontal extent, the digester would be called a *displacement type* and would more resemble an animal's intestines. Digester

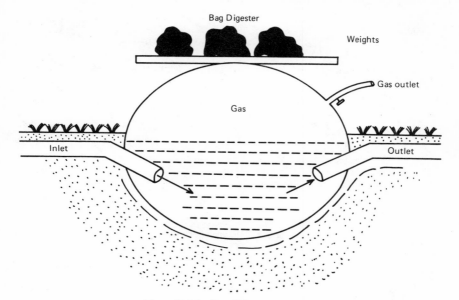

Figure II.3.5. Simple balloon digester.

upsets—due to toxins, oxygen, etc.—in one area can be corrected by reversing the flow and reinoculating the disturbed area. Displacement digesters also assure a definite residence time for all feed material and are easily cleaned.

A promising design of ultimate simplicity has been developed for large-scale algal digestion. This digester is an earthen pond similar to those used for algal growth, but about ten meters deep. Gas collection is provided as shown in Figure II.3.6 where the submerged plastic "tent" funnels the digester gas up to the center collector (Oswald 1976b). A careful choice of materials and design could provide solar heating of the pond to optimal and quite stabile temperatures due to the large mass.

An alternative to the displacement type is the verticle mixing design used by, among others, the Khadi Village Industries in India. Variations of this design are nearly limitless, but the one shown in Figure II.3.7 (Khadi 1975 and ESCAP 1976) demonstrates the principles for a two-stage, cylindrical digester (3 meters round and 4 meters high) with a brick partition separating the two chambers. Both the inlet and outlet are at the bottom to aid stirring and assure even retention. These two-stage digesters are less common than the single-stage, vertical mixing varieties which are easier to build. Single-stage digesters do suffer from several disadvantages:

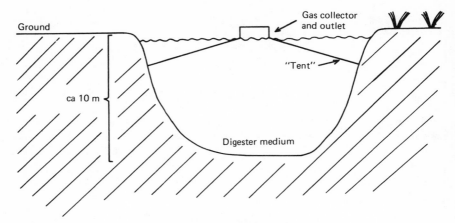

Figure II.3.6. Earthen pond digester.

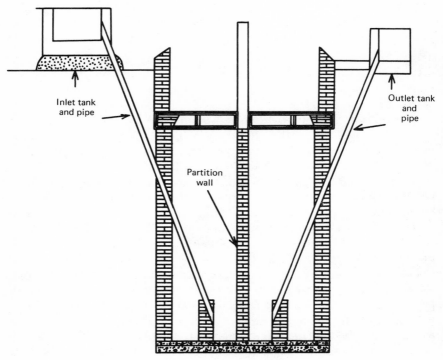

Figure II.3.7. Two-stage vertical digester.

- rapid vertical mixing can prevent proper residence time for some of the feed material;
- digester disruption is more difficult to correct;
- digester cleaning is more difficult;
- oxygen entering with each load can slow digestion.

Each of these problems is more readily avoided in a two-stage digester.

A different concept of the same process is the recently developed "anaerobic upflow packed-bed biological reactor" (ANFLOW) or anaerobic filter (Griffith and Compere 1975). The ANFLOW reactor consists of verticle columns (a 5-meter high by 1.8-meter diameter pilot plant is planned) filled with variously shaped packing materials made from ceramics, stoneware, plastic or aluminum. Anaerobic bacteria and protozoa (manure and sewage sludge) are fed into the *bottom* of the columns and either collect in the packing spaces or are attached to the packing itself using gelatin (cross-linked with glutaraldehyde, ethylchloroformate, or formaldehyde). ANFLOW columns are essentially displacement type digesters and have all the advantages discussed above. In addition, however, the immobilized vertical grid of bacteria provides much more uniform contact between waste and bacteria, shortens liquid retention time for a given waste retention time, eliminates the need for any stirring, and allows the formation of a series of bacterial populations along the column. Clearly, however, very small waste particle size is required, but both suspended and dissolved organic matter will pass. Methane yields are not yet determined, but they are expected to be superior to normal digesters. Whether increased yields or improved digestion justifies the extra economic investment remains to be seen. The geometry is particularly disadvantageous from the heat-loss point of view.

Construction Materials. Anaerobic digesters in industrialized countries are made of stainless steel, plastics, glass, etc. Such construction materials are extremely scarce in developing areas and contribute unnecessarily to costs: stainless steel reactors cost about $10/liter, reinforced concrete costs $3.6/liter, and earth ones cost only $0.07/liter (Oswald 1976b). The popular Khadi Village Industries (India) gobar-gas plant consists of a cement digester and a steel gas holder; of the total $200-400 cost for a 6-meter^3 unit, 50-60% is for the steel-gas holder (Khadi, 1975, and Prasad, Prasad, and Reddy 1974). Not only is such capital unavailable to the normal small farmer (even with favorable Indian Government loans), but steel and cement are sorely needed for competitive uses. Attempts are being made to replace concrete with locally made burnt or sundried bricks, treated wood, plywood, plastic, jute, bamboo, earth embankments, stone, fireclay, pottery, epoxy-painted paper or straw, chipboard, or polyethylene bag, fibreglass, natural or synthetic rubber. Possible replacements for the steel gas collector are polyethylene-lined wooden frame, hardwood barrel, or rubber. In each case cost must be weighed against durability. Similarly the large decrease in digester volume (residence time) with thermophilic systems may, even with the additional complexity of stirring, heating and insulating, provide a net decrease in cost.

Digester Products. Gas phase. Gas evolved from a properly functioning anaerobic digester will contain approximately the mixture of gases shown in Table II.3.6, which in turn represents the products of the several simultaneous fermentation reactions.

The heating value of this gas mixture (alternately called marsh gas, dung gas, gobar gas, and biogas) is typically 24 MJ/meter^3 (650 Btu/ft^3) and decreases linearly with the carbon-dioxide percentage from pure methane's heating value of 37 MJ/meter^3 (1000 Btu/ft^3). Biogas has about half the heating value per volume of natural gas, a third that of propane and 50% more than coal gas. This heating value can be increased by removing the carbon dioxide, but unless the biogas is to be mixed into the main gas grid, compressed or transported long distances, it is neither profitable nor necessary. Recovery of CO_2 for nutrient use in algal ponds might, however, provide an extra incentive. Carbon dioxide and water can be removed by adsorbing the former on monoethanolamine (MEA) and 2% flow; MEA can be reclaimed by heating it in a stripper. Alternatively, carbon dioxide can be removed by bubbling it through lime water which can be recovered by drying and baking the precipitate (Konstrandt 1976).

Depending on feedstock, hydrogen sulfide can be a more difficult problem and in unreasonable quantities can cause an offensive odor, health hazard, and corrosion in combustion chambers or motors. Sulfide in digester gas increases with increased substrate-protein content and is

Table II.3.6 Biogas Composition (New Alchemy 1973).

methane	54–70%
carbon dioxide	27–45%
nitrogen	0.5–3%
hydrogen	1–10%
carbon monoxide	0.1%
oxygen	0.1%
hydrogen sulfide	trace

never a problem when cow dung is digested. It can, if necessary, be adsorbed onto ferric oxide and iron filings where the iron is converted to ferrous sulfide. Fresh air blown through such a scrubber will regenerate it after saturation (Konstandt 1976).

Gas yields per unit digested material are as follows for fats, proteins and hydrocarbons (Konstandt 1976):

hydrocarbons	0.8 liter/gram	50% CH_4 + 50% CO_2
proteins	0.7	70% CH_4 + 30% CO_2
fats	1.2	67% CH_4 + 33% CO_2

Methane gas yields per unit solid *substrate* is of course the more interesting digester parameter, but one must recall that it is a function of the material retention time and digestion rate; the latter is also a function of temperature. A quoted gas yield, therefore, is meaningful only if the temperature, retention time, and material digested are known. Most useful from the energy conversion point of view would be a plot of the gas yield vs temperature when the retention time has been adjusted for maximum yield at that temperature. Not only is such data not available, but the energy and economic costs for digesters and temperature control greatly complicates the optimization. Optimum conditions, which need not be the same energetically and economically, are strongly dependent on ambient conditions and resource base of the country involved. Most important in this connection is the unjustified dismissal of temperature control based on the assumption that *methane* is used to heat the digester. Desired digester temperatures (30-60°C) lie well within the range of low-grade heat and do not require a high quality fuel. Low-grade heat is available as waste heat or from simple solar collectors in carefully integrated systems. Only in such contexts will the full potential of methanogenic systems be realized.

In the absence of adequate yield data, production rates for various substrates digested at 30°C are summarized in Table II.3.7 to at least give an approximation (Imhoff and Thistlethwaite 1971). Accurate temperature corrections are equally uncertain, but for a simple, mesophilic, cow-dung digester at ambient temperatures from 14°C to 30°C, the gas yields vary from 0.16 to 0.44 meter^3/kg dry dung, respectively, (or, 3.8 to 10 MJ/kg dry dung, respectively; Parikh and Parikh 1976). Comparative data at thermophilic temperatures are more sparse, but for a 30-day retention time the yield increases approximately 50%: shredded urban wastes yield 0.46 meter^3/kg volatile solids (57% of total solids) at mesophilic temperatures (37°C) and 0.68 meter^3/kg at thermophilic temperatures (65°C; Cooney and Wise 1975).

Two points are important to notice from the data in Table II.3.7.

Table II.3.7   Gas Yields from Anaerobic Digestion (30°C) of Waste Solids (Imhoff and Thistlethwaite 1971).

Waste	Total Gas Yield (meter^3/kg dry solids)	% Methane	Half-Digestion Period (days)
municipal sewage sludge	0.43	78	8
municipal sewage skimmings	0.57	70	—
municipal garbage only	0.61	62	6
waste paper only	0.23	63	8
municipal refuse (combined, free of ash)	0.28	66	10
abattoirs waste:			
cattle paunch contents	0.47	74	13
intestines	0.09	42	2
cattle blood	0.16	51	2
dairy wastes, sludge	0.98	75	4
yeast wastes, sludge	0.49	85	—
paper wastes, sludge	0.25	60	—
brewery wastes (hops)	0.43	76	2
stable manure (with straw)	0.29	75	19
horse manure	0.40	76	16
cattle manure	0.24	80	20
pig manure	0.26	81	13
wheaten straw	0.35	78	12
potato tops	0.53	75	3
maize tops	0.49	83	5
beet leaves	0.46	85	2
grass	0.50	84	4
broom (1 in. /25 mm/ cuttings)	0.44	76	7
reed (1 in. /25 mm/ cuttings)	0.29	79	18

First, volatile solids in shredded urban waste, as in this example, constitute little more than half the total, and hence care must be taken in how the results are quoted. Secondly, while the total yield is high, the daily gas volume yields per reactor volume (0.21 and 0.31, respectively) are low due to long retention times which in turn increase capital costs per unit of methane produced. This latter fact is partly due to the slow digestibility of the wastes, but an example for feedlot wastes at short retention times represents something of an extreme alternative: 4.5 liters gas per liter of digester per day for a three-day retention time and 8% volatile-solids concentration, 40% of which are digested. This enormous range of digestion parameters depends not only on substrate and operating conditions but also on the desired objective—energy production or waste treatment.

The large number of parameters makes data quoting and digester design extremely difficult. Decisions about temperature and retention times are of course specific to feed material, ambient conditions, availability of capital material and particularly low-grade heat. For simple unheated digesters, however, rough approximations have been given for digester volumes (Golueke and Oswald 1973): 0.28 meter^3 per person, 0.28 meter^3 per cow, and 0.07 meter^3 per chicken. These are for ambient temperatures, whereas shortened retention times for heated digesters will allow proportional decreases in digester volume and may, depending on design, represent considerable capital savings.

Digestion of waste grown algae deserves special mention for reasons discussed in Part II, Chapter 2. Algae has proportionally more fixed nitrogen relative to carbon than is required for optimal anaerobic digestion, but it is uniquely attractive digestion substrate in several other respects. Firstly, algae because it is small and water-borne, requires no mechanically strong and biologically resistant structural material such as ligno-cellulose in wood or crystalline cellulose in plants. Therefore it is not surprising that not only microalgae (such as *Chlorella* and *Scenedesmus*) but also blue-green and other larger algae (e.g., *Microactinum, Euglena, Oscillatoria* and *Spirulina*) have proved to be converted efficiently to methane (Uziel et al. 1975). At 37°C, 60% of dry algae's combustion energy can be converted to biogas at a near optimal loading rate of 2.4 grams volatile solids per fermentor liter per day; the optimum retention time is about 15 days. For a typical heat of combustion of 23 MJ/kg volatile substance (5-6 kcal/gram), this conversion efficiency produces gas yields of 14 MJ/kg volatile solids. Corresponding data for thermophilic algal disgestion are not yet available. Algae's second advantage as a digestion substrate is its constancy and efficiency of production. It can be grown with about 2-3% solar efficiency (total radiation) and provides substrate for a stable, self-sustaining digestion process. Its important potential was discussed in Part II, Chapter 2.

Sludge phase. Following digestion, all nonvolatile products, undigested feed material, and bacteria will settle to the bottom of the digester unless stirring keeps them in suspension with other soluble components, such as ammonia. The addition of new feed material (most systems are semi-continuous digesters), of course, will require the removal of an equal volume of sludge suspension. Some of this sludge's components, such as undigested cellulose, will provide little nutrient value, but is potentially important as a soil conditioner. Generally sludge is excellent both as fertilizer and as a soil conditioner. Not only are no nutrients lost during digestion, but the availability of fixed nitrogen in an assimilable form increases from ¾% for manure to 2% in the digester sludge. Ammonia is particularly important, but the list of nutrients includes potassium and phosphorus, plus trace elements such as boron, calcium, copper, iron, magnesium, molybdenum, sulfur, zinc, etc.

Sludge is the key to all organic-waste recycling schemes, but in an urban context sludge poses a "waste"-disposal problem. Anaerobic digestion has often formed the first stage in urban sewage treatment with the effluent being further treated for nutrient removal. Without sufficient secondary and tertiary treatment, however, the nutrient-rich waters cause a serious pollution problem. Release of effluent to natural lakes and waterways causes excessive algal and plant growth. This folly is the curse of careless wealth and indicates the diseconomy of waste treatment which is not coupled to agricultural production.

Sludge does present certain constraints to its use due to the large bulk of water, but these are easily met on the local level. The three major uses for this dilute, constant flow of nutrients are:

1. Sludge for normal farmland fertilizer (Khadi 1975, Prasad, Prasad, and Reddy 1974, and New Alchemy 1973). In this case the sludge (dilute and water-borne) is spread on the fields by hand, spray or in conjunction with irrigation systems. Such cycles re-use all nutrients (essential in closed or semi-closed systems), and in addition the humus doubles as a soil conditioner to increase its water-holding capacity and improve soil structure. Holding tanks (perhaps doubling as algal ponds) may also be required to have sludge availability and irrigation and fertilization needs coincide. Some chemical conditioners are still required as evidence indicates that continued sludge-use causes soil acidity even though the sludge should be at near neutral pH.

2. Hydroponic plant growth, i.e., the cultivating of plants in a contained nutrient solution instead of soil, can be continuously fed with digester sludge. The high-nutrient transport rates result in high growth rates especially for water hyacinth, *Ipomoea repens,* and grasses, like rye, fescue and canary grass. Fodder crops would be most suitable to such a system.

3. Algal ponds are a variation on the hydroponic scheme and were discussed in Part II, Chapter 2. Such systems are in operation for sewage-water treatment, but in these cases aerobic decomposition and algal growth are combined without previous removal of the waste's stored energy. Harvesting of algae for feed, fodder, or for anaerobic digestion are applications under development in many areas.

Bacterial Conversion of Organic Wastes to Hydrogen

Hydrogen (H_2) is an energy-rich gaseous fuel whose production by microorganisms is complicated by its suitability as a *substrate* and as the only biosphere reductant which can reduce NAD(P). Hydrogen-metabolizing bacteria, the largest group of chemoautotrophs, are invariably facultative which when grown aerobically require only H_2, CO_2 and O_2, roughly in the ratio 6:2:1. Production of cell material requires only the fixation of CO_2 with protons from hydrogen gas activated by hydrogenase and energy from the oxidation of hydrogen (Doelle 1975):

$$CO_2 + H^+ \rightarrow (CH_2O) + H_2O \qquad\qquad \Delta G_0 = +8.2 \text{ kcal}$$
$$H_2 + O_2 \rightarrow H_2O \qquad\qquad \Delta G_0 = -56.5 \text{ kcal}$$

$$CO_2 + 2H_2 + O_2 \rightarrow (CH_2O) + 2H_2O \qquad\qquad \Delta G_0 = -48.3 \text{ kcal}$$

Hydrogen also provides both energy and reducing power for anaerobic fermentation, as with the thermophile *Methanobacterium thermoautotrophicum*. The absence of oxygen prevents the complete oxidation, and energy-rich methane gas is a metabolic product:

$$4H_2 + CO_2 \rightarrow 2H_2O + CH_4 \qquad\qquad \Delta G_0 = -33 \text{ kcal}$$

Hydrogen *production* via anaerobic fermentations occurs naturally by many bacteria, the most important of which are listed in Table II.3.8. Hydrogen evolution by red, green and blue-green (cyanobacteria) algae has already been discussed in Part II, Chapter 2. The difference between photosynthetic and fermentative hydrogen production is that in the former solar energy is converted to hydrogen by one and the same organism—

Table II.3.8   Some Hydrogen-Producing Microorganisms.

Microorganism	Energy Source	Fermentation Balance
Strict Anaerobes:	sugar fermentation	
Clostridium butyricum		4 glucose → 2 acetate + 3 butyrate + $8CO_2$ + $8H_2$
C. pasteurianum		
C. welchii		
C. kluyveri		2 ethanol → butyrate + $2H_2$
Micrococcus lactilyticus		
Desulfovibrio vulgaris		
Facultative Anaerobes:	sugar fermentation	
Escherichia coli		2 glucose → 2 lactate + acetate + ethanol + $2CO_2$ + $2H_2$
Aeromonas hydrophilia		formic acid → CO_2 + $2H_2$
Bacillus macerans		
Photosynthetic Organisms:	light, organic compounds	
Rhodospirillum rubrum		acetate + $2H_2O$ → $2CO_2$ + $4H_2$

either directly under unnatural conditions or indirectly as in heterocystous blue-green algae. Fermentative hydrogen production, however, results from the metabolism of organic substrates produced by other photosynthetic organisms.

An intermediate system, photosynthetic (purple) bacteria, requires both light *and* organic substrates. All of these natural hydrogen-producing systems are listed below, but only the latter two will be discussed here.

Light:

red and green algae $\qquad H_2O \rightarrow$ photosynthetic system $\rightarrow H_2$

blue-green algae (Cyano bacteria) $\quad CO_2 + H_2O \rightarrow$ photosynthetic system $\rightarrow$ organic substrates $\rightarrow H_2$

purple bacteria $\quad$ organic substrates or thiosulfate $\rightarrow$ photosynthetic system $\rightarrow H_2$

Dark:

bacteria $\qquad$ organic substrates $\rightarrow H_2$

Dark Fermentation. Fermentative hydrogen production can convert up to 33% of simple organic compounds' (e.g., glucose or ethanol) combustion energy to hydrogen (Thauer 1976). This relatively low conversion rate (compared to ca 85% for conversion to methane) results from the absence of the following theoretically possible reaction:

$$\text{glucose} + 6H_2O \rightarrow 6CO_2 + 12H_2 \qquad \Delta G_0 = -6.18 \text{ kcal/mole} \\ (25.8 \text{ MJ/mole})$$

This 99% efficient reaction is not naturally catalyzed precisely because it is *too* efficient: not enough energy is released per reaction to synthetize one molecule of ATP, the energy form required by the organism. Furthermore, the existing glucose-metabolism pathways (e.g., the citric acid cycle) thermodynamically cannot convert glucose to CO_2 and H_2 as these dehydrogenation reactions cannot be coupled to proton reduction.

The fermentations which do occur leave an organic molecule, e.g., acetate, in addition to hydrogen and carbon dioxide:

$$\text{glucose} + 2H_2O \rightarrow 2 \text{ acetate} + 2H^+ + 2CO_2 + 4H_2 \\ \Delta G_0 = -51.2 \text{ kcal/mole.}$$

Four molecules of hydrogen gas per glucose is the maximum observed (33% efficiency), but a pressure of over one atm hydrogen reduces the gas production efficiency to 20%. Maximum production requires a partial pressure around 10^{-3} atm.

A practical hydrogen-producing system would require not only constant removal of the gas, but an initial degradation step to produce glucose from cellulose and other organic wastes. The three alternatives for this initial stage were discussed earlier—microbial, enzymatic, or acid hydrolysis.

Photosynthetic Fermentation. Green and purple photosynthetic bacteria offer a more effective system to evolve hydrogen gas by solving two problems (Pfennig 1967 and Stanier, Dondoroff, and Adelberg 1972):

1. Energy stored in the hydrogen gas is provided in large part by the single photosynthetic system (for photophosphorylation or ATP synthesis) and only partly from the fermented substrate.
2. Oxygen inhibition of the hydrogen production system (hydrogenase) does not occur because not only do these bacteria require strictly anaerobic conditions for photosynthesis, but since water is *not* the electron donor for photosynthesis, oxygen is not evolved. Hydrogen production in these organisms results from the reduction of protons by electrons excited in a single photosystem and donated by organic or inorganic substrates.

The inability to use water as the electron donor, as is standard for all other photosynthetic organisms, gives these bacteria a very severe growth limitation, but ecological niches do exist, typically beneath an algal culture in fresh or marine water. There, little oxygen exists, and the uniquely low-energy (red and infrared) light transmitted by the algae is used by the unusual bacterial pigment which absorbs where other pigments do not, as shown in Figure II.3.8. Heavily polluted water is a particularly suitable environment as bacterial action depletes the oxygen tension and suitable electron donors are present. These donors, required because water cannot be used, are the various inorganic (H_2S, thiosulfate, or H_2) or organic compounds, including fermentable organic wastes. Use of such electron donors for the single photosystem is part of the evidence which indicates an early (preoxygen) evolutionary appearance for these organisms. Carbon dioxide can be fixed or released as an oxidation product depending on whether the organic substrate is respectively more or less reduced than the cell material. The CO_2 pathway is summarized as:

$$\text{organic substrate} \pm CO_2 \xrightarrow{\text{light}} \text{organic cell material}.$$

Production of hydrogen gas (i.e., proton photoreduction) by photosynthetic bacteria requires an abundance of substrate and is inhibited by

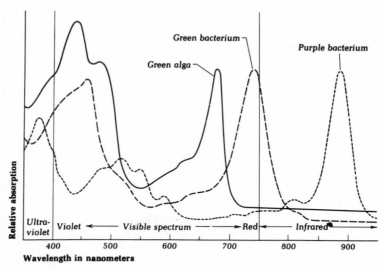

Figure II.3.8. Absorption spectra of green and purple bacteria and green algae (by permission of Stanier et al. 1972).

nitrogen gas and ammonium ions. The latter facts, together with the inability of a nitrogenase deficient (nif⁻) *Rhodopseudomonas capsulata* to evolve hydrogen, indicate that although hydrogenase is present it is *not* the enzyme system involved in hydrogen production. Hydrogenase in these organisms is apparently only used for hydrogen uptake as an electron and energy source in which case dark CO_2 fixation is also possible. Use of the nitrogenase system for hydrogen production is disadvantageous not only for its nitrogen, ammonium and oxygen inhibition, but primarily because energy (ATP) is consumed unnecessarily; hydrogenase catalyzing the same reaction requires no ATP. The quantity of ATP consumed by the nitrogenase system and the biochemical explanation for this requirement are not certain, but estimates range from 6 to 15 ATP per hydrogen molecule evolved. This ATP is the prime use of the photosynthetically stored energy.

One of the most promising organisms, *Rhodospirillum rubrum,* is a nonsulfur, purple photosynthetic bacterium which evolves hydrogen gas via the following light-dependent reactions (Gest, Omerod, and Omerod 1962):

$$C_2H_4O_2 + 2H_2O \rightarrow 2CO_2 + 4H_2 \qquad \text{acetate}$$
$$C_4H_6O_4 + 4H_2O \rightarrow 4CO_2 + 7H_2 \qquad \text{succinate}$$
$$C_4H_4O_4 + 4H_2O \rightarrow 4CO_2 + 6H_2 \qquad \text{fumarate}$$
$$C_4H_6O_5 + 3H_2O \rightarrow 4CO_2 + 6H_2 \qquad \text{malate}$$

For photosynthetic organisms metabolizing energy-rich organic compounds the efficiency, greatest for resting cells, is 73% for fumarate (typical for other substrates) if one does *not* include the solar input. These organisms are efficient for converting energy in organic chemicals to a more useful form, hydrogen gas, but are not useful for energy storage; energy storage occurs in the photosynthetic organism which first produced the organic matter. Quantum yields are not accurately known for these reactions.

Photosynthetic bacteria can be used in a variety of simple transparent fermentors, but a recent alternative uses *R. rubrum* immobilized on a 5% agar solution (Weetall and Bennett 1976). Malate (0.01M) is passed through the reactor, and hydrogen gas is collected in a trap. A laboratory-scale reactor with four grams of cells yielded 10^5 μliters in its 150 hours of operation; the reactor half-life is 115 hours, and costs are approximately two orders of magnitude too high, but the development of such techniques may solve long-term stability problems.

Chromatium is typical of the sulfur-metabolizing bacteria whose oxidation of thiosulfate ($S_2O_3^{2-}$) is accomplished in the light: thiosulfate is the electron donor for chlorophyll whose light-excited electron serves in turn to reduce a proton to hydrogen gas via hydrogenase (Doelle 1975). Sulfur compounds exist as wastes from certain industrial processes, such as paper-pulp wastes, and also occur naturally in some sulfur-rich lakes and springs where hydrogen gas is then produced.

A practical system for hydrogen production requires a constant source of digestible organic compounds. Large quantities are available in the form of cellulose wastes if the cellulose can first be metabolized to the required organic acids. This problem is a variation of the cellulose-degradation problem for sugar production. If organic-acid intermediates are desired rather than sugar, a fungus, such as *Rhizopus arrhizus,* can be employed which aerobically converts carbohydrates to fumaric acid. An anaerobic cellulytic bacteria, e.g., *Bacteriodes succinogenes,* is a better choice as it can be combined with the anaerobic hydrogen-evolving bacteria in a single-reaction vessel. Either of these organisms may serve in a cellulose-waste-to-hydrogen gas converter (Morowitz 1974):

carbohydrate $\xrightarrow{\text{Rhizopus arrhizus}}$ fumaric acid $\xrightarrow{\text{Rhodospirillum rubrum}}$ hydrogen

(100 g., 1.5 MJ)   (70 g.)   (7 g., 1 MJ)

cellulose $\xrightarrow{\text{Bacteriodes succinogenes}}$ succinic acid $\xrightarrow[\text{light}]{\text{Rhodospirillum rubrum}}$ hydrogen

(100 g., 1.5 MJ)   (80 g.)   (9.5 g., 1.4 MJ)

Overall conversion efficiencies can be expected to be about 50%, but will of course depend upon the degradability of the cellulose waste. As with other cellulose processes, bagasse, corn stalks, scrub wood, and other cellulose/starch-containing wastes are suitable substrates. Process costs need to be $0.02/kg to be competitive with the substrate costing about the same (Morowitz 1974).

ENZYMATIC HYDROLYSIS OF CELLULOSE

Cellulase Enzymology

Cellulose (see Part II, Chapter 1, especially Figures II.1.10 and II.1.11), depending upon its crystallinity and degree of association with lignin and hemi-cellulose, can be hydrolyzed to its constituent glucose monomers by a special multienzyme system, cellulase. Each of the many cellulase components serve to disrupt cellulose crystallinity and/or to hydrolyze the $\beta(1\text{-}4)$ glucan linkages between glucose molecules. The component enzymes (Wood 1975) are traditionally grouped functionally into the following classes, although the labels are tentative at best due to the high degree of cooperativity:

1. Endo-β1,4 glucanases (sometimes called C_x) attack $\beta(1\text{-}4)$ glucan linkages randomly along cellulose chains which are soluble, swollen, or partly degraded; crystalline cellulose is not attacked. This class contains numerous enzymes with varying activities relative to different cellulose substrates. Cellobiose and cellotriose, not glucose, are the major products. Five endoglucanases have been isolated from *Sporotrichum pulverulentum* (Almin, Eriksson, and Petterson 1975) with weights varying from 28,000 to 37,000, activities (A_m, bonds broken per molecule per second) from 50 to 710, and K_m's (grams/liter) from two to nine.
2. Exo-β1,4 glucanases encompass two enzyme subgroups:
   a. β1,4 glucan glucohydrolases which remove glucose monomers from the nonreducing cellulose chain end, and
   b. β1,4 glucan cellobiohydrolase (CBH may correspond to "C_1" cellulases) removes cellobiose (glucose dimer) from the nonreducing end of a cellulose chain. CBH has the greatest affinity for cellulose and is essential for degrading crystalline cellulose.
3. β-glucosidase (cellobiase) hydrolyzes cellobiose, cellutriose, and very short-chain cellulose, but its activity decreases rapidly with increasing degree of polymerization.
4. An enzyme oxidizing a glucose hydroxyl group (number 6 position)

to a carboxylic group, thereby breaking a hydrogen bond and promoting swelling.

5. Cellobiose:quinone oxidoreductase reduces quinones or phenoxy radicals in the presence of cellobiose, the latter then being oxidized to cellobiono-δ-lactone and on to cellobionic acid via lactase; the quinones are produced by phenol oxidases during lignin degradation (Westermark and Eriksson 1974).

Some evidence (Wood 1975) suggests the above grouping is artificial and not demanded by the enzyme system. Cellulases have been difficult to fractionate thus clouding the classification problem and delaying the realization that cellulase is a well-integrated synergistic system: Cell-free cellulase fractionated into three β1,4-glucanases, two β-glucosidases, and a β1,4 glucan cellobiohydrolase (or "C_1"), had upon pair-wise recombinations of these three major groups' activities less than 5% of the unfractionated or totally recombined mixture. All components acting in concert are apparently required for effective hydrolytic activity. It can be further shown that each component is limiting with a clear activity maximum for the proper proportions. Activities ascribed in the above categories may result from inadequate separations except for those data based on soluble or very short-chain cellulose when synergism is not observed.

Cellulase Temperature Effects.   All cellulase components are remarkably heat stable: no activity is lost when incubated at 37°C, pH 5.0 for twelve weeks (Wood 1975). At 60°C the endoglucanases are far more stable than glucosidase or the "C_1" cellulases (cellobiohydrolase), the latter two losing 80% of their activity after four hours, while endoglucanase loses 10%. Some evidence (Lee and Blackburn 1975) suggests that a thermophilic cellulolytic *Clostridium* (unidentified "M7" strain) can take advantage of the increase in degradation rate with temperature (activity maximum occurs for pH 6.5, 67°C), but as incubation was for only two hours, long-term stability and degradation of crystalline cellulose were not measured.

Molecular Weights.   Molecular weights vary widely (5300 to 400,000; Wood 1975) depending on the source and component with the glycosidases generally having higher molecular weights than endoglycanases. At least a part of this variation can result from carbohydrate associated with the enzyme which in some cases reaches 50%.

Lignin-Degrading Enzymes.   Most of the world's cellulose is associated with lignin which, like cellulose, is naturally degraded by microorganisms but far more slowly. Among the more effective (known) lignin-degrading organisms are the white-rot type fungi, so named because of their ability to metabolize lignin preferentially over cellulose which is white. One mycelia-forming species, *Sporotrichum pulverulentum,* is used as an enzyme (Eriksson and Petersson 1975) and protein source (von Hofsten 1975), while the mushroom *Pleurotus ostreatus* (oyster mushroom) metabolizes 98% of the lignin in rice to produce delicate human food of high quality (Kurtzman, Jr. 1976). Little is known about the enzymes involved, but lignin degradation must follow a mechanism very different from that of cellulose as there are few hydrolyzable bonds among the polymerized phenyl-propane units. Like cellulose, however, lignin is highly dispersed, and the two are commonly interdigitating and perhaps even bonding. Like cellulase, the lignin-degrading enzymes must also be extracellular but perhaps not dissociated from the fungal hyphae as coenzymes are probably required.

Whatever the enzymology involved, the end products of lignin degradation are CO_2 and H_2O. Intermediate steps include the oxidation of α-carbon side-chain atoms (sometimes with the elimination of two carbons) and the cleavage of aromatic rings, the latter being possible only after two hydroxyl groups are placed ortho or para to each other (Kirk 1975). Such preparation of the aromatic ring for cleavage requires hydroxylation or demethylation, each in turn requiring the coenzyme NADH or NADPH for activity; as no truly extracellular (coenzyme) requiring enzymes are known, the inaccessibility of the lignin greatly complicates degradation. In addition, after cleavage, an army of enzymes is still required for complete degradation: lactonizations, delactonizations, isomerizations, hydrolyses, transfers, dehydrogenations, and decarboxylations.

Virtually nothing is known of the systems involved in lignin degradation, but the complexity of this coenzyme-requiring multienzyme complex—suggests that intact organisms will be required for digesting lignocellulosics.

Although lignin degradation is generally coupled with cellulase activity, both being required for wood degradation, an interesting mutant of *Sporotrichum pulverulentum* has been found (Eriksson 1976) which degrades lignin, but lacks cellulase. Successful application of this strain would produce lignin-free long chain cellulose from wood, useful for fodder or paper.

Cellulase Sources

Of the known microorganisms producing cellulase, *Trichoderma viride* (a mold) best degrades insoluble cellulose, but others include *T.*

lignorum, T. koningii, Chrysosporium lignorum, C. pruinosum, Penicillium funiculosum, P. iriensis and *Fusarium solani* (Sternberg 1976). *Sporotrichum pulverulentum,* a white-rot fungus, may prove superior for lignocellulose.

Cellulase effectiveness depends not only on the quantity and specific activities of the constituent components, but on the proper proportion of cellulase types necessary for significant degradation of the cellulolytic substrate involved. Some types have a superior initial rate of degradation for the reactive amorphous cellulose portion, but lack the ability to attack ordered cellulose. The extent of crystalline cellulose destruction varies by orders of magnitude with species: *T. viride* successfully saccharifies 63% of ball-milled paper (5% solids, pH 4.8, 50°C) in 48 hours with all but 7% of the remainder attributable to lignin (Mandels 1975).

Not only the cellulase-producing species but also the substrate determines the quantity and cellulase types synthesized, cellulase being an induced enzyme. Normally spores are inoculated in a cellulose medium, e.g., shredded and ball-milled waste paper, with cellulase appearing after four days and synthesis ceasing after seven days. The glucose produced, however, represses cellulase formation and can inhibit existing cellulase, although it does stimulate cell growth; 1% glucose is sufficient to halt cellulase synthesis.

The presence of a cellulase-synthesis regulatory mechanism invites the search for derepressed mutants. The *T. viride* used is, however, already probably partly derepressed for cellulase synthesis as it produces three to four times more enzyme than the wild type. Increases of 10-20 times have been induced for organisms hyperproducing other extracellular enzymes, e.g., amylase and invertase from *Neurospora*. However, two differences exist for cellulase:

1. cellulase is an inducible enzyme, requiring the presence of substrate for synthesis;
2. total cellulase-protein weight to achieve a given reaction rate is far greater than, e.g., for invertase. *T. viride* cellulase hyperproducers secrete 1.6 mg protein/ml (Sternberg 1976), whereas substrate cellulose is only 7.5 mg/ml, some substrate being required for respiration and cell growth.

This can be compared with the 0.15 mg protein/ml from the *Neurospora* hyperproducer discussed above. Mutant selection might, however, produce constitutive cellulase producers, independent of cellulose or glucose concentration, and mutants which do not excrete unnecessary enzymes, e.g., amylases, mannanase, etc.

Physical and Chemical Constraints to Cellulose Hydrolysis

The well-integrated complexity of cellulosic materials provide one of Nature's most degradation-resistant substances. Physical proximity of the hydrolytic enzyme and substrate is first among the prerequisites for effective degradation, a diffusional process frustrated by the physical close-packing and large enzyme size (equivalent size: 60 Å diam for a sphere or 30 × 200 Å for an ellipsoid). Physical accessibility of the enzyme to the cellulose is determined by several factors (Cowling and Kirk 1976):

1. Moisture content below a certain threshold (25% for wood and 10% for cotton) prevents degradation by microorganisms. Water is chemically required for hydrolytic cleavage of the glycosidic link, improves accessibility by swelling the cellulose and is required for diffusion of the water-soluble enzyme complex.

2. The relative sizes of the enzyme and the pits or pores in the substrate's structure determine normal (untreated) accessibility. Two types of capillary voids exist in cellulosic fibers:
   a. gross capillaries (e.g., cell lumina) and pit openings ranging from 200 Å to more than 10 microns, and
   b. cell-wall capillaries which absorbed water can be forced open and expand up to 200 Å diam—typically the median diameter is 5 Å for water-swollen wood and 10 Å for water-swollen cotton.

   The latter comprise three orders of magnitude more surface area than the gross capillaries ($2 \times 10^3 cm^2$/gram raw cellulose vs $3 \times 10^6 cm^2$/gram). The cellulase molecule is, as mentioned above, typically 60 Å diam and therefore is excluded from the cell-wall capillaries.

3. Degree of crystallinity greatly affects the speed of degradation; amorphous regions, rare in elementary fibrils, are quickly hydrolyzed.

4. The degree of polymerization varies from 15 to 14,000 glucose units per cellulose molecule, but polymerization plays a role predominantly via the decreased order of short chains; end-wise cleaving enzymes apparently are of less significance than random cleaving ones.

5. Noncellulose materials associated with cellulose fibers can impede degradation by:
   a. preventing physical access, e.g., lignin which penetrates or binds (perhaps) with cellulose;
   b. poisoning the cellulolytic organism, e.g., toxic phenolic compounds, mercury, silver, copper, chromium and zinc salts, or
   c. specifically inhibiting enzymes.

6. Degradation rates can be promoted by minerals (e.g., manganese, cobalt, magnesium, and calcium in the presence of phosphate), vitamins (particularly thiamin), or soluble substrate carbohydrate which stimulate microorganism growth.
7. Chemical substitution of the hydrogen from primary and secondary hydroxyl groups with more reactive (e.g., methyl, ethyl, hydroxy-ethyl and carboxymethyl) groups increases water solubility, decreases crystallinity and thus promotes hydrolysis.

The Process of Enzymatic Conversion of Cellulose to Glucose

The technical processes involved in enzymatically converting cellulose to glucose or other end products isolate and optimize each of the processes which naturally occur within one organism. In the synthetic process the end product is altered according to need, but the individual stages remain the same (Spano 1976):

- collection and separation of raw-cellulose material;
- pretreatment;
- enzyme preparation;
- digestion in the reactor;
- separation of glucose from cellulase and nondigested cellulose, the latter two for recycling;
- purification of glucose for ultimate use (if necessary);
- conversion to final product.

Figure II.3.9 summarizes the overall conversion process, the individual steps of which will now be discussed further.

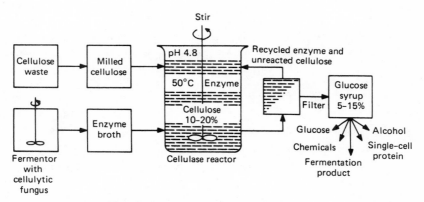

Figure II.3.9. Enzymatic cellulose hydrolysis.

Separation.   Separation is a nontrivial step particularly when discussing urban wastes. Agricultural wastes are a much simpler matter although one might consider separate pretreatments of, for instance, wood wastes, cotton, straw, etc., according to their size and resistance to digestion; separation for these materials can also occur after one pass through the digester. Urban wastes pose a complex problem, requiring not only mechanical separation according to size or metal content, but perhaps also manual selection. Many chemicals common to refuse, especially industrial wastes, can completely inhibit the digestion process.

Pretreatment.   The rate of enzymatic conversion depends strongly upon disruption of the crystalline cellulose and lignocellulose associations described above. Each alternative pretreatment process has the same intent: to increase the physical availability (solubility) of glucan linkages which are easily digested in amorphous cellulose. Disruption is possible chemically or physically.

Chemical pretreatment.   Sodium hydroxide solutions of mercerizing strengths (>20%) cause swelling in crystalline cellulose, perhaps by the saponification of intermolecular ester bonds. Microbial hydrolysis, as shown in Figure II.3.10 for poplar wood and wheat straw, increases linearly with sodium hydroxide concentration up to about a 10% solution—straw always being about 30% more digestible than wood (Millett et al. 1976). For spruce wood, cellulose digestion increased from a few percent without treatment to 80% carbohydrate conversion in cold 2 N sodium hydroxide. The increase due to hydroxide treatment varies greatly with wood species and age, both factors being correlated with lignin content: treatment of silver maple (18% lignin) increased its digestibility from 20 to 41%, while red oak (24% lignin) increased only from 3 to 14% (Millett, Baker, and Satter 1976).

As a 2% residual sodium hydroxide content is tolerated by sheep, extraction of the sodium hydroxide may not be necessary if the treated cellulose is mixed with corn silage or alfalfa hay and used directly as fodder. Ruminant digestion of treated straw (1.5% NaOH for 24 hours) increases by about 25% making it nutritionally equivalent to medium quality hay. For bagasse and rice straw, boiling for three hours in 1% sodium hydroxide is adequate.

- Ammonia (aqueous or gaseous) treatment of lignocellulose, used since 1905, causes swelling by a mechanism similar to that of sodium hydroxide: ammonolysis of the intermolecular glucuronic ester crosslinks. The response is again dependent on species, especially the lignin content; aspen shows the strongest response to ammonia

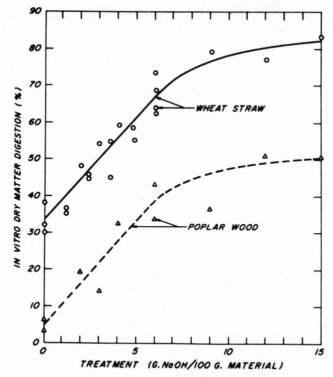

Figure II.3.10. Effect of NaOH pretreatment on cellulose digestibility (by permission of Wilson and Pigden 1964).

treatment (50% digestibility), while rice straw digestibility increased from 29% to 57%.

Treatment generally employs ammonia at room temperature, e.g., soaking in a 5% solution for 30 days (straw) or gas treatment (70 pounds per square inch) for two hours (aspen sawdust). These treatments have the additional advantage of increasing the nutritionally important ammonia content if it is to be used directly for fodder.

- Peracetic acid (acetic anhydride + 35% hydrogen perioxide) is effective for delignifying cornstalks and some sawdusts.
- Sodium chloride in acetic acid particularly aids lignin solubilization; alternatively gaseous CO_2 can be used as the active agent of the process but at a chemical cost of $200/ton.
- Ammonium bisulfate, the conventional pulping procedure, can also produce fodder from wood (e.g., Douglas fir) comparable to medium-quality hay.

- Sulfur-dioxide treatment (120°C, 30 pounds per square inch, for two to three hours) increased cellulase wood digestion from 0-9% for nontreated woods to 50-60% for treated wood as usual hardwoods are superior and after treatment are comparable to high-quality hay. Eight-week feeding trials revealed no palatability problems.
- Steaming (e.g., 165°C at 100 pounds per square inch for two hours) of wood (e.g., aspen) chips delignifies them at a cost of $5/ton (4000 tons/yr) providing a fodder acceptable at 60% of the total ration. Hardwoods again demonstrate a superior response.

Physical pretreatments.   Susceptibility to acid, bacterial or enzymatic hydrolysis increases rapidly with decreasing cellulose particle size, due to the increased surface areas. Physical methods are particularly attractive if the product is to serve any biological role, such as food or fodder. Which method is most satisfactory will, of course, depend on the starting material, but at present the following three methods, while energy-intensive, are the best: vibratory ball milling, pot milling, and hydropulping.

The above methods are all comparable in cost to each other and to alkaline treatment, but based on the percent of saccharification (cellulose conversion to sugar) ball milling is the most successful. Cellulose-source species variations still exist, but for all mechanical pretreatments, a plateau in the ultimate extent of saccharification is reached after about 20 minutes of treatment, a level which ranges from 20% to 80%. Typical particle sizes are 50μm with treatment costs (see Table II.3.9, after Nystrom 1975) per kg cellulose averaging about 10¢ or 20¢/kg glucose for 50% conversion. For completeness the following as yet noncompetitive methods are included:

- Gamma irradiation has an optimal hydrolysis enhancement effect at 10^8 rad giving 78% digestibility for aspen (but 14% for spruce). The radiation decreases the degree of polymerization to 40 but at a prohibitive cost of more than $100/ton.

Table II.3.9   Ball-Milling Costs (Nystrom 1975).

Mesh Size	Micron	kg/hp-hr	Power Cost $/Ton	Maintenance $/Ton	Overhead Cost $/Ton	Total Cost ¢/kg
40	420	7.3	<$2.00	$1.40	$ .20	0.4●
80	117	2.3	4.00	4.40	.40	0.9
100	149	1.8	5.00	6.50	.50	1.3
200	74	0.45	20.00	24.00	1.00	5.0
270	53	0.25	36.45	45.00	1.40	9.0

- High-intensity ultraviolet light in the presence of sodium nitrate as sensitizer.
- High temperature (200°C for 32 hours) gives moderate improvements in hydrolysis.
- Low temperature (−75°C) reduces polymerization especially after repeated treatment.
- High pressures (8000 kg/cm² at room temperature for a half-hour).

Enzyme Preparation. The most common cellulase source is the *Trichoderma viride* QM 9123 or QM 9414 mutant strains grown on mineral salts, 1% cellulose, 0.075% proctose peptone, and 0.0-0.2% Tween 80 (Mandels and Weber 1969). As cellulose is an induced enzyme system with several synergistically related components, the cellulose substrate used for enzyme preparation is varied according to the waste cellulose to be digested. The medium held at 28°C, buffered at pH 4.8, stirred and aerated, is inoculated with a spore suspension and harvested after 10-13 days by glass-wool filtration (Mandels, Hontz, and Nystrom 1974). Acetone precipitation provides more concentrated enzyme solutions. While this single-stage process is the more common, a two-stage enzyme production system has been described (Mitra and Wilke 1975). Industrial-grade cellulases grown on wheat bran are also available.

Cellulase Reactor. The rate and extent of cellulose conversion to glucose will depend on a number of variables whose optima are only tentatively determined:

- starting material and concentration;
- lignin and noncellulose content;
- cellulase activity, concentration and relative composition;
- digester conditions, e.g., agitation speed, temperature and pH;
- glucose (product) concentration; and
- residence time.

Several of these variables are summarized in Figures II.3.11 and II.3.12 which show the percent saccharification and glucose concentration vs reaction time for pulped paper at varying substrate and enzyme concentrations ([S] and [E], respectively). Two days are typically sufficient to reach near maximum conversion, but this time decreases somewhat for increased enzyme concentration (Spano 1976). Note that a large cellulose concentration results in a low degree of decomposition but a high reduced sugar concentration. Since glucose and cellobiose are inhibitory, maximum glucose concentrations are 10-15%.

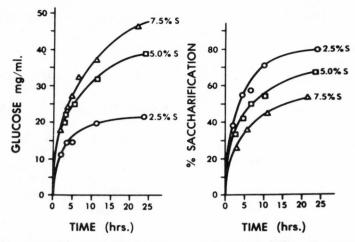

Figure II.3.11. Glucose concentrations and percent saccharification from enzyme hydrolysis as a function of time (by permission of Spano, Madeiros, and Mandels 1976).

Separation, Purification and Final Use Stages. (Wilke and Mitra 1975, and Wilke, Yang, and von Stockar 1976) Enzymatic digestion of cellulose is typically 33-77% complete and results in a digestate containing 1.6-4.6% glucose, nondisintegrated cellulose and cellulase enzyme. The latter is partly adsorbed on spent solids and partly in solution. Separa-

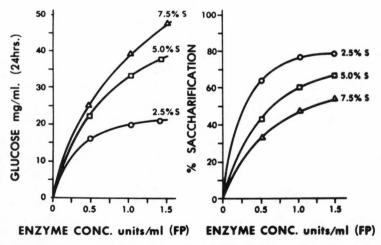

Figure II.3.12. Glucose concentration and percent saccharification from enzyme hydrolysis as a function of enzyme concentration (by permission of Spano, Madeiros, and Mandels 1976).

tion of the three components is expedited by the fact that the enzyme is reversibly adsorbed onto cellulose although not equally for the various cellulase components. This observation allows the separation of liquid and solid effluent phases by a vacuum-driven filter. The liquid phase contains the ca 4% sugar and solubilized enzyme, and the two are separated by passing the liquid through a multistaged adsorption train. Cellulase is adsorbed onto fresh cellulose, while the sugars remain in solution. The cellulose is then returned to the digester, while the sugar solution is ready for use in processing.

Solids from the initial filtering are passed through a series of slightly acidic (pH 4.8-6.0) counter-current mixer filters as shown in Figure II.3.13. There the enzyme is washed free from the spent solids for recycling to the digester, while the solids can be burned, used for further digestion processes or perhaps upgraded to ruminant fodder.

The scheme shown in Figure II.3.13 recovers 35% of the cellulase—less than half the initial predictions—and produces a 6% sugar solution

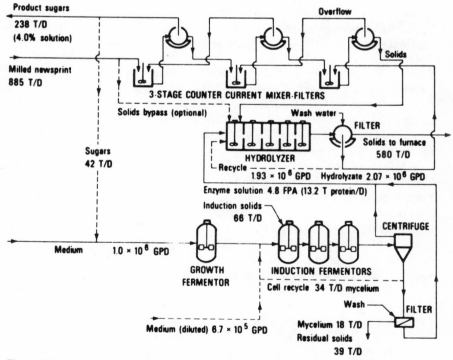

Figure II.3.13. Flow diagram of enzyme hydrolysis process (by permission of Wilke et al. 1976).

(72% glucose, 22% cellobiose and 8% other sugars; Wilke, Yang, and von Stockar 1976). Purification and/or concentration of the sugar solution is, of course, possible, but the economics depend on the starting material and desired end product. Such a sugar solution is an ideal medium for countless fermentations without further treatment. In the case of sugars produced from industrial or urban wastes, poisons and pathogens may limit their uses to a conversion substrate for fuel or in some cases fodder products. It is highly unlikely that purification and drying of the sugar solution to table quality would ever be justified.

Economic Estimates

Economic estimates are extremely difficult for processes such as enzymatic cellulose hydrolysis, which are partly a basic research problem and partly a development problem. Several parameters can be expected to undergo major alteration before it is put into commercial operation. In addition to technical questions are also those of scale and capital intensiveness of the operation. Finally, the starting material plays a dominant economic role which only begins with its cost of credit (for some wastes), but includes pretreatment requirements (and hence costs) and constraints on final-product usage (e.g., hygiene level).

One leading cellulase group estimates glucose production costs at 25¢/kg (Spano 1976), while another's estimate is less than half this (Wilke, Yang, and von Stocker 1976), and yet another's is two times higher (Brandt 1975). Both of the former assume zero cellulose cost or credit, while the latter assumes a cellulose cost and enzyme recovery rate of 20% vs the 35% discussed earlier. The first estimate appears to assume no attempted enzyme recovery.

While these differences are highly significant, the low-cost assumption (for Figure II.3.13) is described in Table II.3.10 to demonstrate the various price factors. For a modern (90% efficient) plant processing 885 tons per day of cost-free newspaper (61% cellulose) and the other assumption shown, enzyme hydrolysis (50% effective) should yield sugar at a cost of $0.11 per kilogram (Wilke, Yang, and von Stocker 1976). A number of points should, however, be mentioned. First, as designed, this process is extremely capital-intensive: fixed capital costs ($23,390,000) for this installation represent nearly 70% of the product cost, and it is one of the most important areas for improvement. Other obvious cost variables are increased conversion efficiencies (increased efficiency from 50-80% would nearly halve the cost to 6.6¢/kg) or increased paper costs (if paper costs $20/ton, the sugar cost would increase to 20¢/kg). Clearly, better cost estimates will follow more repre-

Table II.3.10 Process Cost Analysis Base Case (Cell Recycle Fraction = 0.65) (Wilke, Yang, and von Stockar 1976).

		Hydrolysis	Pretreatment	Enzyme Recovery (34%)	Enzyme Make-Up	Total
Fixed capital cost[1]	($)	6,200,960	2,816,130	2,060,410	12,309,260	23,386,760
Annual investment-related costs[2]	($)	1,482,030	673,060	492,440	2,941,910	5,589,440
Annual utilities costs[3]	($/hr)	122,990	61,500	122,990	122,990	430,470
Annual raw materials costs[3]	($)	239,415	132,290	26,560	446,850	845,115
Annual manufacturing costs	($)	—	—	—	1,312,190	1,312,190
Daily manufacturing costs	($)	1,844,435	866,850	641,990	4,823,940	
Sugar costs	(¢/kg)	5,589	2,626	1,945	14,618	24,778
		2.6	1.2	0.9	6.8	11.5

[1]Plant cost = 2.45 × principal equipment cost.
[2]Depreciation is 10% for pretreatment and fermentation, but 5% for hydrolysis and recovery, interest = 5%, maintenance = 5% × plant cost, no taxes.
[3]Power (10¢/kWh), process steam (72¢/ton), cooling water (2.7¢/meter³), process water (13¢/meter³).

sentative pilot-plant operation and greater understanding of the basic processes involved, but these calculations indicate both the potential and problem. Lowering the high capital costs would extend use of enzyme hydrolysis to many cellulose-rich countries, while increased enzyme recovery and/or cellulose conversion efficiencies probably are the better chances for reducing sugar costs. While the difficulties in adapting enzyme hydrolysis to a low-technology form cannot be overemphasized, the potential and limitations can be expected to become clearer as the enzymological process is studied further.

CHEMICAL CONVERSIONS?

Acid Hydrolysis of Cellulose to Glucose

Cellulose can be hydrolyzed by acid to free glucose molecules at a rate which depends strongly on the reaction temperature. Intermediate hydrolysis stages include chains of two, three and four glucose molecules, but in contrast to enzyme hydrolysis, acid hydrolysis is nonspecific and simply breaks the weakest bonds first, including those *within* glucose molecules. The linkages between glucose units are quickly hydrolyzed for soluble or amorphous cellulose, but crystalline cellulose poses a problem, just as it does for enzyme hydrolysis. This hydrolysis of crystalline cellulose to glucose occurs at a rate (k_1) comparable to the acid-decomposition rate of glucose itself (k_2):

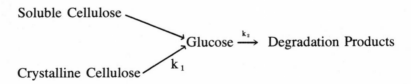

Figure II.3.14 demonstrates this sequence of events for acid (0.8% H_2SO_4) hydrolysis at 180°C of hemicellulose-free wood and shows that the reaction rates and k_1 and k_2 are very similar, 0.0265 and 0.0242, respectively (Millett, Baker, and Satter 1976).

As the decomposition of glucose gives nonuseful products, the reaction must be stopped after hydrolysis of the amorphous cellulose. Optimization of the acid concentration plus hydrolysis time and temperature for maximum net sugar yield is shown in Figure II.3.15 (Grethlein 1975). These data show that the ratio k_1/k_2, i.e., the relative rates of cellulose and glucose decomposition, increases with both acid concentration and temperature. At 230°C the reaction time is 20 seconds and is about the

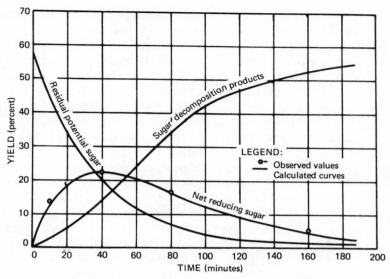

Figure II.3.14. Acid hydrolysis (0.8% H_2SO_4) of wood residue at 180°C (by permission of Saeman 1945).

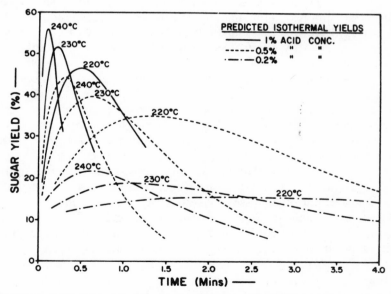

Figure II.3.15. Sugar yield's dependence on time, temperature, and acid concentration (by permission of Grethlein 1975).

minimum time limit for substrate handling. Just over 50% of the potential glucose yield is attained under these near optimal conditions.

Lignification slows the reaction considerably, but unlike enzymatic hydrolysis, the reaction does proceed. Unfortunately, the lignin is decomposed and lost as a potential by-product. As chemical uses for lignin do exist, a pretreatment (as discussed earlier) to remove lignin is preferable if it can be done economically. The short reaction time for optimal sugar yield requires that the cellulose be pretreated anyway (e.g., ball milled to ca 2 mm particles), and the feed slurry (10-30% solids) should be preheated to the reaction temperature.

Impurities not removed in pretreatment will not inhibit the hydrolysis reaction as they do in enzymatic hydrolysis, but as a result may constitute a serious impurity in the glucose hydrolysate. The low pH does, however, require post-hydrolysis treatment with calcium hydroxide to neutralize the acidic slurry. Following hydrolysis the solids are filtered off, and the product is a mixture of glucose plus organic impurities (principally decomposed sugars). Undigested solids, principally crystalline cellulose, can be recycled for additional pretreatment and rehydrolysis.

Economic considerations of the acid hydrolysis process (assuming neither cost nor credit for refuse cellulose) are quite favorable relative to the manufacturing of sugar from molasses, for example. Sugar produced by the hydrolysis of wastes, even when corrected for inflation, has in recent years dropped below that of sugar from blackstrap molasses: from 1971 to 1974 the cost of sugar from molasses increased from 7.5¢/kg to 20¢/kg sugar, mostly due to feedstock cost increases, while sugar from hydrolyzed wastes costs about 5-7¢/kg, depending on scale. Increased scale for this process does decrease costs up to a plant size of about 1000 tons per day (Grethein 1975).

Energy costs are important due to heating requirements and are noticed for example by the doubling of process costs for decreasing the solid/liquid ratio from 0.3 to 0.1. Energy considerations, in fact, offer a sobering view of this process: A simplified calculation shows that the gross heating costs of a 10% cellulose solution about equal the energy in the recovered glucose (assuming 50% recovery). Some recovery of the heat can be made, but other losses make acid hydrolysis from the energetic point of view a net loss. Sugar from this process cannot be considered for any type of fuel production, but may be well justified for other processes which more directly require sugar's properties and high quality.

Fuel Cells

A fuel cell is an electrochemical device in which the energy derived from chemical reactions maintained by a continuous supply of chemical reactants is catalytically converted to electrical energy (Sisler 1961 and 1971, Maugh 1972a). A fuel cell differs from a normal electrochemical cell (e.g., an automobile battery) only in that the reactants are continuously supplied; normal cells sacrifice the electrode which limits the cell life.

Since fuel cells are electrochemical devices which directly convert chemical energy into electricity, the intermediate heat-generation step of normal electric generation processes is eliminated. As a result several advantages accrue (Hammond, Meta, and Maugh 1973, Maugh 1972a):

- Fuel cells can be added in modular fashion for a generation of one to a thousand volts and power levels up to 100 MW.
- They can be used in remote areas away from transmission-line grids.
- They use fuel which can be distributed at 20-30% of the energy cost of electrical transmission.
- They have efficiencies better than gasoline (20%) or diesel generators (20-35%) or even large steam turbine plants (39% for output greater than 100 MW). Fuel-cell efficiency (55%) is furthermore independent of load and scale down to 25 kW.
- They emit no air or noise pollutants and minimal heat.
- They require little installation time, even in remote areas.
- They can be run in reverse to store energy from traditional power plants during off-peak times.
- They require no operating personnel or maintenance.

The primary disadvantage with fuel cells is their cost: $350-450/kW with service life of 16,000 hrs. Costs have been reduced 20-fold, and service life increased fivefold in the last few years, but economic competition requires a halving of the cost and a further doubling of the service life (Maugh 1972a).

Chemical Fuel Cells.   A fuel cell consists of two electrodes (an anode where oxidation occurs and a cathode where reduction occurs) connected through a conducting electrolyte and a resistive load where the electrical work is done. If hydrogen is the fuel, as in Figure II.3.16, it is oxidized at the anode and releases an electron to the electrode and a proton to the electrolyte solution according to the reaction,

$$H_2 \rightleftarrows 2H^+ + 2e^-.$$

The proton from hydrogen migrates to the cathode where oxygen gas is reduced by the electron traveling across the load,

$$1/2O_2 + H_2O + 2e^- \rightarrow 2OH^- \text{ (cathode)}.$$

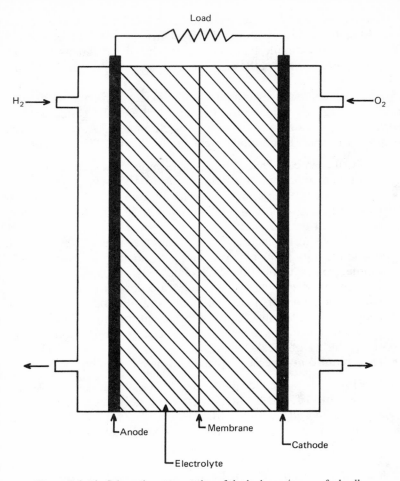

Figure II.3.16. Schematic representation of the hydrogen/oxygen fuel cell.

The proton and hydroxyl ion diffuse and combine in the electrolyte to form water,

$$2H^+ + 2OH^- \rightarrow 2H_2O.$$

Overall, the reaction is simply the same as the combustion of hydrogen with the energy yielded as electrical power instead of heat,

$$H_2 + 1/2O_2 \rightarrow H_2O + 210J.$$

The potential difference (voltage) resulting from such a cell is determined by the cell reaction equilibrium constant of the change in free energy (Moore 1965),

$$E° \text{ (volts)} = -\Delta G°/nF = (RT/nF)\ln K_a.$$

$E°$ in the above equation is the standard electromotive force (emf, with unit activity for products and reactants), $\Delta G°$ is the change in free energy, K_a is the equilibrium constant, and all other numbers are constant. For any fuel cell reaction, the resulting potential can be calculated in advance from the free-energy change. As the hydrogen/oxygen fuel cell is the most developed, it is taken as the reference fuel, but other potential fuels are listed in Table II.3.11.

The voltages indicated in Table II.3.11 depend only on the energy yield of the cell reaction; an effective fuel cell must, however, develop not only a useful potential difference (voltage), it must also release electrons at an adequate rate. This rate at which electrons are released is equivalent to the electrical current and depends upon chemical reaction kinetics. Neither current nor voltage alone adequately describes a fuel cell, but rather it is their product—the power—which describes a cell's ability to do electrical work.

As discussed above, the ability to develop a potential difference depends only on the reactant's thermodynamic properties. Reaction rates, however, can be limited by physical or chemical barriers at any of a number of steps (Lewis 1966):

- rate of reactant transport to the electrodes;
- rate of reactant adsorption at the electrode;
- rate of electron transfer to the electrode;
- reactions on the electrode surface;
- rate of desorption of reaction products;
- rate of transport of products away from the electrode.

A decrease in the net reaction rate results in "polarization" and a concomitant decrease in useful power output. Polarization is charac-

Table II.3.11   Relative Fuel Activities (Lewis 1966).

Fuel	Voltage (theoretical)	Relative Activity
Hydrogen	1.23	100
Formaldehyde	1.13	50
Methanol	1.21	30
Propane	1.10	3
Methane	1.10	2

terized by the "relative activities" of several fuels in Table II.3.11 and indicates the limited usefulness of some energy-rich compounds. The comparison between hydrogen and methane is a useful one: Judged purely on the basis of energy yield upon complete combustion, i.e., the resulting voltage, the two fuels are comparable with only a 10% superiority for hydrogen. The 50-fold difference in reactivity, however, currently excludes methane as a useful fuel cell substrate and results from its four-stage oxidation reaction (Lewis 1966):

$$
\begin{aligned}
CH_4 \quad &+ H_2O \quad \rightarrow CH_3OH \quad + 2H^+ + 2e \\
CH_3OH \quad & \qquad\qquad \rightarrow HCHO \quad + 2H^+ + 2e \\
HCHO \quad &+ H_2O \quad \rightarrow HCOOH \quad + 2H^+ + 2e \\
HCOOH \quad & \qquad\qquad \rightarrow CO_2 \quad\quad + 2H^+ + 2e \\
\hline
CH_4 \quad &+ 2H_2O \rightarrow CO_2 \quad\quad + 8H^+ + 8e
\end{aligned}
$$

Note that the rate-limiting factors listed above apply to each of the four intermediates, whereas hydrogen is not only a smaller molecule, but its oxidation involves only a single, two-electron transfer.

Molecular size is, of course, a limitation both for mass transport in the electrolyte and for adsorption to the electrode surface. Surface problems include not only adsorption but proper orientation and reactivity for electron transfer. Increased temperature is one common solution, but generally some form of catalyst is required to lower the reaction energy barrier. Typically a platinum or paladium catalyst is used for the hydrogen/oxygen fuel cell, but with other fuels, catalyst problems remain the major impediment to fuel-cell development. Among these problems is corrosion (particularly at high temperature) and poisoning due to impurities (e.g., CO).

Biological Fuel Cells.   The extreme temperatures and simple fuels required for modern fuel cells stand in strong contrast to the complex fuels (carbohydrates) used at ambient temperatures by biological organisms. Since the construction of the first biochemical fuel cell over 60 years ago, many approaches have been attempted to exploit the advantages of naturally occurring systems. In each case high energy, typically organic molecules (normal photosynthetic products), are converted to electroactive intermediates which yield electrons for current generation. A fuel cell which is partly biological can use a normal chemical or a biological catalyst of which the major alternatives are sketched in Figure II.3.17.

An indirect "type A" fuel cell is feasible with present technology. Hydrogen is produced in one reactor via bacterial fermentation (see page

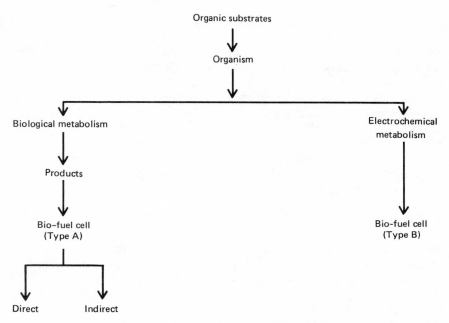

Figure II.3.17. Bio-fuel cell classification scheme.

272 of this chapter) and is fed to the anode of a normal catalytic fuel cell; air at one atm provides oxygen to the cathode. Such a system allows the separate optimization of conditions for fuel production and fuel-cell efficiency. A direct type A fuel cell combines the two reactors with an obvious saving in hardware. In such a fuel cell the bacteria are located on the electrode (anode) and convert an organic substrate dissolved in the electrolyte to a fuel usable by the electrode (Karube et al. 1977, Lewis 1966). Typical substrates and organisms involved include:

urea → ammonia   (*Micrococcus ureae, Bacillus pasteurii*)
carbohydrate → hydrogen   (*Clostridium cellobioparus*)
sulfate → hydrogen sulfide   (*Desulfovilrio desulfuricans*)
carbohydrates → ethyl alcohol   (*Saccharomyces sp., Pseudomonas lindueri*)

Of the above examples, the hydrogen cell is again superior to the others, but even this cell does not compare to the indirect cell cited above: the current density of the latter is 500 times greater than the direct type A cell (18 mA/cm^2 at 0.78 V vs 0.1 mA/cm^2 at 0.3 V; Blanchard and Foley 1971). The discrepancy depends in part on mass transport of metabolites into the cells and the incompatibility of optimal conditions for fermentation and electrode reaction processes. Ammonia as a fuel suffers from low reactivity, alcohol does not oxidize completely, and sulfide is not recycled to sulfate by the cathode. Improvements are conceivable, but hydrogen is likely to remain the preferred fuel.

A microbiological cathode can, in principle, be constituted using photosynthetically-produced oxygen by algae. Variations on this theme could include algal fermentation, e.g., for hydrogen production for the anode and photosynthetic-oxygen evolution for the cathode, or two separated algal cultures—one normal culture for oxygen and one under argon for hydrogen production. Hydrogen production by photosynthetic bacteria may also be possible. A second cathode alternative is the reduction of nitrate to nitrogen gas by *M. dentrificans* (Lewis 1966).

Enzyme-Catalyzed Fuel Cells. In terms of electrical-power generation, the well-established H_2/O_2 fuel cell has the greatest current density (0.1-0.2 A/cm^2 at 1 V) and efficiencies of 70% (Hammond, Metz, and Maugh 1973). Direct biological fuel cells have current densities two orders of magnitude lower. The high-current densities for inorganic fuel cells are of limited economic value, however, due to the high price of the noble metal catalysts, extreme operating conditions (high temperature, high pressure and high fuel concentration) and the requirement for simple, high purity fuels.

Oxidations in the organisms used for the biological fuel cells discussed above have several advantages:

- mild operating conditions (ambient temperature and atmospheric pressure);
- the ability to oxidize complex, photosynthetically-produced organic molecules (e.g., carbohydrates, fats, glucose);
- reaction specificity for decreased susceptibility to poisons;
- ability to utilize fuels in low concentration.

Isolated oxidation-reduction enzymes as replacements for the metal catalysts of so-called indirect "type A" fuel cells have in addition to the above, the advantages of low cost, and potentially unlimited supply.

The absence of any enzyme-catalyzed fuel cell points to the existence of several primary problems: enzyme stability, enzyme immobilization and electrical coupling of the enzyme to electrode and external circuit.

Enzymes can be used in several ways to catalyze fuel cell reactions. For a system in which the enzyme (E) is in free solution with reactant (R) and product (P) molecules, a three-step reaction is required before the electron is transferred to the electrode surface (M):

$$E + R \;\rightleftarrows ER$$
$$ER + M \rightleftarrows M[ER]$$
$$M[ER] \;\;\rightleftarrows M + E + H^+ + e^-$$

The reactivity of such a system would generally be far too slow due to enzyme size and stereospecificity unless a small mediator, or electron-carrying molecule, is added. An effective mediator must be able to accept quickly the electron from the enzyme-active site and diffuse to the electrode when the electron is transferred. Yet another alternative eliminates dependency on diffusion and would attach the enzyme to the electrode via a conducting organic polymer. While most organic molecules are non-conducting, delocalized nonbonding or pi-bond electrons give some organic polymers a conductivity approaching that of a metal. In some cases a cofactor such as NAD^+ is bound close to the enzyme's active site to act as an electron probe connecting the polymer and enzyme.

The potential of enzyme-catalyzed fuel cells is enormous and combines the advantages of biological and inorganic catalyst fuel cells. Progress rests with the development of enzyme-stabilization techniques, synthesis of organic analogues, and understanding of electron-transfer mechanisms. Present studies include several fuels and their corresponding enzyme systems: hydrogen (hydrogenase), glucose (glucose oxidase), ethanol (alcohol dehydrogenase), and urea (urease).

The enormity of the potential usefulness of these fuel cells is matched by the enormity of the fundamental applied and developmental research required to even adequately speculate on their future. What is clear is the worthiness of the attempt.

THERMAL CONVERSIONS

Methanol Production

Methanol (methyl alcohol, CH_3OH) can be chemically produced from any carbonaceous source, though not in a single step. The first stage in this process is the production of synthesis gas (CO, CO_2, H_2) via a process which in principle is simple but in practice varies greatly in difficulty with starting material.

Synthesis gas is normally produced from coal, but can use any organic source via the following gasification or watergas reactions (Reed and Lerner 1973),

$$C + H_2O \;\rightarrow CO + H_2$$
$$C + 2H_2O \rightarrow CO_2 + 2H_2.$$

These reactions are endothermic and therefore require heating, but there is an alternative, exothermic combustion process (particularly for coal) which uses pure oxygen to produce synthesis gas plus methane. Methane can also be used directly to produce synthesis gas, it first being produced as above, biologically, or by using hydrogen gas and any organic source,

$$C + 2H_2 \rightarrow CH_4.$$

Synthesis gas is then formed from synthetic methane or more commonly from natural gas according to the reactions,

$$CH_4 + H_2O \rightarrow CO + 3H_2$$
$$CH_4 + 2H_2O \rightarrow CO_2 + 4H_2$$
$$CO + H_2O \rightarrow CO_2 + H_2.$$

Industrial methanol production can involve any of a host of processes (Mitre 1975), but most commonly synthesis gas (CO, CO_2, H_2) or methane (CH_4) are used as the only feedstocks with low carbon number and which give adequate methanol yields. Synthesis gas can be used to make a wide variety of products (Figure II.3.18) via processes which are in many respects similar. Each process selects for the desired product by optimizing operating conditions (temperature and pressure), reactor design and catalyst. By altering these factors, virtually any carbonaceous substrate can be processed to a nearly limitless number of organic products. Note that for fuel production, methanol purity is of little importance, and in fact the presence of higher alcohols (ethanol, propanol and isobutanol) increases energy content and gasoline solubility.

An important consideration for production of methanol or any other synthesis gas from wastes or biomass is the feedstock's water content. A feedstock with 50% moisture content, for example, will require the consumption of 20% of its energy for drying. Biomass water content is one strong encouragement for employing biological conversion methods, e.g., conversions to ethanol, methane, sugar or possibly methanol.

All but one of the potential methanol synthesis processes are capital-intensive because of the high-pressure and high-temperature reactors involved. The destructive distillation of wood is, however, the modern version of a millenia-old tradition. It now involves the cooking of hardwood in sealed ovens with limiting air, but in the past a wood pile was simply covered with sod to keep it semi-anaerobic while air was allowed to enter only to the center active combustion area. Charcoal is the principal product of these processes, but 2% methanol, 3% acetic acid and several other chemicals are by-products originating from the lignin. Soft woods yield no methanol. A laboratory-phase development of these treatments is the wood-liquification process using hydrogen at 200°C and

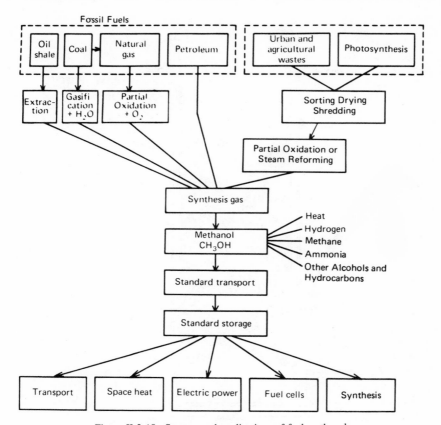

Figure II.3.18. Sources and applications of fuel methanol.

200 atm. This process, like coal liquification, yields 5% methanol, 15% propanol and 40% char.

Figure II.3.19a shows a related process and one potentially suited to small scale as it was in fact widely used on the private automobile; two and a half kilograms of wood was equivalent to a liter of petroleum. The lower Figure II.3.19b shows an extension or modernized version of the same concept, but one designed to convert any type of organic matter first into carbon monoxide and hydrogen and then ultimately to methanol.

Economics of Methanol Production. Methanol can be produced from biomass (or coal) whenever a liquid fuel is required. The industrial processes described, however, suffer from two major drawbacks typical of chemical conversions:

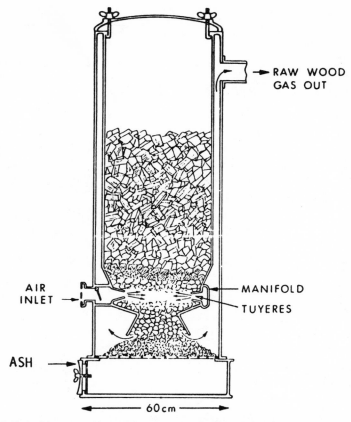

RAW WOOD
GAS OUT

AIR
INLET

MANIFOLD

TUYERES

ASH

← 60 cm →

Figure II.3.19.A. Biomass gasifier: small automobile wood gas generator from 1945 (2 kg wood = 1 liter petrol) (with permission of Reed 1976).

1. biomass contains water whose vaporization requires energy before the desired conversion reaction occurs;
2. industrial conversions are capital-intensive and demonstrate the typical economics of scale.

The first point above indicates one advantage of a microbiological (water-soluble) process, the possibilities of which are described below. The second point indicates the capital intensity of high-temperature/high-pressure processes (biological process at ambient temperatures and atmospheric temperatures show less economy of scale) both for fuels and the required reactors. Practical sizes range from 100 to 25,000 tons of methanol per day with corresponding capital cost estimates of $10- and

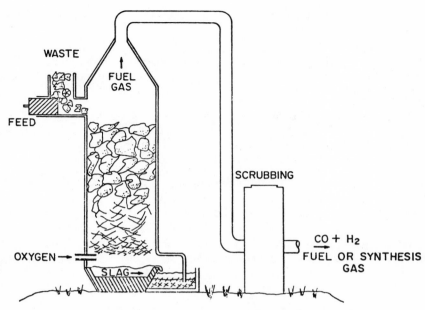

Figure II.3.19.B. Biomass gasifier: schematic purox process module (with permission of Reed 1976).

$480-million, respectively, i.e., a fivefold decrease in capital cost per ton with increased plant size (Reed and Lerner 1974). Manufacturing costs, i.e., costs excluding capital investment, depend little on scale and average $9/ton.

Fuel costs can vary from a negative cost for waste disposal to $20/ton for some coal. A pattern exemplified by methanol synthesis, but recurring in other processes, is the fact that a small plant converting wastes to methanol is competitive with a large-scale coal-converting plant. Such factors allow the use of widely dispersed waste-treatment plants where the saving of transport costs plus a credit for waste disposal become strong economic incentives.

A recent price for methanol is 6¢ per liter (in 23,000 tank cars, 1974 prices), whereas gasoline is 8¢ per liter. As the energy density of methanol is only half that of gasoline, the costs per energy unit are .4¢/MJ and .25¢/MJ, respectively. As fuel methanol would cost 10-30% less than the present pure methanol due to relaxed purity requirements and estimates for the large-scale coal plants project an even lower unit cost of 3.4¢-4.5¢ per liter ($41-54/ton), large-scale methanol production, particularly from wastes, can soon become competitive. While it is not as ideal a fuel as petroleum, methanol has several advantages (discussed in Part II, Chapter 4) which will secure for it an increasingly important role.

Biological Production of Methanol. Biological conversion of organic wastes or biomass to methanol is preferable to the previously discussed chemical synthesis because:

- water content of biomass is advantageous for biological conversion but disadvantageous for chemical processes;
- biological processes do not require the high temperature and pressure reactors necessary for chemical processes.

The first point is a direct energy saving, while the second eliminates the enormous capital investment (and fuel consumption) for chemical processes and hence removes the economics of large scale. A small-scale system would allow conversion of wastes or biomass where they occur.

No known organism excretes methanol, but the more than 100 methane utilizers (Whittenbury, Philips, and Wilkinson 1970) or methylotrophs probably oxidize methane according to the following scheme (Kosaric and Zajic 1974):

$$CH_4 \rightarrow CH_3OH \rightarrow \text{'HCHO'} \rightarrow HCOOH \rightarrow CO_2$$

N-methyl compounds   cell material

The most interesting first step is believed to be catalyzed by a monooxygenase enzyme where XH_2 is a reducing agent obtained from other metabolic pathways (Higgins and Quayle 1970),

$$CH_4 + O_2 + XH_2 \xrightarrow{\text{oxygenase}} CH_3OH + H_2O + X.$$

A practical system for fermentative methanol production (Foo and Hedén 1976) would utilize methane (e.g., from an earlier fermentation), but the fermentation process must be blocked beyond methanol. Two alternatives exist: blocking by chemical inhibition of methanol dehydrogenase or isolating mutants lacking this enzyme or otherwise deficient in methanol utilization. The first method has yielded some tentative results (Foo and Hedén 1976), while the latter is under investigation. Other applications of "leaky mutants" have been developed, but these studies are too premature to make any further speculations.

Pyrolysis

Pyrolysis (alternatively called destructive distillation, thermal decomposition or carbonization) is normally defined as an oxygen-deficient combustion process in which biomass or any other organic matter is irreversibly converted to "pyrogas" (hydrogen, methane, and carbon monoxide and dioxide), oil and char (Lewis and Ablow 1976). Pyrolysis reactors, such as that shown in Figure II.3.20, are normally indirectly heated, but the term has been extended to include *gasification,* a similar process but one in which some of the pyrogas is combusted with a limited amount of oxygen.

Heat is required for this endothermic reaction which normally requires temperatures of 500-1200°C, but at least a portion of this energy is recovered in the products. Energy recoveries as high as 98.6% have been quoted, but these figures ignore external heating. More typical figures are 58% for biomass with 10% water content and 50% for 25% water (Lewis and Ablow 1976).

Typical product figures are difficult to give as they vary dramatically with feedstock, its water content and reaction conditions, but Table II.3.12 shows the conditions and product yields for three agricultural wastes. Oil from pyrolysis typically has a heating value of 35 MJ/kg or 33 MJ/liter, while the gas phase has about 18 MJ/meter^3 (Crentz 1971).

Pyrogas composition and hence its heating value can also vary greatly with reaction temperature. For the system sketched in Figure II.3.20, which recycles the oil for maximum gas production, these pyrogas variations are shown in Figure II.3.21. At low temperatures (400°C) the pyrogas heating value is maximum and its composition is similar to that of anaerobic digester gas (50% methane, 40% carbon dioxide), but the reaction rates are likely to be very slow, perhaps uneconomically slow

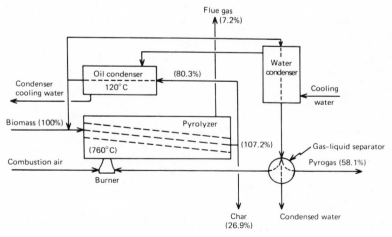

Figure II.3.20. Schematic pyrolysis system (redrawn from Lewis and Ablow 1976).

Table II.3.12   Pyrolysis Products (Crentz 1971).

Waste	Cow Manure	Rice Straw	Pine Bark
Pyrolysis temperature	500–900°C	200–700°C	900°C
Energy/ton waste (dry)	15.6 GJ	13.85 GJ	18.55 GJ
Energy/quantity of products per ton waste			
Gas	5.9 GJ (310 meter^3)	2.7 GJ (170 meter^3)	9.35 GJ (570 meter^3)
Oil	2.2 GJ (66 liters)	1.4 GJ (42 liters)	0.7 GJ (21 liters)
Char	5.4 GJ (320 kg)	6.2 GJ (360 kg)	8.6 GJ (350 kg)
	13.5 GJ	10.3 GJ	18.6 GJ

(Lewes and Ablow 1976). At higher temperatures hydrogen becomes the dominant gas (55%), but the overall heating value is decreased.

The char produced has potential economic value similar to activated charcoal which can be burned, but is also useful as a chemical adsorbant, e.g., for alcohol distillation and pollution control. It contains all of the nonvolatile carbonaceous material and mineral ash.

It should be emphasized that the scheme in Figure II.3.21 is only one of several variations in the pilot-plant stage. Which system is best—and feedstock may also be a factor—remains to be determined. Pyrolysis' best chance for economic success, however, lies in urban centers where the high-capital costs are tractable and pathogens/toxins are problematic for many of the earlier discussed alternatives. Credits for disposal costs ($6-8

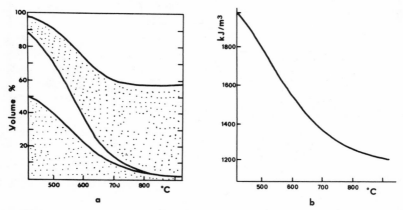

Figure II.3.21. Pyrogas composition (a) and heating value (b) (by permission of Lewis and Ablow 1976).

and \$8-12/ton for landfill and incineration respectively) make the economics in urban centers particularly attractive.

Further developments may improve and simplify the techniques to allow smaller-scale operation. Operation of this process at atmospheric pressure is highly advantageous for low-capital use, but the inability to remove nutrients and metals restricts the most appropriate use to low-quality dry wastes typical of urban centers.

Hydrogenation

Hydrogenation (also called deoxygenation and chemical reduction) was developed for coal at the US Bureau of Mines (Hammond, Metz, and Maugh 1973), but cellulose and lignin have proved to be even easier substrates: as much as 99% of the waste carbon (85% is more common) is converted to a low-sulfur oil with heating value 35 MJ/kg (84% that of the common No. 6 heating oil). Typical oil composition is 80% carbon, 8% hydrogen, 10% oxygen, and less than 0.4% sulfur. The hydrogenation process, summarized as,

$$1 \text{ ton waste } + 5\% \text{ alkaline catalyst} \xrightarrow[240-380°C \text{ for 1 hr}]{100-250 \text{ atm CO } + \text{ steam}} \text{oil}$$
$$\text{or} \quad 2 \text{ barrels oil } (0.32 \text{ meter}^3)$$

results in the removal of oxygen from cellulose by carbon monoxide and steam. It is completely enclosed in a reactor containing typically a sodium-carbonate catalyst at high temperature and pressure. Conditions have less effect on the net yield than on product quality which at room temperature ranges from a soft tar-like solid (for a 250°C process temperature) to a free-flowing liquid (for 380°C process temperature); low-temperature operation has the advantages of:

- lower energy consumption;
- less than one-third the pressure (1500 pounds per square inch at 250°C vs 5000 pounds per square inch at 400°C) saves energy and autoclave-construction costs;
- negligible CO consumption while at higher temperature the watershift reaction proceeds,

$$H_2O + CO \rightleftarrows CO_2 + H_2.$$

On the other hand, high-temperature conversion produces oil with a lower oxygen content and is sometimes required for cellulose with a high-lignin content.

Economic estimates have not yet been made public from the pilot plants, but high-pressure chambers place costs above that of pyrolysis for example. Noneconomic problems include handling and introduction of substrate solids to the high-pressure chamber and finally separation of products. Like its chief competitor, pyrolysis, hydrogenation stands its greatest chance of economic success with the help of an urban waste-disposal credit, but unlike pyrolysis, it produces only a single valuable product—oil.

Energy analysis is not yet possible in detail, but the *direct* process energy inputs are known to require ca 0.8 barrels for the two barrels produced, or a 60% conversion efficiency. The inclusion of incomplete conversion and capital-energy costs (e.g., reactors) will lower this effiency value considerably.

Combustion

Combustion of organic compounds in air is not only the oldest energy conversion process, but also constitutes the initial step for more than 90% of all energy consumed in the US; in the developing world human and animal metabolism are the only major forms of energy conversion other than combustion.

Heat, the direct product of combustion, is the ultimate desired energy form for only a third of the US energy consumption. The bulk of the heat produced provides torque or turning motion via turbines or pistons: 25% of the energy is for transportation (internal and jet combustion) and 25% is for turbine-driven electric generators (about 30% efficient to generated electricity, not end use).

Efficiency is, of course, a major factor in all processes, but the production of mechanical energy by combustion is particularly plagued by the by-product waste, low-grade heat. Ultimate limits to efficiency are set by the laws of thermodynamics (see Part I, Chapter 2), but a general rule is to keep the energy at as high quality as possible. For heat engines this means keeping temperature differentials high; steam-powered electric generation plants are only about 30% efficient due primarily to the turbine stage, not to combustion itself. Internal-combustion engines are even less efficient. Combustion itself can occur over a wide range of efficiencies with losses due to:

- incomplete combustion;
- inert contaminants;
- volatile components which take heat to vaporize, or
- inefficient transfer of heat from flame and combustion products.

Unnecessary inefficiencies occur not only on the primitive level—Indian peasants burning cow-dung cakes at only 10% efficiency—but include sophisticated burners and motors. Inefficiency has been tolerated out of laziness and the accessibility of inexpensive, high-quality fuels.

Considering the prevalence of combustion as an energy-conversion process and the primitiveness common to most of these processes, both in primitive and modern applications, it is not surprising that major advances can be made with nominal effort. The guiding principles are the same as discussed elsewhere, i.e., to maximize the temperature of the hot reservoir or combustion chamber.

Two such devices are shown in Figure II.3.22, one for simple combustion as a heat source and the other to provide a high-temperature reaction chamber (Lloyd and Weinberg 1975). Both are designed on the principle of maximized heat exchange between the incoming fuel/air and the outgoing combustion products without dilution or mixing of fuel and products.

The "flame" or central combustion region is clearly the area of highest temperature, but the attainable temperature and removable heat depend on fuel, usage, and burner geometry. A normal heat engine would use the center region as the hot reservoir and the outside as the cold reservoir.

The efficiency of such a heat engine is limited by the maximum obtainable flame temperature, which for the burners shown above, is limited by the construction materials. Burners built of the metal alloy Inconel 600 have a maximum of ca 1200°C giving a Carnot-cycle efficiency (assuming a 20°C cold reservoir) of 80%.

Carnot efficiencies do not tell the whole story, however, as they as-

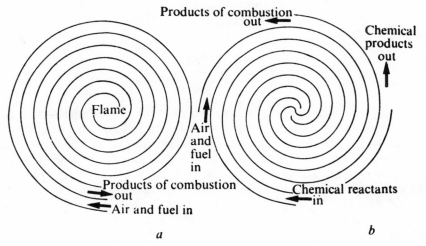

Figure II.3.22. Spiral combustion (a) and reaction chambers (b) (by permission of Lloyd and Weinberg 1975).

sume ideal systems. Heat losses in a practical system with a high Carnot efficiency can result in a low thermodynamic *conversion* efficiency due to incomplete combustion, condensation losses and heat losses in exhaust gases. The "Swiss Roll" design in Figure II.3.22 recirculates exhaust gases to preheat the fuel and provides some startling results (Loyd and Weinberg 1975):

- higher efficiencies are obtained with fuels diluted with a great excess of air. For a normal burner, attainment of 1500°C costs one-third more fuel than for this Swiss roll;
- limits of flammability are virtually eliminated: methane can burn in less than 1% concentrations.

Not only do such burners provide higher efficiencies, they also extend the definition of a fuel to extremely low-energy-content materials, including many waste and ventilation (e.g., from mines) gases.

Fuel Combustion.   The maximum obtainable heats of combustion are shown in Table II.3.13 for various fuels. Note in particular the incomparable superiority of gasoline—49 MJ/kg—well over twice that of wood (with 20% water). Most fuels have a low-energy content because they are already partially oxidized. This fact has led to an empirical relation between the weight percent of carbon, hydrogen, and oxygen in an organic fuel and its heat of combustion (Spoehr and Milner 1949):

$$\text{heat of combustion (kJ/grams)} = \frac{R}{1.89} + 1.68,$$

where

$$R = \text{degree of reduction}$$
$$= \frac{C \times 2.664 + \%H \times 7.936 - \%O}{3.989}$$

In addition to heat loss due to partial oxidation, wood and most other biomass have the additional problem that each kilogram of water added per kilogram fuel lowers the heat of combustion by 2.3 MJ/kg. Typical water contents of various types of biomass are: fresh wood—47%, oven-dried wood—20%, terrestrial plants—60%, aquatic plants—95%, and solid animal wastes—20%. As self-sustaining combustion requires water contents below 10-15%, drying of biomass can be a serious problem. Solar drying is both a a traditional and energetically sound method.

For urban centers, the high concentrations of low-energy wastes (14-16 MJ/kg) must be evaluated as a combustion fuel both economically and

Table II.3.13   Heats of Combustion of Solid Energy Sources (after Edwards 1975).

Source	Heat of Combustion
Solid Fuels:	MJ/kg (dry):
Carbon	32.5
Charcoal	28–30
(10–15% H_2O)	
Coal	14.5–32.5
(liginite-bituminous)	(29.)
Wood (20% H_2O)	20
Garret pyrolytic char	21
from urban refuse (20%)	
Urban refuse (as is)	10–11.5
Glucose	14–15.5
Crops	13.5–16
Liquids:	MJ/kg:
Coal tar fuels	37–41
No. 4 petro. fuel oil	31–34
Garret pyrolytic oil	24
Gasoline	49
Methanol	21–24
Ethanol	28–30
Acetone	30
N-Butanol	35
Gases:	MJ/meter^3:
Hydrogen	10–12
Methane	37
Carbon monoxide	12
Natural gas	33–45
Oil gas	11–41
Producer gas	4.658–5.6
(CO, N_2, H_2, CO_2)	
Garret pyrolytic gas (27%)	20

energetically against the gas/liquid fuels derived from more capital-intensive treatments, such as pyrolysis. Waste combustion does pose a number of problems due to its nonuniformity, pollutants, corrosive elements, and of course its low heat content. These problems have been solved best in a number of European countries by combining state-owned waste collection/disposal services with power generation. In the US, utilities are privately owned, and such effective coordination has not been practiced. Combustion of urban wastes is still currently viewed from a disposal point of view with the emphasis on effective incineration. Increased oil costs for power generation may well improve the attractiveness of wastes as a commercial fuel, but the supply must also be dependable to justify the investments. A recent study by the US Environmental

Protection Agency estimates that only 5% of the wastes' energy content would be required to process, transport, and fire the wastes. Most efforts are for minor modifications of existing steam-producing systems to accept only slightly prepared wastes.

If a conversion is necessary, then the fuel is combusted to raise the temperature of one body (t_2) in the presence of a cooler body (at temperature t_2), such as atmospheric air or cooling water. The two bodies then constitute a heat engine, whether it is a refrigerator or a steam engine. The maximum thermodynamic efficiency can then be rewritten in a form involving only the temperatures of the two reservoirs (Moore 1965),

$$\text{eff} = \frac{t_2 - t_1}{t_2}.$$

Efficiencies can never reach 100% (the second law of thermodynamics), and the amount of heat lost is a measure of the inefficiency. For these reasons, Carnot-cycles, as these heat-engine processes are called, should be avoided whenever possible in energy conversions, and when necessary, the generated heat (hot reservoir) should be held at a maximum temperature. The 50% efficiency point for a 20°C cold reservoir is, for example, not reached until the hot reservoir is over 300°C (313°C).

MECHANICAL CONVERSION

Leaf Protein (green-crop fractionation)

Tradition has dictated that Man consume only a small portion of the nutrients produced by photosynthesizing green plants. His attention was fixed on the seeds and tubers for good reason, however, as there resides the greatest concentration of palatable and easily digestible food. The remainder, if not used for fodder or nonfood purposes, was discarded.

Domesticated animals, particularly ruminants, are good converters of low-quality plant products to high-quality protein but at a very low efficiency: approximately 10% of the protein fed any farm animal can be recovered as edible meat or milk. Low efficiency is acceptable if the plant has no other value, but alternative food-quality products do exist: mechanical disruption for example allows the extraction of protein from otherwise unpalatable plant fibers and leaves (Pirie 1975b or c). The advantages of this so-called leaf protein or green-crop fractionation are:

- Several times as much protein can be extracted from a forage crop by mechanical fractionation as by ruminant conversion (40-60% vs 10-30%, respectively).

- Crops are harvested at a less-mature stage for protein extraction than for silage or haymaking, thus decreasing disease and pest risks.
- Early harvesting of perennials allows multiple harvesting, increases annual average canopy density (increases annual photosynthetic conversion efficiency and yield), and prevents erosion.
- Drying costs are reduced to less than half.
- Plants unsuitable even for ruminants can be used.
- Harvesting is independent of the weather, in contrast to hay.
- Protein extraction can be done simply on small scale, and all products can be safely used locally (on a mixed farm).
- The product, suitable for human consumption, is 60-65% protein, and the fats are highly unsaturated.

The Process. A block-flow diagram of the protein extraction and fractionation process from green leaves or forage is summarized in Figure II.3.23. Following harvesting of the leaves or forage, the process involves the following steps:

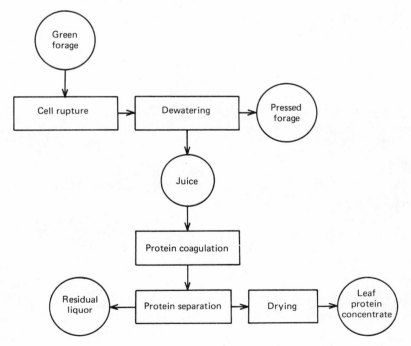

Figure II.3.23. Flow diagram for mechanical protein extraction.

1. Cell rupture and juice extraction. The protein rich juice is liberated from the cells by a mechanical rubbing process, such as with the "screw expeller" shown in Figure II.3.24. This mechanical device rubs leaf against leaf and at the same time maintains pressure long enough (7 seconds) to allow the juice to flow away without being reabsorbed by the fiber pulp. Effective extraction is possible with up to 1 cm thick pulp pressed at pressures of ca 4 kg/cm^2 and at a process rate of five tons of crop per hour. Alternative pulping/pressing devices are considered, and the optimum arrangement may vary with crop.
2. Protein coagulation: Extracted juice contains about 10% solids of which a third is protein and a third is sugar. As a juice it can be used directly as fodder (e.g., for pigs), but a more effective use is to remove the protein by heat coagulation. Slow heating, first to 50°C and then up to 70°C, results in two coagulated fractions, but rapid heating (one to two seconds) with injected steam produces a hard easily-filtered protein curd. Rapid heating to 100°C prevents some undesirable reactions (by inactivating chlorophyllase) and may partially sterilize the product (Pirie 1975b or c).
3. Protein separation: Protein curd can be easily separated from the residual liquor by simple filtration.

The Products.   The three products from protein extraction consist of a fiber fraction, a leaf-protein concentrate and a residual liquor. The fiber

Figure II.3.24. Screw expeller for protein juice extraction (by permission of Pirie 1975b).

Table II.3.14 Composition of Leaf Protein as Usually Prepared (from Pirie 1975b).

True protein	60–70%
Lipid	20–30%
Starch	5–10%
β-carotene	1–2 mg per gram
Fiber	<2%
Water-soluble compounds	<1%
Ash	<3%
Acid-insoluble ash	<1%

fraction constitutes the bulk of the original dry weight (74%) and protein (57%) and hence constitutes an ideal fodder for ruminants (Bray 1976). Not only can otherwise unsuitable plants with toxins and liquor be made palatable for the animals, but drying to produce a ton of fodder requires the removal of only a third as much water (five to ten tons water/ton dry forage vs two to three tons/ton dry fiber). The ease of drying and storage of the fiber fraction on the farm has been the principal interest in protein fractionation.

The protein concentrate fraction, whose contents are listed in Table II.3.14, contains 33% of the original forage protein, 13% the original dry matter and is 60-70% high-quality protein. The fats are highly unsaturated, and hence drying and proper storage is required to prevent rancidity. Leaf protein, normally green, slightly acid and with a cheese-like texture, can be made colorless and tasteless, but the natural product (from lucerne, for example) is acceptable in many communities and is superior in quality to the best seed protein, sesame. Its use as human or animal food is a question of psychology, not quality or safety.

The residual liquor contains 10% of the original protein, peptides, free amino acids, and 1-3% carbohydrate. If concentrated, it resembles molasses, but otherwise can be used for fertilizer or for microbial growth.

Crops for Protein Extraction. The least suitable crop for protein extraction and fractionation is one which is itself edible by the intended consumer. Thus, a forage crop should not be fractionated unless the protein concentrate is used for human consumption *or* unless the crop-drying/storage problem warrants it.

Any crop, harvested prematurely, is appropriate for fractionation. The only exception might be those leaves rich in tannin or phenolic compounds (they inhibit protein extraction) or leaves which are fibrous and dry. Some observed yields of leaf protein are, for instance, 2 tons/ha-yr (3 tons expected) from winter wheat, fodder radish, and mustard in the UK, nearly 2 tons/ha-yr from irrigated lucerne in New Zealand, and 3 tons/ha-yr (5 tons potential) from cowpea in India (Pirie 1975c).

Normal crop breeding selects for high seed yield and not total protein yield. If desired, different crops could be bred for leaf-protein extraction, but a more reasonable approach is to use wastes from present agricultural crops. Among those plants successfully yielding protein from their wastes or by-products are (Pirie 1975c): brans (*Phaseolus spp.* and *Vicia faba*), jute (*Corchorus spp.*), peas (*Pisum sativum*), ramie (*Boehmeria nivea*), potato (*Solanum tuberosum*), and sugar beet (*Beta vulgaris*).

Of these, potato haulms, for example, could yield 600 kg protein per hectare (Pirie 1975b), whereas the potatoes typically yield less than 400 kg protein per hectare (Leach 1976). Likewise the haulms of peas harvested for freezing and canning could yield more protein than the peas themselves.

Trees, like weeds, are probably of little interest. Two exceptions include coppiced trees and water plants (weeds). The latter are a major and expensive nuisance in many overly fertilized waterways, and where a single, easily harvested species exists, removal can provide protein, nutrient removal and elimination of the need for herbicides.

Economics. Economic studies in the developed world have been based largely on the savings in crop-drying costs. These savings, together with the increased value of the protein concentrate, yield a decreased production cost per ton of dry forage from £55.20 to £46.45 (Bray 1976). In addition, the protein-concentrate production cost is £79.45 per ton, but sells for nearly twice as much as dry forage, i.e., £95/ton vs £54/ton. One particularly important factor is that the crop is 38% and 22% of the cost for the fractionated forage and protein fractions, respectively, which could be drastically reduced if crop wastes rather than a forage crop were used. Disposal of the residual liquor is nearly 20% of the leaf-protein cost, but would ideally have a positive value if used for fermentation or further processing. Clearly, economic considerations have meaning only in a specific context, but both the economics and simplicity of operation show uniquely high promise.

4

Fuels

Fuels are low-entropy, high-quality, stored-energy forms which can in turn be converted to a useful energy flow at a desired time and place. Ease of storage, transport, and conversion are the most important criteria for high-quality fuels. The energy-flow forms desired include, of course, heat and light, but also high-quality forms like very high-temperature heat, mechanical energy (motion, pumping) and electrical energy. As was discussed in Part I, Chapter 2, electricity should only be produced when absolutely necessary and then preferably via processes which avoid heat generation, as in steam plants or combustion engines. Economically practical alternatives do not currently exist, but this reemphasizes the need for the development of fuel cells for which hydrogen is the ideal fuel. Flexibility for present and future uses should be considered in all fuel-use planning.

Only those less familiar fuels which can also be biologically produced (presently or potentially) are discussed below: these include two gases, hydrogen (H_2) and methane (CH_4), and two liquids, methyl and ethyl alcohol (CH_3OH and C_2H_5OH, respectively). Table II.4.1 summarizes some of the physical data for these and for a few other fuels.

HYDROGEN

Hydrogen, the chemically simplest fuel, is a gas at room temperature and a liquid below $-253°C$. While not a fuel which occurs naturally in any abundance, hydrogen can always replace hydrocarbons (oil or natural gas) in addition to being uniquely well-suited to fuel cells (for electricity, see Part II, Chapter 3) and low-temperature catalytic heaters (heat). Chief among hydrogen's advantages are (Maugh 1972b, Gregory 1973):

- Little or no pollution—in oxygen the only combustion product is water, while in air oxides of nitrogen are the sole additional products. These occur in amounts far lower than with oil or even natural gas; no carbon or sulfur oxides nor particulates can result.

Table II.4.1 Fuel Data (from Dryden 1975).

	HEAT OF COMBUSTION		Boiling/ Melting Point °C	Density kg/meter^{-3}	Ignition Temperature °C	Lower/Upper Flammable Limit, %
	kJ/gram	kJ/cm^3				
Gas						
Hydrogen H_2	124.7	0.0010	$-252.5/-259.14$	0.178	570	4/74
Methane CH_4	61.1	0.0044	$-164 /-182.48$	0.718	650–750	5/15
Liquid						
Methanol CH_3OH	20.1	15.9	$64.5/-97.8$	800	464	6.7/36.5
Ethanol C_2H_5OH	26	20.5	$78.4/-116$	790	392 (95%)	3.28/18.95
Hydrogen H_2	124.7	8.7	$-252.5/-259.14$	70	—	—
Gasoline C_8H_{18}	44.3	30.9	$99/-107$	660–690	400–510	0.95/6.0
Solid						
Wood $C_{.32}H_{.46}O_{.22}$	17.5	14.2	—	350–500	—	—
Coal $C_{.28}H_{.42}$	32.2	41.8	—	1200–1500	—	—
Charcoal	26 (13–29)	7.–15.	—	280–570	—	—
Carbon	29.5	55.–62.	—	1800–2100	—	—
Hydrides (VH_2)	4.7	28.4	—	—	—	—

- Low-ignition energy (0.02 MJ or 7% of natural gas) makes hydrogen suitable for low-temperature catalytic burners which produce no pollutants.
- Rapid combustion increases hydrogen's efficiency as a motor fuel (50% greater for gasoline engines).
- Combusts with a wide range of air mixtures.
- Highest energy density (by weight) of all fuels: twice that of methane and as a liquid nearly three times that of hydrocarbon fuels.
- Hydrogen gas can be transmitted at the same rate (energy/time) as natural gas in existing pipelines.
- Electrical generation efficiencies of 80% are possible with hydrogen/oxygen fuel cells, independent of scale.
- Low viscosity (a third that of natural gas).
- Heat capacity for liquid hydrogen is 30 times that of normal jet fuel and can cool the surface of hypersonic aircraft.

The fact that hydrogen is not a commonly used fuel indicates that it is not without its disadvantages:

- Low energy density by volume for both hydrogen gas ($0.001 \ kJ/cm^3$ or less than a fourth that of methane) and liquid ($8.7 \ kJ/cm^3$ or less than a third that of gasoline).
- Explosive due to low combustion-threshold energy.
- High diffusibility.
- Hydrogen production requires conventional fuels.
- Liquification is energy- and capital-intensive.
- Storage and transport of liquid hydrogen requires great care and is expensive, nearly a 100-fold increase for both.
- Hydrogen presently costs about twice as much as methane.

Hydrogen is currently used more as a chemical substrate than fuel; 40% of the US annual production (one-third of the world's total) is, for example, used for ammonia (fertilizer) synthesis. As a result of this specialized use, no attempts at hydrogen production, storage, transport, or use as a fuel have been made on a scale which would permit an adequate comparison with the petroleum technology. Use of hydrogen as a fuel stands against the enormous investment and traditions of a petroleum and natural gas economy and would require a series of similar developments to realize that role.

Production

More than half of today's hydrogen is made by catalytic cracking of hydrocarbons, thus placing its cost well above its oil source and explaining why hydrogen is only used for chemical purposes. Several other alternatives for hydrogen production do however exist. First, via electrolysis, essentially a battery runs backwards, water can be split into hydrogen and oxygen using electricity. Electrolyzers have a theoretical electrical conversion efficiency of 120% (heat from the surroundings is also absorbed), but realistic potential efficiencies are about 85%, and typical present efficiencies are 60-70% (Maugh 1972b). Hydrogen costs clearly exceed electricity costs (per unit energy), but conversion can be justifiable as a load-leveling system to allow coal-based power stations to always run at full power. The excess during off-peak periods (often 50% of maximum demand) is stored as hydrogen and reconverted when required.

A second consideration can also justify electrolytic hydrogen production: electrical transmission and distribution account for 45% of its delivered cost ($2.22 of $4.89 per 10^6 Btu (1.05 GJ) vs $0.47 per 10^6 Btu for methane). Conversion to hydrogen followed by reconversion to electricity (via fuel cell) can therefore become economically preferable for electrical transmissions greater than 400 km above ground, or 32 km below ground (Maugh 1972b).

Electrolysis will not yield a generally usable fuel unless cheap nonpetroleum-based electricity is available: electricity at 0.4-0.7¢/kWh would yield hydrogen for $1.50-$2.50/10^6 Btu (vs natural gas at $0.50-$1.00/10^6 Btu). Current electrical costs are 0.6-0.9¢/kWh.

A second alternative for hydrogen production is via the endothermic chemical decomposition of water at 2500°C, far higher incidentally than the expected 900°C nuclear-reactor temperatures. Multistage decomposition reactions have claimed theoretical efficiencies of greater than 50% at 800°C and can use reactor or focused solar heat if the corrosive intermediates can be contained at high temperature and pressure (Hammond, Metz, and Maugh 1973). The final class of alternatives, photolysis, includes such high-technology approaches as the use of ultraviolet radiation from a fusion-reactor plasma for photochemical reactions, but of greater interest are the biological systems discussed elsewhere: algae (see Part II, Chapter 2), bacteria (see Part II, Chapter 3), and semi-synthetic enzyme systems (see Part II, Chapter 5).

Storage

Hydrogen gas can be stored by all systems presently used for natural gas (pressure tanks and underground porous rock), but the decreased energy density requires a three-fold volume increase. This represents a considerable cost and material investment and excludes the practicality of compressed-gas storage, most importantly for automobile use.

Energy density of hydrogen increases 10,000-fold when liquified (below $-253°C$) and stored in well-insulated containers. Such systems are in common use and have only two drawbacks: container materials are expensive and damage (e.g., car accidents if used for transport) carries a high fire-risk.

A promising chemical storage method with high-energy density (three times that of liquid hydrogen by volume), ease and safety takes advantage of hydrogen's high diffusibility and reactivity. With this method, the pressurized gas penetrates into a lattice of solid metals (e.g., vanadium) or alloys which form simple hydrides (e.g., VH_2 or $FeTiH_{1.6}$) with hydrogen. These hydrides are commonly compounds of the form AB_5H_X, where A is a lanthanide rare earth element, B is nickel or cobalt and X is as large as six hydrogen atoms (Powell et al. 1976). The hydride formation is exothermic and is reversed simply by heating to release hydrogen gas. High metal costs and weight remain the limit to applicability, but this system remains the most promising for transport use.

Transmission

Hydrogen gas can be transmitted via existing natural-gas pipelines at only a slight increase in cost as the decrease in energy density (volume) is nearly offset by the increased diffusibility and flow rate (Gregory 1973). While liquid hydrogen will not be transmitted by pipeline in the near future, it is conceivable, but insulation costs probably make transport of storage containers more economical.

Use

Use of hydrogen for combustion requires no new technology although burners must be modified. In car motors efficiency is increased 50% due in large part to the low combustion temperature and rapid combustion speed (Maugh 1972b). All energy is released quickly at the beginning of the cycle with less heat lost to the engine wall. The only pollutant, nitric oxide, is about 0.01 gram/km or 20-fold better than the postponed US emission standards. Water, the major combustion product, is the ideal recycled element of this fuel cycle, especially for home use where the total absence of pollutants eliminates the need for a chimney and adds thereby 30% to home-heating efficiency.

Aside from being pollution-free, hydrogen's greatest advantage is as an efficient generator of electricity via fuel cells. These were discussed in Part II, Chapter 3.

Aviation is expected to be the first large-scale hydrogen-use sector before the new century. The increased weight for cryogenic containers is more than offset by the three-fold increase in energy density for liquid hydrogen. Nitric oxides are drastically reduced, and the liquid can be used to cool the aircraft surfaces at hypersonic speeds. Economics, not technology, remains the decisive factor (Gregory 1973).

METHANE

Methane (CH_4) is the principal component of natural gas which is most commonly obtained from subterranean fields, both in association with crude oil and without. Its organic-matter origins relate it directly to a process of greater long-term interest, the anaerobic microbial digestion of organic wastes. The resulting fermentation products, so-called biogas (also "gobar" or marsh gas, see Part II, Chapter 3), are commonly 60% methane, with the remainder being mostly carbon dioxide. Small and variable amounts of hydrogen sulfide (H_2S) often contaminate biogas, which if abundant can be corrosive, offensive, and potentially harmful. If necessary, hydrogen sulfide and carbon dioxide can be removed by an alkali scrub or molecular sieve adsorption. A 2% flow of monoethanolamine, MEA, is often used and later recycled by heating.

Transport and storage of natural gas or methane is a problem due to the low energy density typical of all gases. For a gas, however, methane has an unusually high energy content (4.4 J/cm^3 or 4.4 $GJ/meter^3$), four times that of hydrogen, by volume. Transport is usually via low or high gas pressure pipelines (35-7000 MPascal, respectively), or for longer distances methane is cooled to $-164°C$ and shipped/stored as a liquid at room temperature (Dryden 1975). Liquification is, however, an expensive and energy-intensive process. Simple compression in standard steel cylinders is also not generally profitable except for special applications. Purified methane, for instance, compressed at one-ton pressure and stored in two cylinders (0.25 × 1.2 meter), has fueled a standard car 120 km (ESCAP 1975). Pipeline transport and gas storage in low-pressure tanks or underground aquifers is most common. Methane's explosiveness in 5-14% mixtures with air demands respect and care in its use.

Pipeline methane from a low-pressure reservoir, the digester itself, or a floating-dome type, can be combusted directly in various burners. These burners, though, first need to be optimized for the biogas. An unmixed burner has a soot-free flame which is weak, rather low-temperature (200°C), and inefficient due to incomplete combustion, whereas a flame which has been mixed with the stoichiometric combustion requirement of 9.5 $meters^3$ air per $meter^3$ gas, burns with a short, stout flame of 550°C. Efficiencies in even simple but carefully designed burners approach 60% (ESCAP 1975). Similar burners equipped with a glow-mantle provide equally troublefree lighting.

A high-quality fuel-gas, like methane, is also well-suited to the invaluable combustion engine. Four-stroke, low-compression petrol or kerosene engines ignited by electric spark perform well with either pure methane or the 60% methane (plus carbon dioxide) mixture in biogas. Power output is reduced by only 10-15% at a use-rate of 500-750 liters of gas (60% methane) per horsepower hour (ESCAP 1975). Diesel motors, using a little oil for ignition, also run well on biogas; a simple fuel injector and carburetor are the only required alterations. Highest motor efficiencies are achieved at a 13.2% gas mixture.

Aside from questions of economics and fuel oil availability, methane or biogas is said to have advantageous side effects from its clean burning characteristics. Engine life is increased fourfold, maintenance is reduced by half, the lubricating-oil life is extended, and carbon-monoxide production is substantially reduced (ESCAP 1975). The cleanliness of the exhaust gases, principally carbon dioxide, would ideally allow its recycling for algal growth (see Part II, Chapter 2). The potential use of methane in electric-current-producing cells is disscussed in Part II, Chapter 3.

METHANOL

Methanol (methyl or "wood" alcohol, CH_3OH) is the partially oxidized product of methane (CH_4) and hence is the simplest alcohol. It is a colorless, odorless, water-soluble liquid which boils at 64.6°C and auto-ignites at 467°C (see also Table II.4.1 for other properties). As a liquid fuel it is surpassed only by gasoline which has twice the energy density by both weight and volume.

As a general-purpose fuel methanol is suited to virtually all energy needs, particularly those requiring a clean flame as in many industrial operations and home heaters. Among its advantages are (Reed and Lerner 1973, Mitre 1974):

- No particulates are released, and existing soot is burned off.
- NO_x in product gases is less than from natural gas and much less than from oil.
- CO in product gases is less than from either oil or gas.
- No sulfur compounds are emitted.
- Negligible amounts of aldehydes, acids, and unburned hydrocarbons are emitted.

In addition, the low energy density is partially offset by methanol's high combustion efficiency.

Production, transportation and end-use alternatives for methanol were shown in Figure II.3.18 in Part II, Chapter 3, where production was discussed. Transport of methanol, unlike gaseous fuels or liquid hydrogen, is identical to that of gasoline with the only exceptions being methanol's water solubility and its tendency to corrode aluminum, magnesium and some gasket materials. With some exceptions, existing facilities can be used if provisions to exclude water are made.

Special end uses suitable to methanol, other than simple combustion, are almost more varied than the petroleum it is to replace:

1. Fuel cells—a methanol/air fuel cell using tungsten carbide and charcoal electrodes (with sulfur-acid electrolyte) has generated electricity continuously for more than 30,000 hrs (Reed and Lerner 1973).
2. Synthesis of ethanol from methanol and synthesis gas (CO and H_2) is possible with a cobalt catalyst. Ethanol can be dehydrated to ethylene, which in turn is a chemical base for production of plastics and many other petrochemicals.
3. Single-cell protein (bacteria) in good yield and purity can be grown on methanol with little risk that pathogenic bacteria can survive on the same substrate (IFIAS 1974).
4. Amino acids, drugs, and other chemicals may be produced by mutated methanol-oxidizing bacteria (IFIAS 1974).
5. Methanol as a motor fuel or fuel additive: Methanol's great potential as a fuel for internal combustion engines, such as in the modern automobile, has been exploited to only a limited extent in race cars and some yet aircraft. Interest in methanol use in internal-combustion engines results from its special properties as a fuel (Gregory 1975, Ingamells and Lindquist 1975):
   a. Methanol blends well with other hydrocarbons.
   b. Methanol has a high octane number, and even a 10% blend increases gasoline's (90%) quality by two or three octane numbers.
   c. Power output can be increased by as much as 10% with rich mixtures of methanol because its high heat of vaporization cools and increases the density of the intake air.
   d. Energy-conversion efficiency with a 10% methanol blend increases ca 2%. A 6%-efficiency increase is observed in methanol-burning turbines.
   e. Nitrogen-oxide ˙emissions for methanol are half those of gasoline, and if a fuel-injection system is used, all pollutants are reduced except for unburned fuel. Aldehyde emission is unchanged.
   f. Higher compression ratios (10-12) and leaner mixtures (air:methanol::14:1) are possible with methanol than with gasoline.

Methanol is, of course, not without its disadvantages as a motor fuel:

a. Methanol (pure or blended with gasoline) is hydroscopic.
b. Small amounts of water (less than 0.1%) in a methanol-gasoline blend cause a phase separation with methanol-water droplets settling to the container bottom.
c. Heat of combustion for methanol is half that of gasoline.
d. Heat of vaporization of methanol is four times that of gasoline and hence requires preheating.
e. Engines fueled by pure methanol will not start below 18°C without a volatile starting fuel (e.g., ether). Cold starts are not a problem with blends up to 40% methanol.
f. Low concentrations of methanol increase the vapor pressure of the blend above that of either component. Vapor locks due to gasoline in the carburetor boiling are frequent if volatile gasoline components aren't removed during refining.
g. Drivability in new cars (particularly with emission-control devices) is aggravated by methanol-gasoline blends (over 10%), as judged by frequency of stalling, hesitation and surging.
h. Aldehyde emissions are not reduced for methanol fuels.
i. Methanol is corrosive. It dissolves the lead-tin gas-tank liner, resulting in clogging by lead hydroxide and tank leakage. Magnesium and magnesium-aluminum alloys, common in engine blocks, fuel pumps and fuel tanks, in addition to some gaskets, are susceptible to corrosion by methanol.
j. Cost per kilometer is three-fold greater for methanol fuel than gasoline.

A close analysis of the above methanol characteristics shows that the advantages of methanol as an additive are disproportionately large. Economy, however, is less clear as methanol prices exceed petroleum threefold. Economic questions, like many of the problems with methanol use in cars, reflect the traditional role of petroleum: internal-combustion engines, carburetors, storage and transport facilities are all optimized for petroleum not methanol. The low average lifetime for a car suggests that redesigning new cars made to use methanol as a fuel will not be the greatest problem. Patchwork adaptations to existing private cars may invite more problems than the $100 conversion cost, but evaluations for cars yet unbuilt should include the $500 for emission-control devices eliminated through the use of methanol; this factor has been deemed worth 4¢/liter extra fuel cost.

The limits to methanol use are its cost and constancy of supply. At present, world consumption is just over 25,000 tons per day with the chemical industry as essentially the sole user (Burke 1975). Chemical use requires a purity far greater than is necessary for fuel use, and impurities, typically higher alcohols, would both lower the price and increase the energy content of an unrefined "methyl fuel." Conversion of cars and factories to methanol use requires, of course, confidence that its availability is assured. This availability is an economic question.

Economics

Chemical-grade methanol synthesized from natural gas increased in retail price from 3¢ to 10¢/liter between 1972 and 1975 (Burke 1975). Although methanol is the simplest and cheapest alcohol, even larger-scale coal-conversion operations (ca 5000 tons/day) cannot hope to lower prices below present levels. Some estimates range as high as 15¢/liter for plants with capital costs of $400-million (5000 tons/day). These costs, if compared with gasoline on an energy-content basis, are 30¢/liter or double the present gasoline cost. Estimates for conversion of Persian Gulf methane (presently flared off and wasted) to methanol are comparable at 20 ¢/liter, but this should decrease somewhat for fuel-grade purity. Conversions of wastes to methanol are more promising as they have the economic advantages of employing a substrate with a negative value, but practicality rests on pretreatment and supply problems.

Toxicity

Methanol is not much more toxic than gasoline, but as an alcohol the chances of ingestion are far greater. Blindness and death can result from either inhalation or ingestion of methanol: the maximum allowable exposure to methanol vapors is 200 parts per million, while ingestion of 10 ml can cause sickness and 30 ml death (Mitre 1974).

ETHANOL

Ethanol or ethyl alcohol (C_2H_5OH) is a liquid fuel resulting from distillation of the yeast-fermentation product (see Part II, Chapter 3). Grain fermentation has been used for thousands of years, traditionally on the most primitive level but in modern times on an advanced industrial scale.

Ethanol's most common use as an inebriating beverage accounts both for its long history and to some extent for its unpopularity as a fuel. As a liquid fuel its chief disadvantage is its low energy density relative to that of petroleum (about half).

Liquid fuels are preferred over gaseous fuels principally for their thousand-fold greater energy densities and consequently their storage ease. Ethanol has the additional advantage of being the only liquid fuel easily produced biologically. Storage is, of course, important to compensate for seasonal variations in substrate availability but also for both transport of fuel and fueling of transport vehicles and other mobile or remote combustion engines.

Ethanol has been used for 50 years as a combustion-engine fuel additive, but only in Brazil has it been used nationwide for automobile use. In normal internal-combustion engines, an up to 20% ethanol blend can be used without any adjustment although 2-8% is more common. Recent experiments in engines designed for use of pure (95%) ethanol show that in spite of its low combustion energy density, ethanol delivers as much or more power per liter than gasoline (Hammon 1977). Due in part to its relatively high ideal compression ratio (10), the pure-ethanol engine delivers 15-18% more power per liter than a petroleum but with a compensatory 18% greater volume rate of consumption. Careful tuning for a lean fuel mixture can in fact give the ethanol engine in practice not only slightly better mileage than the gasoline engine, but also lower pollution: carbon monoxide and nitrogen oxides (NO_x) are reduced by half. Lead pollution is avoided completely as the gasoline additive, tetraethyl lead, is not needed.

Diesel motors, typical of larger transport vehicles and stationary engines for irrigation and power generation, cannot use pure ethanol directly. A special dual-carburetion system does however allow a 50-50 diesel fuel-ethanol mixture.

Alcohol use for heating and cooking is of course possible, but as a high-quality fuel its use should be restricted to uses with high-quality requirements. Transport and storage considerations may however extend ethanol's use.

Ethanol's potential misuse cannot be ignored in practice as it is in this brief discussion and may require an additive to reduce its palatability.

5

Long-Range Research
for Semi-Synthetic Systems

Scientific research has narrowed the distinction among, for example, biology, chemistry and physics. In the present context, biology need not be confined to living organisms as these organisms can be viewed as complex factories manufacturing equally complex functional systems from simple abundant materials—water and air—and solar energy. Purified enzymes can replace rare-element catalysts, and plant pigments can replace photoelectric cells. A few examples of synthetic uses of biological components are considered here.

IMMOBILIZED ENZYMES AND CELLS

Enzymes. Enzymatic hydrolysis of cellulose discussed earlier employed the soluble enzyme, cellulase. Two factors should be noted when considering other enzymatic processes. First, enzyme-related costs, such as the initial enzyme supply, loss make-up, and enzyme separation/recovery costs, were shown to be a major economic burden in the overall process. Secondly, most cellulose (crystalline and semi-crystalline cellulose) is not soluble, and hence the hydrolyzing enzyme must be transported in solution to the substrate rather than vice versa. This second point makes cellulose a rather unique substrate, but then few substrates in the context of this discussion are as abundant.

A multitude of soluble substrates exist which while of rather small abundance are nonetheless of considerable importance. Table II.5.1 shows several important conversions catalyzed by immobilized enzymes. Most of these are relevant to the modern food industry and are thus rather specialized for this discussion, but conversions of starch to glucose, inter-conversions of various sugar forms (dextrose to fructose, sucrose to dextrose and fructose), and oxygen scavenging can have wide application.

Any suitable *soluble* substrate has the advantage over cellulose in that

Table II.5.1   Applications for Immobilized Enzymes (Weetall 1975).

Enzyme Category	Application	Specific Enzymes
Proteases	protein hydrolysis	papain, ficin, bromelain, trypsin, chymotrypsin, pepsin, pronase, amino-peptidase
	cheese manufacturing	rennin, pepsin
	beer, ale, and other malt beverages, chill-proofing and haze removal	papain, ficin, bromelain
	treatment of fish press water to low viscosity	ficin, papain, trypsin, pepsin
	microbial hydrolysis	trypsin, lysozyme, nucleases
	rendering	papain, ficin, bromelain
	viral inactivation or destruction	trypsin, RNase, lysozyme
	hydrolysis of pectin esters in fruit juice	pepsin, ficin, bromelain
	amino acid production from sludge and other protein-containing wastes	pronase, peptidases
Carbohydrases	hydrolysis of cheese whey, whey products and lactose in milk	lactase
	hydrolysis of raffinose in beet sugar	α-galactosidase
	convert sucrose to dextrose and fructose	invertase
	convert dextrose to fructose	glucose isomerase
	hydrolysis of starches	amylases (alpha and beta)
	dextrose production	amyloglucosidase
	hydrolysis of celluloses	cellulase
	hydrolysis of poly-saccharides such as gum and mucilages	amylases
	preparation of sub-stituted sugars and pentoses	amylases
Lipases	hydrolysis of lipids	lipases
	rendering	lipases
Esterases	nucleotide production	RNase, DNase, alkaline phosphatase nucleotide phosphorylase
	solubilization of tea cream	tannase

continued

Table II.5.1—*continued*

Enzyme Category	Application	Specific Enzymes
Miscellaneous	hydrolysis of pectins, clarification of fruit juices, extract purees, concentrates, vinegar, cider, etc.	pectinases
	glucose removal from eggs prior to drying	glucose oxidase, catalase
	oxygen scavanging	glucose oxidase
	sterilization or cold pasteurization	catalase
	steroid transformations	transferases, hydroxylases, carboxylases, decarboxylases, dehydrogenases
	fine chemical production	—
	desulfunation of gas	—
	resolution of racemic mixtures of amino acids	aminoacylases
	penicillin production	penicillin amidase

the expensive enzyme can be immobilized and the substrate allowed to flow past (Messing 1975, and Weetall 1975). In addition to conserving expensive enzyme for continuous reuse, other advantages include:

- the enzyme catalyzed process can be stopped by simply removing the enzyme matrix;
- the enzyme's lifetime, i.e., its stability, is often increased as a result of immobilization;
- the product solution is free from enzyme, thus eliminating undesired reactions and the need for complex separation steps.

Each of the above factors is important not only for decreasing conversion costs, but also for decreasing the complexity of the process and the equipment required. Process complexity in some cases is little more than pouring a suitable substrate in the top of a vertical column containing the immobilized enzyme and collecting the product at the bottom.

Immobilization represents an expense, of course, and the enzyme must be expensive to justify it. Several alternative immobilization techniques are shown in Figure II.5.1 of which some require only a very simple matrix such as cellulose (Messing 1975):

- Adsorption of the enzyme in or on a carrier matrix. Charcoal, the first carrier, was first used in 1916, but now glass, organic polymers (e.g., cellulose), mineral salts, metal oxides, and many other porous

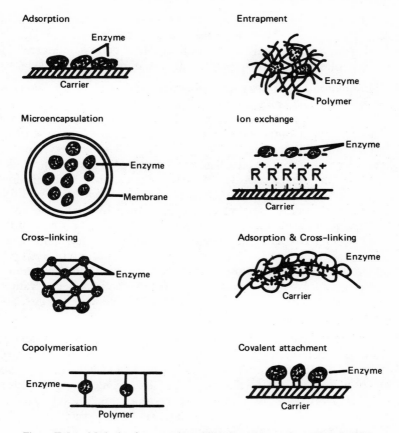

Figure II.5.1. Methods of enzyme immobilization (redrawn from Weetall 1975).

compounds have been used. Adsorption can be ionic, covalent, or in several cases held via as yet not understood mechanisms.
- Cross-linking enzymes via bi-functional molecules (i.e., compounds such as glutaraldehyde which have two binding sites). Cross-linking can be done free in solution to form a gelatinous polymer or together with a carrier matrix.
- Encapsulation within a membranous vesicle or entrapment within a tangled polymeric matrix.

Cell Immobilization.   A further extension of enzyme immobilization—which at first glance might be considered a retrograde step as it is simpler than immobilization—is the encapsulation or immobilization of whole bacterial cells. That is, instead of a single purified enzyme, one immobilizes a cell population. Immobilization of cells—biological "factories"—is often preferred to single enzyme immobilization because:

1. Elimination of the enzyme purification stages is an enormous savings in time, complexity, and capital material.
2. The enzyme(s) is in its natural cellular environment while only the more resilient outer membrane is subjected to the strains of synthetic immobilization.
3. Some processes are multistaged and require a series of several enzymes. In some cases several enzymes can be immobilized together, but utilizing whole cells extends the range of possibilities. More importantly, several enzymes—which by definition are not consumed during the reactions they catalyze—require additional factors called coenzymes which *do* need to be regenerated after each reaction. That is the coenzyme donates a portion of itself (e.g., a proton, H^+) which must be replaced by some other reaction not directly involved in the enzyme-catalyzed reaction. Regeneration of coenzymes is extremely difficult with immobilized enzymes, but is possible with immobilized cells.

Immobilized bacterial cells should not be confused with fermentation processes. Fermentation occurs with growing and dividing cells which metabolize a high-energy substrate ("food") and happen to have a "waste product" which is valuable, e.g., the conversion of glucose to alcohol. Immobilized cells are in their resting, nondividing phase, an ambivalent state between the normal concepts of living and dead. They are not "fed" as in fermentation, but rather are provided with an intermediate metabolite which is a particular enzyme's substrate.

The only commercial application to date is the production of amino acids via the resolution of a racemic mixture to optically active, L-amino acids (Chibata et al 1974), e.g., L-aspartic acid from fumarate. Even this seemingly esoteric conversion has significance for medicine and nutrition. It exploits the versatile intestinal bacteria, *Escherichia coli*. An additional process recently shown to be feasible is the conversion of lactose to the simple sugars glucose and galactose (Ohmiya, Ohashi, and Kobayashi 1977). Lactose is also a sugar from milk and cheese whey, but is not tolerated by those who have a genetic defect for the relvant digestive enzyme, lactase.

Immobilized bacteria can also be used for fuel production: the photosynthetic bacterium *Rhodospirillum rubrum* was immobilized in agar and successfully converted malate ($C_4H_6O_5$) to hydrogen plus CO_2 (Weetall and Bennett 1976).

Encapsulated/immobilized bacterial cells represent a whole new dimension of enzyme engineering by extending the range of potentially exploitable enzymatic reactions. Use of mutated cells with abnormally high concentrations of desired enzymes or with unusual blockages in enzymatic pathways will further extend the applications.

ENZYMATIC HYDROGEN PRODUCTION

Hydrogen production from intact organisms has been discussed previously (see Part II, Chapters 1, 2, and 3), but the simplest biological system would eliminate the organism and include only its organelles and enzymes. The required components for the semi-synthetic system in Figure II.5.2 would be:

- *chloroplasts* acting like a photocell to produce a voltage (charge separation, or reduction potential) from collected solar energy;
- *ferredoxin* to carry the electron (i.e., an electrical "current") from the chloroplast surface membrane to
- *hydrogenase*, the enzyme which catalyzes the production (or dissolution) of hydrogen gas from (to) a proton (H^+) and an electron.

As implied above, hydrogen gas should be evolved if one shines light on a reactor containing the appropriate elements (discussed further in Part II, Chapter 1). This seemingly trivial experiment has successfully produced hydrogen gas (Benemann, Berenson, and Kaplan 1973), but this description belies a long list of serious problems. This list begins with the critical and instable enzyme involved, hydrogenase.

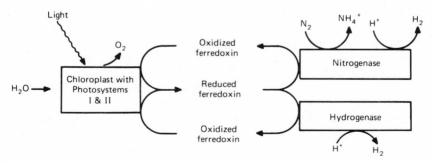

Figure II.5.2. Semi-synthetic system for solar hydrogen and ammonia production.

Hydrogenase is a single polypeptide containing a four-iron cluster which is nearly identical to that of a bacterial ferredoxin. The similarity is not surprising as both have the same electron-transfer function, but hydrogenase has the additional ability to bind a proton in a so-called enzyme hydride complex (EH^-; Krasna 1976). Such a function is built into the protein structure surrounding the iron-sulfur active center and explains in part the tenfold greater molecular weight (60,000 vs 6000 for bacterial ferredoxin) of hydrogenase.

The enzyme-hydride complex is the intermediate form in the reaction catalyzed by hydrogenase,

$$2H^+ + 2e^- \xrightleftharpoons{\text{Hydrogenase}} H_2.$$

Note that the electrons (e^-) are provided by the electron-transfer enzyme, ferredoxin, and that the equilibrium of the above reaction is not altered by hydrogenase, it merely promotes the equilibration process. This is in contrast to the hydrogen production via nitrogenase discussed previously for which only the forward reaction (gas production) is possible, a reaction which in that case requires energy (ATP).

The hydrogenase reaction can also be contrasted with platinum or palladium catalysts which, while only promoting the equilibration process, do so via a so-called homolytic cleavage,

$$H_2 + Pt \rightleftharpoons HPtH \rightleftharpoons 2H^+ + 2e^- + Pt.$$

Hydrogenase is a heterolytic reaction as described above, but the details of the process are not known. Electron-spin resonance studies show that the enzyme can transfer both the required electrons by using all three of its available oxidation states, but the alternative—use of only two oxidation states and ferredoxin at each stage of hydrogen production (cleavage)—is more probable; diatomic hydrogen is not bound (Erbes, Burris, Orme-Johnson 1975).

Not only protons bind to hydrogenase. Carbon monoxide (CO) is a competitive inhibitor which not only binds, but also deforms the iron-sulfur active site. Much more serious, and perhaps the ultimate impediment to the exploitation of hydrogenase, is its extreme oxygen sensitivity. This sensitivity varies widely with the organism, and the inhibition can be either reversible (e.g., *Chromatium* hydrogenase) or irreversible (e.g., *Clostridium*). In no case is the inhibition mechanism known, but practical improvements can be attained by immobilization of the enzyme on glass beads. As discussed on page 330 of this chapter, immobilization primarily conserves valuable enzymes in flow-through reactors, but in this application the volatile gas product creates no analogous separation problem.

Instead, the hope is that covalent bonding of the enzyme to some fixed support (e.g., glass beads) will increase enzyme stability without decreasing its activity. While the oxygen-inactivation mechanism is not understood, hydrogenase stability is known to vary with source organism (Gilitz and Krasna 1975), suggesting that oxygen inactivates by attacking somewhere other than the active site. Immobilization of the hydrogenase enzyme at the site of oxygen attack would then increase stability, but would leave the active site free.

Preliminary immobilization attempts are promising but hardly definitive. Figure II.5.3 summarizes some of the data (Lappi et al. 1976) and shows that immobilization on glass slows inactivation (to 50% activity) by several orders of magnitude: from one to two minutes for soluble hydrogenase to several days when bound to succinyl carbodimide. Inactivation follows two different rates and suggests that immobilization has merely selected a stabile isoenzyme. Immobilization has, however, the second side-effect of reducing hydrogenase activity by more than 90% via a mechanism which is *not* time-dependent and varies with the glass support used. It can result in part from enzyme inactivation, but stearic blockage of the active site seems to dominate: activity increases for smaller sized substrates (i.e., electron donor) less affected by blockage. Ferredoxin, the normal substrate, is much less effective than, for example, the smaller electron donor, methyl viologen (Lappi et al. 1976, and Berenson and Benemann 1977). This stearic exclusion of ferredoxin increases for those glasses which best protect against oxygen inactivation,

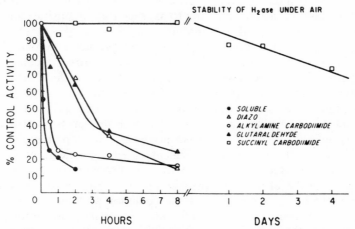

Figure II.5.3. Inactivation of bound hydrogenase with time in air (by permission of Lappi et al. 1976).

suggesting that the two sites may be identical or physically very near. Ferredoxin and hydrogenase were immobilized together on glass, but proved to be inactive; bound ferredoxin alone has however 5-20% the activity of the soluble form (Berenson and Benemann 1977).

Too little is known yet to decide if hydrogenase can be employed for practical hydrogen production. If the active center is the site of oxygen inactivation, then lability is inherent to the active enzyme and oxygen cannot be allowed. However, the wide range of oxygen sensitivities for hydrogenases from different sources suggests otherwise. Recent dramatic progress in synthetic compounds modeling the active sites of ferredoxin and hydrogenase promises an additional path for understanding and perhaps replacing these enzymes (Holm 1976).

APPLICATIONS OF STABILIZED CHLOROPLASTS

Chloroplasts are the plant organelles which absorb and convert solar energy to chemical potential energy. It is therefore the membrane enclosed biological photocell (potential, ca one volt) which powers all later syntheses via the connecting enzymatic "wire," ferredoxin. Viewed as an isolated organelle, chloroplasts are therefore the basic solar-energy conversion unit. Clearly, conversion can also be considered at the molecular level, e.g., chlorophyll, and such attempts with a more versatile and stable pigment, bacteriohodopsin, will not be discussed.

As discussed early in Part II, Chapter 1, the conversion efficiency of red light is ca 37%, and only the realities of the solar spectrum and later losses lower maximum theoretical efficiencies to ca 13%. These efficiencies and the chloroplast's adaptation to average light intensities and spectral properties will be hard to improve.

Two difficult and not necessarily soluble problems remain before chloroplasts can be considered for practical use. The first concerns photosaturation. While chloroplasts have high efficiency at low light intensities, they have lower efficiency in stronger light and can even self-destruct at high intensities. Typical applications would probably involve thin layers of chloroplasts and thus exacerbate this problem which less frequently occurs in Nature. Furthermore, *in vivo* repair mechanisms (Part II, Chapter 1) would be lacking in *in vitro* applications. Genetic breeding, as discussed earlier, may solve this problem, or perhaps algae grown under artificial intense light may provide chloroplasts with a high saturation threshold. One remaining manipulation possibility is a class of non-natural electron acceptors which can replace the rate-limiting reaction connecting the two photosystems (Saha et al. 1971). Much remains unknown about the factors affecting photosystem unit size, limiting reaction rates, quantum efficiency and photoinhibition/destruction.

The second problem, chloroplast instability, is also exacerbated by strong light, but even under ideal dark conditions (in 0.4 M sucrose, pH 6.5-8.0, 0.4°C) chloroplasts retain activity for only several hours. Storage in serum albumin (1-2%) can extend the dark lifetime to 24 hours at 25°C and three days at 0°C (Wasserman and Fleisher 1968, Lien and San Pietro 1976).

Stabilization efforts are complicated by the lack of understanding of the degradation mechanisms. The known sensitivity to degradative enzymes, particularly membrane surfaces, has motivated attempts to stabilize the outer membrane. Three main approaches have been tried (Lien and San Pietro 1976):

1. chloroplast fixation with inter- and intramolecular cross-linkages (e.g., with glutaraldehyde; Packer 1976);
2. microencapsulation of the chloroplast in semi-permeable membrane vesicles which retain enzymes and cofactors while allowing free passage of substrates and products (Butler and Kitajima 1975);
3. immobilization on glass beads (Nugent 1972).

Long-term chloroplast stability is such a recent research goal that the present lifetimes should not be discouraging.

BACTERIORHODOPSIN—A NONCHLOROPHYLL PHOTOSYSTEM

Where virtually all photosynthetic organisms employ chlorophyll as the light absorber and photoreaction center, the halophilic bacterium, *Halobacterium halobium,* contains instead a variant of the visual pigment, rhodopsin (Oesterhelt 1976). The normal function of bacteriorhodopsin, as this purple protein is called, is to transport protons from one side of the cell membrane to the other and thus create an electric field gradient (potential) at the expense of light energy. This gradient is normally used to drive biochemical reactions, but its adaptability as an ion exchanger (e.g., for desalination) or a photovoltaic cell is of greater interest here.

Bacteriorhodopsin is a single polypeptide (molecular weight 27,000) which normally is organized into a two-dimensional hexagonal lattice structure in purple membranes. These membranes contain phospholipids and carotenoids (25% by weight) in addition to the bacteriorhodopsin and are integrated as patches on the normal cell membrane. Each purple-membrane patch is about 0.5 μm in diameter, 4.8 nm thick, and contains about 10^5 bacteriorhodopsin molecules oriented in the same direction (Oesterhelt 1976).

The unidirectional orientation of bacteriorhodopsin molecules in purple membranes permits a concerted effect when "pumping" protons from one membrane face to the other. No mechanism yet satisfactorily accounts for this proton transfer process, but it has a quantum efficiency (quanta absorbed photons/protons transferred) approaching one. At normal light intensities the purple membrane absorbs one quantum per second at 570 nm and hence transfers protons at approximately the same rate. However, as absorption occurs at 560-570 nm, only a few percent of the incident energy is even potentially absorbed.

Potential Applications

Photovoltaic cell. A synthetic photovoltaic cell has been constructed using bacteriorhodopsin. As the purple membranes themselves are too small to work with directly, a millipore filter impregnated with phospholipids was used to bind bacteriorhodopsin-containing vesicles (proteoliposomes) on one side. This filter complex is used to cover a one centimeter aperture in a Teflon partition of a two-compartment cell. The maximum photovoltage produced was 150 mV, while a three-cell "battery" consisting of two bacteriorhodopsin-filters produced twice this potential (Drachev et al 1975, Skulachev 1976).

pH gradient. The same scheme can be used to create a pH gradient by electrically shunting the two sides of the cell. This pH gradient, i.e., the creation of one alkaline and one acidic cell, could be used to drive certain chemical reactions (Skulachev 1976).

ATP generator. ATP-ase, the enzyme system(s) which catalyze the attachment or removal of a pyrophosphate moiety, are believed to operate also in conjunction with transmembrane proton transfers. As ATP is required to supply the energy for many biochemical reactions, its supply in large quantities will soon be demanded by the enzyme engineering industry. The scheme in Figure II.5.4 is amenable to phosphorylation by incorporating ATP-ase and bacteriorhodopsin in the vesicles. Such a system could conceivably be developed for totally synthetic photophosphorylation.

Ion exchanger (e.g., desalinator). The electric field gradient caused by the proton pumping can instead create an alkali-ion gradient if the membrane separating the two pools can be made permeable to only a

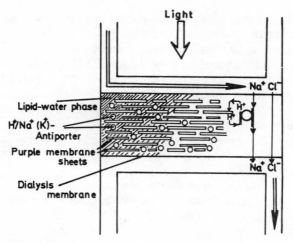

Figure II.5.4. Light-driven bacteriorhodopsin proton pump (by permission of Oesterhelt, 1976).

certain ion(s). If the membrane cannot be made selectively permeable, a carrier or ionophore which is lipid-soluble can selectively carry specific cations through the membrane; chloride has a high permeability and would probably follow across without the aid of a carrier.

A sketch of such a light-driven, quantum desalinator is shown in Figure II.5.4 (from Oesterhelt 1976). Purple-membrane sheets of macroscopic sizes would need to be constructed, possibly by applying purple membranes on mica sheets, orienting them with electromagnetic fields, and finally stabilizing the array with a gel or by cross-link binding.

A FINAL COMMENT ON RESEARCH

The examples of this final chapter are but a few of the ideas conceived by researchers, who increasingly are under pressure to consider applications of their basic investigations. Much of this fantasizing is stimulated by both real needs and the remarkable elegance and flexibility with which biological organisms—and their constituent parts—perform complex functions. In those instances where, for example, immobilized cells are already commercially functioning as miniature modular factories, the traditional industro-chemical factory performing analogous self-controlled functions would require an astounding complexity. Clearly the bioengineering age has only begun, but the final example mentioned above, the photo-enzymatic desalinator, is usefully illustrative of an important point.

If one assumes that rhodopsin in the desalinator described will absorb 10% of the incident sunlight (because like any photopigment it has a limited absorption bandwidth) then a solar radiation of 5×10^{-2} J/cm^2-sec will result in the absorption of about 10^{20} quanta/$meter^2$-sec (600 nm light is about 210 kJ/Einstein). This implies that two sunlight hours would be required to remove one mole of salt, such as desalinating from 1 M to 0.01 M NaCl. This desalination rate, 0.5 liters/hour-$meter^2$, does not compare favorably with a primitive flat glass plate distillator which can yield several liters/$meter^2$-hour. In addition, relative construction costs of the two alternative stills would separate their water purification costs by orders of magnitude.

This rather trivial example simply points out the economic nonfeasibility of an elegant and undoubtedly feasible biological fantasy. Such examples are easy to find, and in fact one needn't fear for economic's inability to weed out those realizable biological dreams which are nonprofitable. Perhaps a more subtle example should have been chosen, such as tracing a microbiological or genetic development to an unlikely, but potentially disastrous result. But the example of using an enzymatic distillator to provide water for village dwellers has value in its simplicity: The biological sciences mustn't rest on the "can do" pedestal once occupied by the physical sciences. The world today is both blessed and cursed by the realized fantasies of previous decades. The sometimes intoxicating elegance of recent and future biological discoveries must not be allowed to automatically justify their exploitation. Exploitation has—in spite of its often pejorative uses when referring to technology—positive connotations: combined careful research developments from the physical and biological sciences can exploit, for example, solar energy to viably energize future societies, but Ockham's razor applies not only to logic but to technology: in a market unbiased by scientific prestige, simplicity will prevail over *unnecessary* complexity. A simple solar distillator of flat glass may still have its place.

APPENDIX:
Conversion Factors

To convert from:	To:	Multiply by:
Length		
centimeters (cm)	inches	0.394
feet (ft)	centimeters	30.5
inches (in)	centimeters	2.54
kilometers (km)	miles	0.621
meters (m)	feet	3.28
meters (m)	yards	1.094
miles (mi)	kilometers	1.609
millimeters (mm)	inches	0.0394
yards (yd)	meters	0.914
Area		
acres	hectares	0.405
acres	sq. meters	4047
hectares (ha)	acres	2.47
hectares (ha)	sq. meters	10,000
sq. centimeters (cm^2)	sq. inches	0.155
sq. feet (ft^2)	sq. meters	0.0929
sq. inches (in^2)	sq. centimeters	6.45
sq. kilometers (km^2)	sq. miles	0.386
sq. kilometers (km^2)	hectares	100
sq. meters (m^2)	sq. feet	10.76
sq. yards (yd^2)	sq. meters	0.836
Volume		
barrels (petroleum, bbl)	liters	159
cubic centimeters (cm^3)	cubic inches	0.0610
cubic feet (ft^3)	cubic meters	0.0283
cubic inches (in^3)	cubic centimeters	16.39
cubic meters (m^3)	cubic feet	35.3
cubic meters (m^3)	cubic yards	1.308
cubic yards (yd^3)	cubic meters	0.765
gallons (gal) US	liters	3.79
gallons (gal) Imp.	liters	4.545
gallons (gal) Imp.	gallons, US	1.20
Weight		
grams (g)	ounces, avdp.	0.0353
kilograms (kg)	pounds	2.205
ounces avdp. (oz)	grams	28.3
pounds (lb)	kilograms	0.454
tons (long)	pounds	2240
tons (long)	kilograms	1016
tons (metric)	pounds	2205
tons (metric)	kilograms	1000
tons (short)	pounds	2000
tons (short)	kilograms	907

continued

To convert from:	To:	Multiply by:
Pressure		
atmosphere	grams/sq.cm	1033
atmosphere	pounds/sq.in	14.7
pounds/sq.in (psi)	grams/sq.cm	70.3
Energy		
British thermal units (Btu)	kilojoules	1.054
calories (cal)	joules	4.19
ergs	joules	1×10^{-7}
kilojoules (kJ)	Btu	0.948
joules (J)	calories	0.239
kilowatt-hours (kWh)	megajoules	3.6
megajoules (MJ)	kilojoules	1000
gigajoules (GJ)	megajoules	1000
terajoules (TJ)	gigajoules	1000
Energy Density		
Btu/gal	joules/cm^3	0.27
Btu/ft^3	kJ/m^3	36.5
Power		
horsepower (hp)	Btu/min	42.4
horsepower (hp)	horsepower (metric)	1.014
horsepower (hp)	kilowatts	0.746
kilowatts (kW)	horsepower	1.341
watts (W)	Btu/hour	3.41
watts (W)	joules/sec	1
Miscellaneous		
liter petrol	megajoules	35
kilogram oil	megajoules	43.2
barrel oil equivalent	gigajoules	6.1
ton coal equivalent	gigajoules	29.3
ton coal equivalent	barrels oil equivalent	4.8
pounds/acre	kilograms/hectare	1.1

REFERENCES

Abeliovich, A., and Azov, Y. 1976. Toxicity of ammonia to algae in sewage oxidation ponds. *Applied and Environmental Microbiology* **31:** 801–06.

Almin, K. E., Eriksson, K-E., and Pettersson, B. 1975. Extracellular enzyme system utilized by the fungus *Sporotrichum pulverlentum* for the breakdown of cellulose. *European Journal of Biochemistry* **51:** 207–11.

Andersen, K., and Shanmugam, K. T. 1977. Energetics of biological nitrogen fixation. *Journal of General Microbiology* **103:** 107–22.

Anderson, L. L. 1972. "Energy Potential from Organic Wastes" (TN 23.U71 No. 8549). Washington, D.C.: U.S. Department of the Interior.

ASAMW. 1976. Personal communication. Stockholm, Sweden: The Association of Swedish Automobile Manufacturers and Wholesalers.

Avron, M. 1976. Personal communication. Rehovot, Israel: The Weismann Institute of Science.

Bardach, J. E., Ryther, J. H., and McLarney, W. O. 1972. *Aquaculture.* New York: Interscience-Wiley.

Barnaby, F. 1977. Nuclear power and proliferation. *New Scientist* **75:** 168–70.

Bassham, J. A. 1977. Increasing crop production through more controlled photosynthesis. *Science* **197:** 630–38.

Benemann, J. R., and Weare, N. M. 1974. Hydrogen evolution by nitrogen-fixing *Anabaena* cultures. *Science* **184:** 174–75.

Benemann, J. R., and Weissman, J. C. 1976. "Biophotolysis: Problems and Prospects" (pp 413–26) In *Microbial Energy Conversion* (ed. Schlegel, H. G., and Barnea, J.). Göttingen, W. Germany: Erich Goltze KG.

Benemann, J. R., Berenson, J. A., Kaplan, N. O., and Kamen, M. D. 1973. Hydrogen evolution by a chloroplast-ferredoxin-hydrogenase system. *Proceedings of the National Academy of Sciences of the United States of America* **70:** 2317–20.

Benemann, J. R., Koopman, B., Weissman, J. C., and Oswald, W. J. 1976. "Biomass Production and Waste Recycling with Blue-Green Algae" (pp 399–412) In *Microbial Energy Conversion* (ed. Schlegel, H. G., and Barnea, J.). Göttingen, W. Germany: Erich Goltze KG.

Benemann, J. R., Weissman, J. C., Koopman, B. L., and Oswald, W. J. 1977. Energy production by microbial photosynthesis. *Nature* **268:** 19–23.

Berenson, J. H., and Benemann, J. R. 1977. Immobilization of hydrogenase and ferredoxins on glass beads. *Febs Letters* **76:** 105–07.

Bergersen, F. J. 1977. "Factors Controlling N_2-Fixation by Rhizobia" (Preprint from Conference on Biological Nitrogen Fixation in the Farming Systems of the Tropics). Canberra, Australia: Division of Plant Industry, CSIRO.

Bergersen, F. J., and Hipsley, E. H. 1970. The presence of N_2-fixing bacteria in the intestines of man and animals. *Journal of General Microbiology* **60:** 61.

Bishop, N. 1975. "Perturbation of Algal Genetics and Its Influence on Photohydrogen Production" In *Proceedings of Workshop on Biophotolysis and Nitrogen Fixation.* Knoxville, Tennessee: University of Tennessee.

Bishop, N. I., Frick, M., and Jones, L. W. 1977. "Photohydrogen Production in Green Algae" (pp 3–22) In *Biological Solar Energy Conversion* (ed. Mitsui, A., Miyachi, S., San Pietro, A., and Tamura, S.). New York: Academic Press.

Björkman, O., and Berry, J. 1973. High-efficiency photosynthesis. *Scientific American* **229:** 80–93.

Blanchard, G. C., and Foley, R. T. 1971. The operation of an ion-membrane fuel cell with microbially-produced hydrogen. *Journal of the Electro-Chemical Society,* **July:** 1232–35.

Bolin, B. 1978. Personal communication. Stockholm, Sweden: Stockholm University.

Bolin, B., and Arrhenius, E. 1977. Nitrogen—an essential life factor and a growing environmental hazard. *AMBIO* **6:** 96–105.

Bradley, D. J. 1974. "Water Supplies: The consequences of Change" (pp 81–98) In *Human Rights in Health* (Ciba Foundation Symposium No. 23, ed. Elliott, K., and Knight, J.), Amsterdam—London—New York: Elsevier.

Brandt, D. 1975. "Remarks on the Process Economics of Enzymatic Conversion of Cellulose to Glucose" (pp 275–77) In *Cellulose as a Chemical and Energy Resource* (Biotechnology and Bioengineering Symposium No. 5, ed. Wilke, C. R.). New York: Interscience-Wiley.

Bray, W. J. 1976. Green crop fractionation. *New Scientist* **70:** 66–8.

Brill, W. J. 1977. Nitrogen fixation. *Scientific American* **236:** 68–81.

Burke, D. P. 1975. CW report: methanol. *Chemical Week,* 24 September.

Burris, R. H. 1977. "Energetics of Biological Nitrogen-Fixation" (pp 278–90) In *Biological Solar Energy Conversion* (ed. Mitsui, A., Miyachi, S., San Pietro, A., and Tamura, S.). New York: Academic Press.

Burris, R. H., and Orme-Johnson, W. H. 1976. "Mechanism of Biological Nitrogen Fixation" (pp 208–33) In *Nitrogen Fixation,* Vol. I (ed. Newton, W. E., and Nyman, C. J.). Pullman, Washington: Washington State University Press.

Butler, W. L., and Kitajima, M. 1975. "Repackaging the Photochemical Apparatus of Photosynthesis" In *Solar Energy: Biological Conversion Systems* (Imperial College, London). London: UK Section, International Solar Energy Society.

Caldwell, M. 1975. "Primary Production of Grazing Lands" (Chapter 3) In *Photosynthesis and Productivity in Different Environments,* International Biological Programme Vol. 3 (ed. Cooper, J. P.). London: Cambridge University Press.

Calvin, M. 1974. Solar energy by photosynthesis. *Science* **185:** 375–81.

Carter, A. P., Leontief, W., and Petri, P. 1976. *The Future of the World Economy.* New York: United Nations Publishing Service.

Chibata, I., Tosa, T., Sato, T., Mori, T., and Yamamoto, K. 1974. "Continuous Enzyme Reactions by Immobilized Microbial Cells" (pp 303–13) In *Enzyme Engineering,* Vol. 2 (ed. Pye, E. K., and Wingard, Jr., L. B.). New York and London: Plenum Press.

Chollet, R., and Ogren, W. L. 1975. Regulation of photorespiration in C_3 and C_4 species. *The Botanical Review* **41:** 137–79.

Cook, E. 1971. The flow of energy in an industrial society. *Scientific American* **224:** 135–44.

Cooney, C. L., and Ackerman, R. A. 1975. Thermophilic anaerobic digestion of cellulosic waste. *European Journal of Applied Microbiology* **2:** 65–72.

Cooney, C. L., and Wise, D. L. 1975. Thermophilic anaerobic digestion of solid waste for fuel gas production. *Biotechnology and Bioengineering* **17:** 1119–35.

Cooper, J. P. 1975. "Control of Photosynthetic Production in Terrestrial Systems" (Chapter 27) In *Photosynthesis and Productivity in Different Environments,* International Biological Programme Vol. 3 (ed. Cooper, J. P.). London: Cambridge University Press.

Cousins, W. J. 1975. "Gasification: A Versatile Way of Obtaining Liquid Fuels and Chemicals from Wood" (pp 49–54) In *Potential for Energy Farming in New Zealand* (ed. Hawcroft, N.). Wellington, New Zealand: Department of Science and Industrial Research.

Cowling, E. G., and Kirk, T. K. 1976. "Properties of Cellulose and Lignocellulosic Materials as Substrates for Enzymatic Conversion Processes" (pp 95–124) In *Enzymatic Conversion of Cellulosic Materials* (Biotechnology and Bioengineering Symposium No. 6, ed. Gaden, Jr., E., Mandels, M. H., Reese, E. J., and Spano, L. A.). New York: Interscience-Wiley.

Crentz, W. L. 1971. "Agricultural Wastes—An Energy Resource of the Seventies" (paper presented at World Farm Foundation Symposium, 8–10 December 1971). Anaheim, California.

Crutzen, P. J., and Ehhalt, D. H. 1977. Effects of nitrogen fertilizers and combustion on the stratospheric ozone layer. *AMBIO* **6:** 112–17.

Cysewski, G. R., and Wilke, C. R. 1977. Rapid ethanol fermentations using vacuum and cell recycle. *Biotechnology and Bioengineering* **19:** 1125–43.

De Bivort, L. H. 1975. World agricultural development strategy and the environment. *Agriculture and Environment* **2:** 1–14.

Delwiche, C. C. 1970. The nitrogen cycle. *Scientific American* **223:** 136–47.

Delwiche, C. C. 1977. Energy relations in the global nitrogen cycle. *AMBIO* **6:** 106–11.

de Maré, L. 1977. "Resources and Needs: Assessment of the World Water Situation" (Report to the UN Water Conference, 1977). Lund, Sweden: Lund Institute of Technology.

Dickinson, W. C., Clark, A. F., Day, J. A., and Wonters, L. F. 1976. The shallow solar pond energy conversion system. *Solar Energy* **18:** 3–10.

Dobereiner, J. 1977. Biological nitrogen fixation in tropical grasses. *AMBIO* **6:** 174–77.

Dobereiner, J., Day, J. M., and von Bulow, J. F. W. 1975. "Associations of Nitrogen Fixing Bacteria with Roots of Forage Grass and Grain Species" (Paper presented at International Winter Wheat Conference, June 1975). Zagreb, Yugoslavia.

Doelle, H. W. 1975. "Photosynthesis and Photometabolism" (Chapter 3) In *Bacterial Metabolism,* second ed. New York: Academic Press.

Doxiadis, C. A. 1976. *Action for Human Settlements.* Athens, Greece: Athens Publishing Center.

Drachev, L. A., Kondrashin, A. A., Samuilov, V. D., and Skulachev, V. P. 1975. Generation of electric potential by reaction center complexes from *Rhodospirillum rubrum. Febs Letters* **50:** 219–22.

Dryden, I. G. C. 1975. *The Efficient Use of Energy.* Surrey, England: IPC Science and Technology Press Ltd.

Dudzik, M., Harte, J., Levy, D., and Sandusky, J. 1975. "Stability Indicators for Nutrient Cycles in Ecosystems" (Lawrence Berkeley Laboratory Report LBL-3264). Berkeley, California: University of California.

Eckholm, E. P. 1976a. *Losing Ground: Environmental Stress and World Food Prospects.* New York: Norton Press.

Eckholm, E. P. 1976b. "The Other Energy Crisis: Firewood" (World Watch Paper No. 1). Washington, D.C.: World Watch Institute.

Edwards, V. H. 1975. "Potential Useful Products from Cellulosic Materials" (pp 321–38) In *Cellulose as a Chemical and Energy Resource* (Biotechnology and Bioengineering Symposium No. 5, ed. Wilke, C. R.). New York: Interscience-Wiley.

Ehrlich, P. R., Ehrlich, A. H., and Holdren, J. P. 1973. *Human Ecology: Problems and Solutions.* San Francisco, California: W. H. Freeman and Company.

Eisa, H. M., Zeggio, V. J., and Jensen, H. M. 1971. Scrubbed diesel exhaust for carbon dioxide enrichment of greenhouse vegetables. *Horticulture Science* **6:** 477–79.

Elliott, K. 1976. Preface (pp 1–2) to *Acute Diarrhoea in Childhood* (Ciba Foundation Symposium No. 42, ed. Elliott, K. and Knight, J.). Amsterdam—London—New York: Elsevier.

Ensign, J. C. 1976. "Biomass Productivity from Animal Wastes by Photosynthetic Bacteria" (pp 455–82) In *Microbial Energy Conversion* (ed. Schlegel, H. G., and Barnea, J.). Göttingen, W. Germany: Erich Goltze KG.

Erbes, D. C., Burris, R. H., and Orme-Johnson, W. H. 1975. On the iron-sulfur cluster in hydrogenase from *Clostridium pasteurianum* W5. *Proceedings of the National Academy of Sciences of the United States of America* **72:** 4795–99.

Eriksson, E., Hallberg, R., Rodhe, H., and Rosswall, T. 1976. "Biogeochemical Cycles of Nitrogen, Phosphorus, and Sulfur'' In *Environmental Issues: SCOPE Report No. 8* (ed. Holdgate, M. W. and White, G. F.). Stockholm, Sweden: Royal Swedish Academy of Sciences.

Eriksson, K-E. 1976. Personal communication. Stockholm, Sweden: Swedish Forest Products Research Laboratory.

Eriksson, K-E., and Pettersson, B. 1975. Extracellular enzyme system utilized by the fungus *Sporotrichum pulverulentum* for the breakdown of cellulose. *European Journal of Biochemistry* **51:** 193–206.

ESCAP. 1975. ''Workshop on Biogas Technology and Utilization.'' Bangkok, Thailand: Economic and Social Commission for Asia and the Pacific.

ESCAP. 1976. ''Biogas Newsletters'' (August 1976). Bangkok, Thailand: Economic and Social Commission for Asia and the Pacific.

Evans, H. J. 1975. *Enhancing Biological Nitrogen Fixation*. Washington, D.C.: National Science Foundation.

Evans, H. J., and Barber, L. E. 1977. Biological nitrogen fixation for food and fiber production. *Science* **197:** 332–39.

FAO. 1973. *Annual Fertilizer Review 1972*. Rome, Italy: Food and Agriculture Organization.

Finn, R. K. 1975. ''Prospects for Fermentation Alcohol from Hydrolyzed Cellulose'' (pp 353–55) In *Cellulose as a Chemical and Energy Resource* (Biotechnology and Bioengineering Symposium No. 5, ed. Wilke, C. R.). New York: Interscience-Wiley.

Foo, E. L., and Hedén, C-G. 1976. ''Is Biocatalytic Production of Methanol a Practical Proposition?'' (pp 267–80) In *Microbial Energy Conversion* (ed. Schlegel, H. G., and Barnea, J.). Göttingen, W. Germany: Erich Goltze KG.

Galtung, J. 1978. ''Self-Reliance: Concepts, Practice and Rationale'' In *Self-Reliance* (ed. Galtung, J., O'Brian, P., and Preiswerk, R.) St. Saphorin, Switzerland: Georgi Publishing Company.

Gates, D. M. 1962. *Energy Exchange in the Biosphere*. New York: Harper and Row.

Gest, H., Ormerod, J. G., and Ormerod, K. S. 1962. Photometabolism of *Rhodospirillum rubrum*. *Archives of Biochemistry and Biophysics* **97:** 21–33.

Gifford, R. M. 1974. A comparison of potential photosynthesis, productivity, and yield of plant species with differing metabolism. *Australian Journal of Plant Physiology* **1:** 107–17.

Gitlitz, P. H., and Krasna, A. I. 1975. Structural and catalytic properties of hydrogenase from *Chromatium*. *Biochemistry* **14:** 2561–65.

Goedheer, J. C., and Kleinen Hammans, J. W. 1975. Efficiency of light conversion by the blue-green alga *Anacystis nidulans*. *Nature* **256:** 333–34.

Golley, F. B. 1972. ''Energy Flux in Ecosystems'' (pp 69–90) In *Ecosystem Structure and Function* (ed. Bohen). Portland, Oregon: Oregon State University Press.

Golueke, C. G., and Oswald, W. J. 1973. An algal regenerative system for single-family farms and villages. *Compost Science,* May–June, 12–15.

Govindjee, and Govindjee, R. 1974. The primary events of photosynthesis. *Scientific American* **231:** 68–82.

Greenbaum, E. 1977. The photosynthetic unit of hydrogen evolution. *Science* **196:** 879–80.

Gregory, D. 1973. The hydrogen economy. *Scientific American* **228:** 13–20.

Gregory, D. P. 1975. ''Methanol as a Future Fuel.'' Chicago, Illinois: Institute of Gas Technology.

Grenon, M. 1975. ''Coal Resources and Constraints'' In *Second Status Report of the IIASA Project on Energy Systems* (ed. Häfele, W.). Laxenburg, Austria: International Institute for Applied Systems Analysis.

Grethlein, H. E. 1975. ''Acid Hydrolysis of Refuse'' (pp 303–18) In *Cellulose as a Chemical and Energy Resource* (Biotechnology and Bioengineering Symposium No. 5, ed. Wilke, C. R.). New York: Interscience-Wiley.

Griffith, B., and Compere, A. 1975. ''ANFLOW Bio-Conversion'' (pp 24–26) In *Oak Ridge National Laboratory Review,* Fall, 1975. Oak Ridge, Tennessee: Oak Ridge National Laboratories.

Hall, D. O. 1976. "Agricultural and Biological Systems" (Chapter 9) In *Solar Energy: A United Kingdom Assessment*. London: UK Section, International Solar Energy Society.

Hammond, A. L. 1977. Alcohol: a Brazilian answer to the energy crisis. *Science* **195:** 564–66.

Hammond, A. L., Metz, W. D., and Maugh, T. H. 1973. *Energy and the Future*, Washington, D.C.: American Association for the Advancement of Science.

Hardy, R. W. F., and Havelka, U. D. 1975. Nitrogen fixation research: a key to world food? *Science* **188:** 633–43.

Hardy, R. W. F., and Havelka, U. D. 1977. "Possible Routes to Increase the Conversion of Solar Energy to Food and Feed by Grain Legumes and Cereal Grains" (pp 299–322) In *Biological Solar Energy Conversion* (ed. Mitsui, A., Miyachi, S., San Pietro, A., and Tamura, S.). New York: Academic Press.

Hepner, L. 1976. "Feasibility of Producing Basic Chemicals by Fermentation" (pp 531–54) In *Microbial Energy Conversion* (ed. Schlegel, H. G., and Barnea, J.). Göttingen, W. Germany: Erich Goltze KG.

Higgins, I. J., and Quayle, J. R. 1970. Oxygenation of methane by methane-grown *Pseudomonas methanica* and *Methanomonas methanooxidans*. *Biochemical Journal* **188:** 201–08.

Hirsch, F. 1976. *Social Limits To Growth*. Cambridge, Massachusetts: Harvard University Press.

Hollaender, A. 1977. *Proceedings of a Conference on Genetic Engineering for Nitrogen Fixation* (ed. Hollaender, A.). New York: Plenum Press.

Holm, R. H. 1976. "Synthetic analogs and the Active Sites of Iron-Sulfur Proteins and Enzymes and Their Relevance to Di-nitrogen Fixation" (Chapter 2) In *Nitrogen Fixation* Vol. I (Proceedings of an International Symposium, ed. Newton, W. E., and Nyman, C. J.). Pullman, Washington: Washington State University Press.

Holmgren, H. 1976. Personal communication. Stockholm Sweden: Department of Physics, Stockholm University.

ICMR. 1974. "Manatee Research" (A Workshop Report 7–13 February 1974). Georgetown, Guyana: International Centre for Manatee Research, National Science Research Council.

IFIAS. 1974. "Socio-economic and ethical implications of enzyme engineering," Solna, Sweden: International Federation of Institutes for Advanced Study.

IFIAS. 1975. "Energy Analysis and Economics." (A Workshop Report). Solna, Sweden: International Federation of Institutes for Advanced Study.

IFIAS. 1976. "Soil Resources of the Earth, Their Utilization and Preservation" (A Workshop Report, Samarkand, Uzbekistan, USSR, 14–21 June 1976). Solna, Sweden: International Federation of Institutes for Advanced Study.

Illich, I. 1973. *Tools for Conviviality*. New York: Harper & Row, Perennial Library.

Illich, I. 1974. *Energy and Equity*. New York: Harper & Row.

Imhoff, K., and Thistlethwaite, D. K. B. 1971. *Disposal of Sewage and Other Water-borne Wastes* (2nd edition). London: Newnes-Butterworths.

Ingamells, J. C., and Lindquist, R. H. 1975. "Methanol as a motor fuel or a gasoline-blending component." Warrendale, Pennsylvania: Society of Automotive Engineers, Inc.

Ingestad, T. 1977. Nitrogen and plant growth: maximum efficiency of nitrogen fertilizers. *AMBIO* **6:** 146–51.

IRRI, 1974. *Annual Report*. Los Baños, Laguna, Philippines: International Rice Research Institute.

Jones, J. W., and Kok, B. 1966. Photoinhibition of chloroplast reactions. *Plant Physiology* **41:** 1044–49.

Karube, I., Matsunaga, T., Tsuru, S., and Suzuki, S. 1977. Biochemical fuel cell utilizing immobilized cells of *Clostridium butyricum*. *Biotechnology and Bioengineering* **19:** 1727–33.

Kemp, C. C., and Szego, G. C. 1975. *The Energy Plantation*. Warrenton, Virginia: Inter-Technology Corporation.

Khadi. 1975. "Gobar Gas: Why and How." Bombay, India: Khadi and Village Industries Commission.

Kikuchi, W. K. 1976. Prehistoric Hawaiian fishponds. *Science* **193:** 295-99.

Kira, T. 1975. "Primary Production of Forests" (Chapter 2) In *Photosynthesis and Productivity in Different Environments* (ed. Cooper, J. P.). London: Cambridge University Press.

Kirk, T. K. 1975. "Lignin-degrading Enzyme Systems" (pp 139-50) In *Cellulose as a Chemical and Energy Resource* (Biotechnology and Bioengineering Symposium No. 5, ed. Wilke, C. R.). New York: Interscience-Wiley.

Knelson, J. H., and Lee, R. E. 1977. Oxides of nitrogen in the atmosphere: origin, fate, and public health implications. *AMBIO* **6:** 126-30.

Kok, B. 1968. "Photosynthesis" (pp 335-79) In *Physiology of Plant Growth and Development* (ed. Wilkins, M. B.). London—NewYork: McGraw Hill.

Kok, B. 1973. "Photosynthesis" (pp 22-30) In *Proceedings of the Workshop on Bio-Solar Conversion* (5-6 September 1973). Bethesda, Maryland: National Science Foundation, RANN.

Kok, B., Forbush, B, and McGloin, M. 1970. Cooperation of charges in photosynthetic O_2 evolution—1. *Photochemistry and Photobiology* **11:** 457-75.

Kolm, H., Oberteuffer, J., and Kelland, D. 1975. High-gradient magnetic separation. *Scientific American* **233:** 46-54.

Konstandt, H. G. 1976. "Engineering, Operation and Economics of Methane Gas Production" (pp 379-98) In *Microbial Energy Conversion* (ed. Schlegel, H. G., and Barnea, J.). Göttingen, W. Germany: Erich Goltze KG.

Kosaric, N., and Zajic, J. E. 1974. Microbial oxidation of methane and methanol. *Advances in Biochemical Engineering* **3:** 89-126.

Kovda, V. A. 1974. *Biosphere, Soils, and Their Utilization* (Report for 10th International Congress of Soil Science. Moscow, USSR: Academy of Sciences of the USSR.

Krasna, A. I. 1976. "Enzymology of Hydrogenase" In *Proceedings on Solar Energy for Nitrogen Fixation and Hydrogen Production* (ed. Beck, R. W.). Knoxville, Tennessee: University of Tennessee.

Krasna, A. I. 1977. "Catalytic and Structural Properties of Hydrogenase and Its Role in Biophotolysis of Water" (pp 53-60). In *Biological Solar Energy Conversion* (ed. Mitsui, A., Miyachi, S., San Pietro, A., and Tamura, S.). New York: Academic Press.

Kugelman, I. J., and Chin, K. K. 1971. "Toxicity, Synergism and Antagonism in Anaerobic Waste Treatment Processes" (Chapter 5) In *Anaerobic Biological Treatment Process* (ed. Gould, R. F.). Washington, D.C.: American Chemical Society.

Kurtzman, Jr., R. H. 1976. "Solid State Fermentation of Lignin by *Pleurotus ostreatus*" (Abstract No. 001) In *Abstracts of 172nd American Chemical Society Meeting* (Division of Microbial and Biochemical Technology). Washington, D.C.: American Chemical Society.

Lappi, D. L., Stolzenbach, F. E., Kaplan, N. O., and Kamen, M. D. 1976. Immobilization of hydrogenase on glass beads. *Biochemical and Biophysical Research Communications* **69:** 878-84.

Leach, G. 1976. *Energy and Food Production*. Surrey, England: IPC Sciences and Technology Press Ltd.

Lee, B. H., and Blackburn, T. H. 1975. Cellulase production by a thermophilic *Clostridium* species. *Applied Microbiology* **30:** 346-53.

Lehninger, A. L. 1975. *Biochemistry*. New York: Worth Publishers, Inc.

Leigh, J. 1976. A chemical fix for nitrogen? *New Scientist* **69:** 385-87.

Lemon, E., Stewart, D. W., and Showcroft, R. W. 1971. The sun's work in a cornfield. *Science* **174:** 371-78.

Lewis, C. W. 1976. Personal communication. Glasgow, Scotland: Energy Analysis Unit, University of Strathclyde.

Lewis, F. M., and Ablow, C. M. 1976. "Pyrogas from Biomass" (pp 341-56) In *Capturing the Sun through Bioconversion* (Proceedings of a Conference, 10-12 March 1976). Washington, D.C.: The Bio-Energy Council.

Lewis K. 1966. Biochemical fuel cells. *Bacteriological Reviews* **30:** 101-13.

Lien, S., and San Pietro, A. 1976. "An inquiry into biophotolysis of water to produce hydrogen" (sponsored by NSF-RANN). Bloomington, Illinois: Department of Plant Sciences, Indiana University.

Lipinsky, E. S. 1977. "Sugar cane versus corn versus ethylene as sources of ethanol" (unpublished paper). Columbus, Ohio: Batelle.

Lloyd, S. A., and Weinberg, F. J. 1975. Limits to energy release and utilization from chemical fuels. *Nature* **257:** 367-70.

Loll, U. 1976. "Engineering, Operation and Economics of Biodigestion" (pp 361-78) In *Microbial Energy Conversion* (ed. Schlegel, H. G., and Barnea, J.). Göttingen, W. Germany: Erich Goltze KG.

Long, S. P., Incoll, L. D., and Woolhouse, H. W. 1975. C_4 photosynthesis in plants from cool temperate regions. *Nature* **257:** 622-24.

Loomis, R. S., and Gerakis, P. A. 1975. "Productivity of Agricultural Ecosystems" (Chapter 6) In *Photosynthesis and Productivity in Different Environments,* International Biological Programme Vol. 3 (ed. Cooper, J. P.). London: Cambridge University Press.

Magee, P. A. 1977. Nitrogen as a health hazard. *AMBIO* **6:** 123-25.

Mandels, M. 1975. "Microbial Sources of Cellulose" (pp 81-110) In *Cellulose as a Chemical and Energy Resource* (Biotechnology and Bioengineering Symposium No. 5., ed. Wilke, C. R.). New York: Interscience-Wiley.

Mandels, M., and Weber, J. 1969. "Production of Cellulase" (pp 391-414) In *Cellulases and Their Applications.* Washington, D.C.: American Chemical Society.

Mandels, M., Hontz, L., and Nystrom, J. 1974. Enzymatic hydrolysis of waste cellulose. *Biotechnology and Bioengineering* **16:** 1471-93.

Mathiesen, H. 1970. Environmental changes and biological effects in lakes (from Danish). *Vatten* **2:** 149-73.

Maugh, T. H. 1972a. Fuel cells: dispersed generation of electricity. *Science* **178:** 1273-74B.

Maugh, T. H. 1972b. Hydrogen: synthetic fuel of the future. *Science* **178:** 849-52.

McBride, B. C., and Wolfe, R. S. 1971. "Biochemistry of Methane Formation" (Chapter 2) In *Anaerobic Biological Treatment Processes* (Advances in Chemistries Series 105, ed. Gould, R. F.). Washington, D.C.: American Chemical Society.

McCann, D. J., and Saddler, H. D. W. 1976. Photobiological energy conversion in Australia. *Search* **7:** 17-23.

McElroy, M. B. 1976. Personal communication. Cambridge, Massachusetts: Harvard University, Earth and Planetary Science Department.

McHale, J., and McHale, M. C. 1975. *Human Requirements, Supply Levels and Outer Bounds.* Aspen, Colorado: Aspen Institute for Humanistic Studies.

McLarney, W. O., and Todd, J. 1974. Aquaculture. *Journal of New Alchemy Institute* **2:** 79-118.

Menz, K. M., Moss, D. N., Cannel, R. Q., and Brun, W. A. 1969. Screening for photosynthetic efficiency. *Crop Science* **9:** 692-94.

Mesarovic, M., and Pestel, E. 1974. *Mankind at the Turning Point.* New York: E. P. Dutton and Co., Inc./Reader's Digest Press.

Messing, R. A. 1975. "Introduction and General History of Immobilized Enzymes" (pp 1-10) In *Immobilized Enzymes for Industrial Reactors* (ed. Messing, R. A.). New York: Academic Press.

Meynell, G. G., and Meynell, E. 1965. *Theory and Practice of Experimental Bacteriology.* London: Cambridge University Press.

Miller, D. L. 1975. "Ethanol Fermentation and Potential" (pp 345-52). In *Cellulose as a Chemical and Energy Resorce* (Biotechnology and Bioengineering Symposium No. 5, ed. Wilke, C. R.). New York: Interscience-Wiley.

Miller, D. L. 1976. "Fermentation Ethyl Alcohol" (pp 307-12) In *Enzymatic Conversion of Cellulosic Materials* (Biotechnology and Bioengineering Symposium No. 6, ed. Gaden., Jr., E. L., Mandels, M. H., Reese, E. T., and Spano, L. A.). New York: Interscience-Wiley.

Millett, M. A., Baker, A. J., and Satter, L. D. 1976. "Physical and Chemical Pretreatments for Enhancing Cellulose Saccharification" (pp 125-53) In *Enzymatic Conversion of Cellulosic Materials* (Biotechnology and Bioengineering Symposium No. 6, ed. Gaden, Jr., E. L., Mandels, M. H., Reese, E. T., and Spano, L. A.). New York: Interscience-Wiley.

Millington, R. J. 1976. "Phosphorus bio-economics" (preprint from author). Canberra, Australia: CSIRO, Division of Land Use Research.

Minchin, F. R., and Pate, J. S. 1973. The carbon balance of a legume and the functional economy of its root nodules. *Journal of Experimental Botany* **24:** 259-71.

Mitchell, R. 1977. "A program for concentration of single cell protein by high gradient magnetic separation" (unpublished paper). Cambridge, Massachusetts: Harvard University.

Mitra, G., and Wilke, C. R. 1975. Continuous cellulase production *Biotechnology and Bioengineering* **27:** 1-13.

Mitre. 1974. "Survey of alcohol fuel technology." McLean, Virginia: The Mitre Corporation.

Moore, W. J. 1965. "The Second Law of Thermodynamics" (Chapter 3) In *Physical Chemistry,* 4th edition. Englewood Cliffs, New Jersey: Prentice-Hall.

Morowitz, H. J. 1974. "Biological production of hydrogen from agricultural wastes" (unpublished paper). New Haven, Connecticut: Yale University, Department of Molecular Biophysics and Biochemistry.

Moss, D. N., Krenzer, Jr., E. G., and Brun, W. A. 1969. Carbon dioxide compensation points in related plant species. *Science* **164:** 187-88.

Myers, J., and Graham, J. R. 1975. Photosynthetic unit size during the synchronous life cycle of *Scenedesmus. Plant Physiology* **55:** 686-88.

NAS. 1976a. *Energy for Rural Development.* Washington, D.C.: United States National Academy of Science.

NAS. 1976b. *Making Aquatic Weeds Useful: Some Perspectives for Developing Countries.* Washington, D. C.: United States National Academy of Science.

NAS. 1977. *Energy and Climate.* Washington, D.C.: United States National Academy of Science.

New Alchemy. 1973. "Bio-gas digesters." Woods Hole, Massachusetts: The New Alchemy Institute.

Nugent, N. A. 1972. "An inquiry into biological energy conversion" (NSF/RANN Report). Knoxville, Tennessee: University of Tennessee.

Nutman, P. S. 1975. *Symbiotic Nitrogen Fixation in Plants* (International Biological Programme Vol. 7). Cambridge, England: Cambridge University Press.

Nystrom, J. 1975. "Discussion of Pretreatments to Enhance Enzymatic and Microbiological Attack of Cellulosic Materials" (pp 221-24) In *Cellulase as a Chemical and Energy Resource* (Biotechnology and Bioengineering Symposium No. 5; ed. Wilke, C. R.). New York: Interscience-Wiley.

Odum, H. T. 1971. *Environment, Power and Society.* New York: Interscience-Wiley.

Odum, H. T. 1973. Energy, ecology and economics. *AMBIO* **2:** 220-27.

Odum, H. T., and Odum, E. 1976. *Energy Basis for Man and Nature.* New York: McGraw-Hill.

Oesterhelt, D. 1976. Bacteriorhodopsin as a light-driven ion exchanger? *Febs Letters* **64:** 20-22.

Ohmiya, K., Ohashi, H., Kobayashi, T., and Shimizu, S. 1977. Hydrolysis of lactose by immobilized microorganisms. *Applied and Environmental Microbiology* **33:** 137-46.

Oort, A. H. 1970. Energy cycle of the earth. *Scientific American* **223:** 54-63.

Orme-Johnson, W. H. 1976. "Enzymology of nitrogenase and hydrogenase" (Paper at Workshop on

Solar Energy for Nitrogen Fixation and Hydrogen Production, 18-20 September 1975). Knoxville, Tennessee: University of Tennessee.

Oswald, W. J. 1973. Productivity of algae in sewage disposal. *Solar Energy* 15: 107-17.

Oswald, W. J. 1976a. "Photosynthetic single-cell protein" (Preliminary manuscript from NSF/MIT Study of Protein Resources). Boston, Massachusetts: Department of Nutrition, Massachusetts Institute of Technology.

Oswald, W. J. 1976b. Personal communication. Berkeley, California: University of California, Berkeley.

Oswald, W. J., and Golueke, C. G. 1968. "Large-Scale Production of Algae" (pp 271-305) In *Single-Cell Protein* (ed. Mateles, R. I., and Tannenbaum, S. R.). Cambridge, Massachusetts: MIT Press.

Packer, L. 1976. Problems in the stabilization of the *in vitro* photochemical activity of chloroplasts used for H_2 production. *Febs Letters* 64: 17-19.

Pagan, J. D., Child, J. J., Scowcroft, W. R., and Gibson, A. H. 1975. Nitrogen fixation by *Rhizobium* cultured on a defined medium. *Nature* 256: 406-07.

Pape, M. 1976. "The Competition between Microbial and Chemical Processes for the Manufacture of Basic Chemicals and Intermediates" (pp 515-29) In *Microbial Energy Conversion* (ed. Schlegel, H. G., and Barnea, J.). Göttingen, W. Germany: Erich Goltze KG.

Parikh, J. K., and Parikh, K. S. 1976. "Potential of Bio-Gas Plants and How to Realize It" (pp 555-91) In *Microbial Energy Conversion* (ed. Schlegel, H. G., and Barnea, J.). Göttingen, W. Germany: Erich Goltze KG.

Peters, G. A. 1977. "The *Azolla-Anabaena azollae* symbiosis: (pp 231-58) In *Genetic Engineering for Nitrogen Fixation* (ed. Hollaender, A.). New York: Plenum Press.

Pfenning, N. 1967. Photosynthetic bacteria. *Annual Review of Microbiology* 21: 286.

Pimentel, D., Hurd, L. E., Bellotti, A. C., Forster, M. J., Oka, I. N., Sholes, O. D., and Whitman, R. J. 1973. Food production and the energy crisis. *Science* 182: 443-49.

Pimentel, D., Dritschilo, W., Krummel, J., and Kutzman, J. 1975. Energy and land constraints in food protein production. *Science* 190: 754-61.

Pine, M. J. 1971. "The Methane Fermentations" (pp 1-10) In *Anaerobic Biological Treatment Processes* (ed. Gould, R. F.). Washington, D.C.: American Chemical Society.

Pirie, N. W. 1975a. "The *Spirulina* Algae" (Chapter 5) In *Food Protein Sources* (International Biological Programme Vol. 4, ed. Pirie, N. W.). London, England: Cambridge University Press.

Pirie, N. W. 1975b. Leaf protein: a beneficiary of tribulation. *Nature* 253: 239-41.

Pirie, N. W. 1975c. "Leaf Protein" (Chapter 14) In *Food Protein Sources* (International Biological Programme Vol. 4, ed. Pirie, N. W.). London, England: Cambridge University Press.

Porter, G. S., and Archer, M. D. 1976. *In vitro* photosynthesis. *Interdisciplinary Science Reviews* 1:(2) 119-43.

Postgate, J. R. 1977. Consequences of the transfer of nitrogen fixation genes to new hosts. *AMBIO* 6: 178-80.

Powell, J. R., Salzano, F. J., Yu, W. S., and Milan, J. S. 1976. A high-efficiency power cycle in which hydrogen is compressed in metal hydrides. *Science* 193: 314-17.

Prasad, C. R., Prasad, K. K., and Reddy, A. K. N. 1974. Bio-gas plants: prospects, problems, and tasks. *Economic and Political Weekly* 9: 1347-64.

Reddy, A. K. N. 1976. "A conceptual framework for environmentally sound and appropriate technologies." (Report from United Nations Environmental Program FP/0402-75-02(826).) Nairobi, Kenya: UNEP.

Reed, T. B. 1976. "When the Oil Runs Out" (pp 366-8) In *Capturing the Sun through Bioconversion* (Proceedings of a Conference, 10-12 March 1976). Washington, D.C.: The Bio-Energy Council.

Reed, T. B., and Lerner, R. M. 1973. Methanol: a versatile fuel for immediate use. *Science* **182:** 1299–1304.

Reed, T. B., and Lerner, R. M. 1974. "Sources and methods for methanol production" (paper at THEME Hydrogen Conference, Miami Beach, Florida, 18 March 1974).

Richmond, A. 1975. Personal communication. Beer Sheeba, Israel: Arid Zone Institute, Ben Gurion University.

Ryther, J. H. 1975. "Preliminary results with a pilot-plant waste recycling marine aquaculture system" (unpublished paper presented at International Conference on the Renovation and Reuse of Wastewater through Aquatic and Terrestrial Systems, 15–21 July 1975). Bellagio, Italy: Rockefeller Foundation Study & Conference Center.

Saeman, J. F. 1945. The kinetics of wood hydrolysis. *Industrial and Engineering Chemistry* **37:** 43.

Safrany, D. R. 1974. Nitrogen Fixation. *Scientific American* **231:** 64–81.

Saha, S., Ouitrakul, R., Izawa, S., and Good, N. E. 1971. Electron transport and photophosphorylation in chloroplasts as a function of the electron acceptor. *Journal of Biological Chemistry* **246:** 3204.

Sands, R. A., and Dunham, W. R. 1974. Spectroscopic studies on two-iron ferredoxins, *Quarterly Reviews of Biophysics* **7:** 443–504.

Santillán, C. 1974. "Cultivation of the *Spirulina* algae for human consumption and for animal feed" (unpublished paper at International Congress of Food Science and Technology, Madrid, Spain, September 1974).

Schneider, S. H., and Mesirow, L. E. 1976. *The Genesis Strategy.* New York: Plenum Press.

Schneider, T. R. 1973. Efficiency of photosynthesis as a solar energy converter. *Energy Conversion* **13:** 77–85.

Schumacher, E. F. 1973. *Small is Beautiful.* London, England: Sphere Books Limited.

Schumacher, E. F. 1976. Patterns of human settlements. *AMBIO* **5:** 91–97.

Schwimmer, M., and Schwimmer, D. 1968. "Medical Aspects of Phycology" (Chapter 15) In *Algae, Man and the Environment* (ed. Jackson, D. F.). Syracuse, New York: Syracuse University Press.

Scrimshaw, N. S. 1975. "Single-Cell Protein for Human Consumption—An Overview" (Chapter 2) In *Single-Cell Protein*, Vol. II, (ed. Tannenbaum, S. R., and Wang, D. I. G.). Cambridge, Massachusetts: MIT Press.

Shanmugam, K. T., and Valentine, R. C. 1975. Molecular biology of nitrogen fixation. *Science* **187:** 919–24.

Shelef, G., Schwartz, M., and Schechter, H. 1973. "Prediction of Photosynthetic Biomass Production in Accelerated Algal-Bacterial Wastewater Treatment Systems" (pp 181–90) In *Advances in Water Pollution Research* (ed. Jenkins, S. H.). London, England: Pergamon Press.

Shelef, G., Moraine, R., Meydan, A., and Sandbank, E. 1976. "Combined algae production—wastewater treatment and reclamation systems" (pp 427–42) In *Microbial Energy Conversion* (ed. Schlegel, H. G., and Barnea, J.). Göttingen, W. Germany: Erich Goltze KG.

Shoji, K. 1977. Drip Irrigation. *Scientific American* **237:** 62–68.

SIPRI. 1977. *World Armaments: The Nuclear Threat.* Stockholm, Sweden: Stockholm International Peace Research Institute.

Sirén, G. 1978. Personal communication. Stockholm, Sweden: Forestry School.

Sisler, F. D. 1961. Electrical energy from biochemical fuel cells. *New Scientist* **10:** 110–11.

Sisler, F. D. 1971. "Biochemical Fuel Cells" (pp 1–11) In *Progress in Industrial Microbiology*, Vol. 9 (ed. Hockenhall, D. J. P.). London, England: J. A. Churchill.

Skulachev, V. P. 1976. Conversion of light energy in electric energy by bacteriorhodopsin. *Febs Letters* **64:** 23–25.

Slesser, M. 1975. Accounting for energy. *Nature* **254:** 170–72.

Smith, B. E., Thorneley, R. N. F., Yates, M. G., Eadys, R. R., and Postgate, J. R. 1976a.

"Structure and Function of Nitrogenase from *Klebsiella pneumoniae* and *Azotobacter chroococcum*" (pp 150–76). In *Nitrogen Fixation*, Vol. I (ed. Newton, W. E., and Nyman, C. J.). Pullman, Washington: Washington State University Press.

Smith, R. L., Bouton, J. H., Schank, S. C., Quesenberry, K. H., Tyler, M. E., Milam, J. R., Gaskins, M. H., and Little, R. C. 1976b. Nitrogen fixation in grasses inoculated with *Spirillum lipoferum*. *Science* **193**: 1003–05.

Soeder, C. J. 1976. "Primary Productivity of Biomass in Fresh Water with Respect to Microbial Energy Conversion" (pp 59–68) In *Microbial Energy Conversion* (ed. Schlegel, H. G., and Barnea, J.). Göttingen, W. Germany: Erich Goltze KG.

Spano, L. A. 1976. "Enzymatic Hydrolysis of Cellulosic Materials" (pp 157–77) In *Microbial Energy Conversion* (ed. Schlegel, H. G., and Barnea, J.). Göttingen, W. Germany: Erich Goltze KG.

Spano, L. A., Medeiros, J., and Mandels, M. 1976. "Enzymatic Hydrolysis of Cellulosic Wastes to Glucose" (pp 541–66) In *Capturing the Sun through Bioconversion* (Proceedings of a Conference, 10–12 March 1976). Washington, D.C.: The Bio-Energy Council.

Spoehr, H. A., and Milner, H. W. 1949. The chemical composition of *Chlorella:* effect of environmental conditions. *Plant Physiology* **24**: 120–49.

Stanhill, G. 1977. An urban agro-ecosystem: the example of nineteenth century Paris. *Agro-Ecosystem* **3**: 269–84.

Stanier, R. Y., Dondoroff, M., and Adelberg, E. A. 1972. "Photosynthetic Bacteria" (Chapter 17) In *General Microbiology*, 3rd edition. London, England: Macmillan.

Stephens, G. R., and Heichel, G. H. 1975. "Agricultural and Forest Products as Sources of Cellulose" (pp 27–42) In *Cellulose as a Chemical and Energy Resource* (Biotechnology and Bioengineering Symposium No. 5, ed. Wilke, C. R.). New York: Interscience-Wiley.

Sternberg, D. 1976. "Production of Cellulase by Trichoderma" (pp 35–53) In *Enzymatic Conversion of Cellulosic Materials* (Biotechnology and Bioengineering Symposium No. 6, ed. Gaden, Jr., E. L., Mandels, M. H., Reese, E. T., and Spano, L. A.). New York: Interscience-Wiley.

Stewart, W. D. P. 1973. Nitrogen fixation by photosynthetic microorganisms. *Annual Reviews of Microbiology* **27**: 283–316.

Stewart, W. D. P. 1977. Present-day nitrogen fixing plants. *AMBIO* **6**: 166–73.

Stewart, W. D. P., and Rowell, P. 1975. Effects of L-methionine-DL-sulphoximine on the assimilation of newly fixed NH_3, acetylene reduction, and heterocyst production in *Anabaena cylindrica*. *Biochemical and Biophysical Research Communications* **65**: 846–56.

Stewart, W. D. P., Rowell, P., and Tel-Or, E. 1975. Nitrogen fixation and the heterocyst in blue-green algae. *Biochemical Society Transactions* **3**: 357–61.

Stuart, T. S., and Gaffron, H. 1972. Gas exchange of hydrogen-adapted algae as followed by mass spectroscopy. *Plant Physiology* **50**: 136–40.

Sverdrup, H. U., Johnson, M. W., and Fleming, R. H. 1942. *The Oceans, Their Physics, Chemistry, and General Biology*. New York: Prentice-Hall.

Talley, S. N., Talley, B. J., and Rains, D. W. 1977. "Nitrogen fixation by *Azolla* in rice fields" (pp 259–82) In *Genetic Engineering for Nitrogen Fixation* (ed. Hollaender, A.), New York: Plenum Press.

Talling, J. F. 1975. "Primary Production of Freshwater Microphytes" (Chapter 10) In *Photosynthesis and Productivity in Different Environments* (ed. Cooper, J. P.). Cambridge—London—New York: Cambridge University Press.

Tel-Or, E., and Packer, L. Submitted for publication. Hydrogenase in isolated heterocysts of blue-green algae. *Biochemical and Biophysical Research Communications*.

Thauer, R. 1976. "Limitation of microbial H_2 formation via fermentation" (pp 201–04) In *Microbial Energy Conversion* (ed. Schlegel, H. G., and Barnea, J.). Göttingen, W. Germany: Erich Goltze KG.

Times Atlas. 1973. *The Times Atlas of the World*. Edinburgh, Scotland: The Times/John Bartholomew & Son Ltd.

Tolbert, N. E. 1977. "Regulation of Products of Photosynthesis by Photorespiration and Reduction of Carbon" (pp 243–63) In *Biological Solar Energy Conversion* (ed. Mitsui, A., Miyachi, S., San Pietro, A., and Tamura, S.). New York: Academic Press.

Troughton, J. H., and Cave, I. D. 1975. "The Potential for Energy Farming in New Zealand" (pp 11–16) In *The Potential for Energy Farming in New Zealand* (ed. Hawcroft, N.). Wellington, New Zealand: Department of Science and Industrial Research.

UN. 1975. "Africa and The Fertilizer Industry" (limited distribution paper). Geneva, Switzerland—New York, New York: UN Economics and Social Council, Economics Commission for Africa, United Nations.

UNEP/Unesco/ICRO. 1977. Fifth International Conference on Global Impacts of Applied Microbiology, 21–26 November 1977. Bangkok, Thailand: Mahidol University.

Uziel, M., Oswald, W. J., and Golueke, C. G. 1975. "Integrated algal bacterial systems for fixation and conversion of solar energy" (paper presented at Annual Meeting of the American Association for the Advancement of Science, 29 January, 1975, New York).

von Hofsten, B. 1975. "Cultivation of a thermotolerant basidiomycete on various carbohydrates" (paper presented at International Symposium on Food from Waste, 7–8 April 1975). Weybridge, England: National College of Food Technology.

WAES. 1977. *Energy Global Prospects 1985–2000* (Workshop on Alternative Energy Strategies, ed. Wilson, C.). New York: McGraw-Hill.

Wasserman, A. R., and Fleisher, S. 1968. The stabilization of chloroplast function. *Biochimica et Biophysica Acta* **153:** 154–69.

Weetall, H. H. 1975. Immobilized enzymes and their application in the food and beverage industry. *Process Biochemistry* **10:** 3–12, 22, 30.

Weetall, H. H., and Bennett, M. A. 1976. "Production of Hydrogen Using Immobilized *Rhodospirillium rubrum*" (p 299) In *Abstracts of the Fifth International Fermentation Symposium* (ed. Dellweg, H.). Berlin: Verlag Versuchs—und Lehranstalt für Spiritusfabrikation und Fermentationstechnologie.

Weissman, J. C., and Benemann, J. R. 1977. Hydrogen production by nitrogen-starved cultures of *Anabaena cylindrica*. *Applied Environmental Microbiology* **33:** 123–31.

Westermark, U., and Eriksson, K-E. 1974. Cellobiose: Quinone oxidoreductase, a new wood-degrading enzyme from white-rot fungi. *Acta Chemica Scandinavica* **B28:** 209–14.

Whittenbury, R., Philips, K. C., and Wilkinson, J. F. 1970. Enrichment, isolation and some properties of methane-utilizing bacteria. *Journal of General Microbiology* **61:** 205–18.

Wilcox, H. A. 1976. "Ocean Farming" (pp 255–76) In *Capturing the Sun through Bioconversion* (Conference Proceedings). Washington, D.C.: Washington Center for Metropolitan Studies.

Wilke, C. R., and Mitra, G. 1975. "Process Development Studies on the Enzymatic Hydrolysis of Cellulose" (pp 253–74) In *Cellulose as a Chemical and Energy Resource* (Biotechnology and Bioengineering Symposium No. 5, ed. Wilke, C. R.). New York: Interscience-Wiley.

Wilke, C. R., Yang, R. D., and von Stockar, U. 1976. "Preliminary Cost Analyses for Enzymatic Hydrolysis of Newsprint" (pp 155–75) In *Enzymatic Conversion of Cellulosic Materials* (Biotechnology and Bioengineering Symposium No. 6, ed. Gaden, Jr., E. L., Mandels, M. H., Reese, E. T., and Spano, L. A.). New York: Interscience-Wiley.

Williams, R. W. M. 1975. "Costs of Production in Energy Farming of Trees" (pp 99–102) In *The Potential of Energy Farming in New Zealand* (Information Series No. 117). Wellington, New Zealand: Department of Scientific and Industrial Research.

Wilson, D. 1972. Variation in photorespiration in *Lolium*. *Journal of Experimental Botany* **23:** 517–24.

Wilson, R. K., and Pigden, W. J. 1964. % dry matter digestion vs. concentration of NaOH treatment. *Canadian Journal of Animal Science* **44:** 122.

Windsor, M., and Cooper, M. 1977. Farmed fish, cows, and pigs. *New Scientist* **75:** 740–42.

Wolverton, B. C., and McDonald, R. C. 1976. Don't waste water weeds. *New Scientist* **75:** 318–20.

Wood, T. M. 1975. "Properties and Mode of Action of Cellulases" (pp 111–38) In *Cellulose as a Chemical and Energy Resource* (Biotechnology and Bioengineering Symposium No. 5, ed. Wilke, C. R.). New York: Interscience-Wiley.

Yount, J. L., and Crossman, Jr., R. A. 1970. Eutrophication control by plant harvesting. *Journal of Water Pollution Control Federation* **42:** 173–83.

GLOSSARY
FOR PART I

absolute zero. In physics, the beginning or zero point in the scale of absolute temperature, equivalent approximately to $-273.1°C$ or $-459.6°F$. According to the kinetic theory of heat, it represents the temperature at which all thermal motion ceases.

amino acids. The 20-25 different naturally-occurring, nitrogen-containing building blocks of protein (see also *protein*).

ammonia. A "fixed" nitrogen form (NH_3) suitable as fertilizer for plant uptake and conversion to protein.

anaerobic digestion. A type of bacterial degradation of organic matter which occurs only in the absence of air (oxygen).

aquaculture. Cultivation of fish, algae, water plants, and other water-borne organisms.

basic needs. Requirements for human survival in good health, i.e., food, shelter, clothing, health, and education.

biogas. (also Gobar gas, marsh gas)—The fuel gas product of anaerobic digestion, predominantly methane (CH_4) and carbon dioxide (CO_2).

biophotolysis. The light-stimulated breakdown of water into energy-rich hydrogen and oxygen gases.

black-body radiation. The standard radiation pattern characterized only by the body's temperature.

C_3 vs. C_4 plants. A division of the plant kingdom into those which first build a three- versus a four-carbon molecule from absorbed carbon dioxide. This esoteric subdivision is relevant because of secondary characteristics of great practical importance.

canopy cover. A numerical description of the extent to which branches and leaves have developed and cover the sunlit area.

carbohydrates. A class of easily digested, natural compounds (e.g., sugar and starch) which contain only carbon, hydrogen and oxygen.

catalyst. A substance (e.g., platinum) which accelerates a chemical reaction but itself is neither consumed nor altered.

cellulose. Long chains of sugar molecules which are not digestible by humans or non-ruminant animals.

centrifugation. A process employing high-speed rotation to separate particles suspended in water (e.g., algal cells) according to their weight and shape.

chloroplast. A small portion of a plant cell which contains the light-absorbing pigment, chlorophyll, and converts light energy to chemical energy.

combustion. The energy (heat) releasing, burning process occurring when oxygen combines with energy-rich organic compounds (e.g., wood, food, peat) and results predominantly in carbon dioxide (CO_2) and water.

desertification. The often irreversible increase in desert size due to removal of trees and other vegetation and the resulting decreased ability to retain soil, nutrients and water.

ecology. Study of the interrelations of different living systems.

ecological food chain. The stages (species) through which food energy is consumed, from sun to green plants, to animal 1, to animal 2 . . . , and finally to low-energy products, e.g., water and carbon dioxide.

ecological niche. A set of conditions (temperature, physical location, point in food chain, etc.) which suits the growth requirements for a particular organism and gives it a unique survivability.

energy. A physical quantity describing the ability to perform work. Unambiguous use of the term is a complex task (a part of thermodynamics), but normally chemical or fuel energy means the heat of combustion, i.e., the energy released when a fuel is burned or metabolized with oxygen.

energy analysis. An accounting system similar to economics but based on energy flows, reservoirs, and sinks.

energy consumption. A common misnomer, shortcutting the complete physical description of the energy conversion—not consumption—which occurred. Strictly speaking, energy is neither produced nor consumed.

energy intensity. Describes the relative extent to which energy is required for a process, e.g., manufacturing, farming, or a society.

energy quality. A loose, qualitative term describing the usefulness of an energy source and normally related to temperature (as of heat energy), energy density, ease of conversion and entropy.

entropy. A mathematically defined quantity which describes the degree of disorder or randomness of a physical system.

enzymes. A biological catalyst; always a protein (see also *amino acids, proteins*).

eutrophication. Lack of oxygen in a body of water; extreme pollution with organic matter can result in the "death" of a waterway.

feedback. The coupling of an effect or product back to its source to continuously regulate (increase or decrease) the latter's activity.

First Law of Thermodynamics (see *thermodynamics*).

flocculation. The separation of e.g., algal cells from a dilute water solution by causing the cells to clump together.

fuel cells. A battery-like apparatus for converting a chemical fuel (e.g., hydrogen and oxygen gases together) to electrical current.

greenhouse effect. An increase in the earth's temperature due to the insolating effect of gases (CO_2) or particles (dust) in the atmosphere.

heat engine. A device for producing mechanical (motion) energy directly from two heat reservoirs of different temperature.

hydrocarbon. A chemical compound containing hydrogen, oxygen, and carbon.

hydrogenase. An enzyme for interconverting hydrogen gas and hydrogen ions (protons) plus electrons.

hydrogenation. A catalytic high temperature and pressure conversion of organic (e.g., plant) material to high-energy fuel oil.

integration. The combination of isolated components (e.g., of technology) into an interwoven system.

intercropping. Planting different crops in alternate rows to increase the effective land area.

leaching. The washing down through the soil of nutrients and other chemicals resulting ultimately in contamination of ground water, rivers or lakes.

legume. A class of plants whose roots are infected by bacteria which synthesize ammonia for the plant's use, thus reducing the need for nitrogen fertilizers.

lignin. Highly resistent long-chain molecules (polymer of aromatic alcohols) which constitute up to 25% wood's weight.

mutation. Accidental or induced changes in an organism's genetic material.

nitrogen fertilizer. Chemical compounds (e.g., nitrates, urea, ammonia) containing nitrogen in a form suitable for plant uptake and conversion to protein.

nitrogen fixation. The chemical or biological process by which inert nitrogen gas is complexed into a more reactive form, e.g., as with hydrogen.

nitrogenase. The enzyme system in certain microorganisms which converts nitrogen gas and hydrogen ions (protons) to ammonia.

ozone. A molecule of three oxygen atoms (O_3) whose presence in the earth's atmosphere results in the absorption of potentially damaging (mutagenic) ultraviolet sunlight.

photoinhibition and *photosaturation*. Increasing light intensity (brightness) increases photosynthesis to a point beyond which it causes no increase (saturation). Further light increases can cause a decrease (inhibition) of photosynthesis.

photosynthesis. The green plant (plus some bacteria) mediated conversion of light energy to chemical energy (ultimately sugar).

photovoltaic. The conversion of light to electrical potential, e.g., by the photocell.

plasmid. A small portion of one organism's genetic material which when inserted into a second organism's cell nucleus performs as it did in the original organism.

protein. A class of plant and animal molecules with both structural and functional (muscle, enzymes) roles. They are long chains of up to 25 different amino acids whose sequence and number uniquely determine the protein's character and function.

pyrolysis. A high-temperature and pressure-destructive conversion of organic matter to a mixture of char, fuel oil, and fuel gas.

Second Law of Thermodynamics (see *thermodynamics*).

self-reliance. The *ability* to meet basic needs locally; it implies maximal development of local resources but neither isolation nor complete self-sufficiency.

symbiosis. A mutally beneficial interdependency between two organisms, e.g., a leguminous plant and its infecting bacteria. Symbiosis can also refer to the net effect of two factors (chemicals, organisms, etc.) which exceeds the sum of the effects of each of the two factors in isolation.

technological fix. A new technology created to deal with a problem those technical aspects may be but one of many facets, and which itself may have been technologically caused.

thermodynamics. A branch of physics which deals specifically with the relations among heat, temperature, energy, entropy, etc.

   First Law of Thermodynamics. Energy is conserved, i.e., it can be neither created nor destroyed.

   Second Law of Thermodynamics. The disorder (entropy) of a closed system always increases
      during an irreversible process.

thermophilic. Organisms or processes which thrive at high temperature.

transpiration. The evaporative loss of water through plant pores (stomates) through which carbon dioxide is also absorbed.

waste. An ambiguous concept whose specific meaning can be expressed for only a specific system in economic or energy terms.

Index